"This is an excellent and very timely text, presenting the modern tools of high-dimensional geometry and probability in a very accessible and applications-oriented manner, with plenty of informative exercises. The book is infused with the author's insights and intuition in this field, and has extensive references to the latest developments in the area. It will be an extremely useful resource both for newcomers to this subject and for expert researchers."

– Terence Tao, University of California, Los Angeles

"Methods of high-dimensional probability have become indispensable in numerous problems of probability theory and its applications in mathematics, statistics, computer science, and electrical engineering. Roman Vershynin's wonderful text fills a major gap in the literature by providing a highly accessible introduction to this area. Starting with no prerequisites beyond a first course in probability and linear algebra, Vershynin takes the reader on a guided tour through the subject and consistently illustrates the utility of the material through modern data science applications. This book should be essential reading for students and researchers in probability theory, data science, and related fields."

– Ramon van Handel, Princeton University

"This very welcome contribution to the literature gives a concise introduction to several topics in 'high-dimensional probability' that are of key relevance in contemporary statistical science and machine learning. The author achieves a fine balance between presenting deep theory and maintaining readability for a non-specialist audience – this book is thus highly recommended for graduate students and researchers alike who wish to learn more about this by-now-indispensable field of modern mathematics."

– Richard Nickl, University of Cambridge

"Vershynin is one of the world's leading experts in the area of high-dimensional probability, and his textbook provides a gentle yet thorough treatment of many of the key tools in the area and their applications to the field of data science. The topics covered here are a must-know for anyone looking to do mathematical work in the field, covering subjects important in machine learning, algorithms and theoretical computer science, signal processing, and applied mathematics."

– Jelani Nelson, Harvard University

"High-Dimensional Probability is an excellent treatment of modern methods in probability and data analysis. Vershynin's perspective is unique and insightful, informed by his expertise as both a probabilist and a functional analyst. His treatment of the subject is gentle, thorough, and inviting, providing a great resource for both newcomers and those familiar with the subject. I believe, as the author does, that the topics covered in this book are indeed essential ingredients of the developing foundations of data science."

– Santosh Vempala, Georgia Institute of Technology

"Renowned for his deep contributions to high-dimensional probability, Roman Vershynin is to be commended for the clarity of his progressive exposition of the important concepts, tools, and techniques of the field. Advanced students and practitioners interested in the mathematical foundations of data science will enjoy the many relevant worked examples and lively use of exercises. This book is the reference I had been waiting for."

– Rémi Gribonval, IEEE & EURASIP Fellow, Directeur de Recherche, Inria, France

"High-dimensional probability is a fascinating mathematical theory that has grown rapidly in recent years. It is fundamental to high-dimensional statistics, machine learning, and data science. In this book, Roman Vershynin, who is a leading researcher in high-dimensional probability and a master of exposition, provides the basic tools and some of the main results and applications of high-dimensional probability. This book is an excellent textbook for a graduate course that will be appreciated by mathematics, statistics, computer science, and engineering students. It will also serve as an excellent reference book for researchers working in high-dimensional probability and statistics."

– Elchanan Mossel, Massachusetts Institute of Technology

"This book on the theory and application of high-dimensional probability is a work of exceptional clarity that will be valuable to students and researchers interested in the foundations of data science. A working knowledge of high-dimensional probability is essential for researchers at the intersection of applied mathematics, statistics, and computer science. The widely accessible presentation will make this book a classic that everyone in foundational data science will want to have on their bookshelf."

– Alfred Hero, University of Michigan

"Vershynin's book is a brilliant introduction to the mathematics which is at the core of modern signal processing and data science. The focus is on concentration of measure and its applications to random matrices, random graphs, dimensionality reduction, and suprema of random process. The treatment is remarkably clean, and the reader will learn beautiful and deep mathematics without unnecessary formalism."

– Andrea Montanari, Stanford University

High-Dimensional Probability

An Introduction with Applications in Data Science

High-Dimensional Probability offers insight into the behavior of random vectors, random matrices, random subspaces, and objects used to quantify uncertainty in high dimensions. Drawing on ideas from probability, analysis, and geometry, it lends itself to applications in mathematics, statistics, theoretical computer science, signal processing, optimization, and more. It is the first text to integrate theory, key tools, and modern applications of high-dimensional probability. Concentration inequalities form the core, and it covers both classical results such as Hoeffding's and Chernoff's inequalities and modern developments such as the matrix Bernstein inequality. It then introduces powerful methods based on stochastic processes, including such tools as Slepian's, Sudakov's, and Dudley's inequalities, as well as generic chaining and bounds based on VC dimension. A broad range of illustrations is embedded throughout, including classical and modern results for covariance estimation, clustering, networks, semidefinite programming, coding, dimension reduction, matrix completion, machine learning, compressed sensing, and sparse regression. Hints for many of the exercises are given at the back of the book.

ROMAN VERSHYNIN is Professor of Mathematics at the University of California, Irvine. He studies random geometric structures across mathematics and data sciences, in particular in random matrix theory, geometric functional analysis, convex and discrete geometry, geometric combinatorics, high-dimensional statistics, information theory, machine learning, signal processing, and numerical analysis. His honors include an Alfred Sloan Research Fellowship in 2005, an invited talk at the International Congress of Mathematicians in Hyderabad in 2010, and a Bessel Research Award from the Humboldt Foundation in 2013. His "Introduction to the non-asymptotic analysis of random matrices" has become a popular educational resource for many new researchers in probability and data science.

CAMBRIDGE SERIES IN STATISTICAL AND PROBABILISTIC MATHEMATICS

This series of high-quality upper-division textbooks and expository monographs covers all aspects of stochastic applicable mathematics. The topics range from pure and applied statistics to probability theory, operations research, optimization, and mathematical programming. The books contain clear presentations of new developments in the field and also of the state of the art in classical methods. While emphasizing rigorous treatment of theoretical methods, the books also contain applications and discussions of new techniques made possible by advances in computational practice.

A complete list of books in the series can be found at www.cambridge.org/statistics.
Recent titles include the following:

21. *Networks*, by Peter Whittle
22. *Saddlepoint Approximations with Applications*, by Ronald W. Butler
23. *Applied Asymptotics*, by A. R. Brazzale, A. C. Davison and N. Reid
24. *Random Networks for Communication*, by Massimo Franceschetti and Ronald Meester
25. *Design of Comparative Experiments*, by R. A. Bailey
26. *Symmetry Studies*, by Marlos A. G. Viana
27. *Model Selection and Model Averaging*, by Gerda Claeskens and Nils Lid Hjort
28. *Bayesian Nonparametrics*, edited by Nils Lid Hjort *et al.*
29. *From Finite Sample to Asymptotic Methods in Statistics*, by Pranab K. Sen, Julio M. Singer and Antonio C. Pedrosa de Lima
30. *Brownian Motion*, by Peter Mörters and Yuval Peres
31. *Probability (Fourth Edition)*, by Rick Durrett
33. *Stochastic Processes*, by Richard F. Bass
34. *Regression for Categorical Data*, by Gerhard Tutz
35. *Exercises in Probability (Second Edition)*, by Loïc Chaumont and Marc Yor
36. *Statistical Principles for the Design of Experiments*, by R. Mead, S. G. Gilmour and A. Mead
37. *Quantum Stochastics*, by Mou-Hsiung Chang
38. *Nonparametric Estimation under Shape Constraints*, by Piet Groeneboom and Geurt Jongbloed
39. *Large Sample Covariance Matrices and High-Dimensional Data Analysis*, by Jianfeng Yao, Shurong Zheng and Zhidong Bai
40. *Mathematical Foundations of Infinite-Dimensional Statistical Models*, by Evarist Giné and Richard Nickl
41. *Confidence, Likelihood, Probability*, by Tore Schweder and Nils Lid Hjort
42. *Probability on Trees and Networks*, by Russell Lyons and Yuval Peres
43. *Random Graphs and Complex Networks (Volume 1)*, by Remco van der Hofstad
44. *Fundamentals of Nonparametric Bayesian Inference*, by Subhashis Ghosal and Aad van der Vaart
45. *Long-Range Dependence and Self-Similarity*, by Vladas Pipiras and Murad S. Taqqu
46. *Predictive Statistics*, by Bertrand S. Clarke and Jennifer L. Clarke
47. *High-Dimensional Probability*, by Roman Vershynin

High-Dimensional Probability

An Introduction with Applications in Data Science

Roman Vershynin
University of California, Irvine

CAMBRIDGE
UNIVERSITY PRESS

University Printing House, Cambridge CB2 8BS, United Kingdom

One Liberty Plaza, 20th Floor, New York, NY 10006, USA

477 Williamstown Road, Port Melbourne, VIC 3207, Australia

314–321, 3rd Floor, Plot 3, Splendor Forum, Jasola District Centre, New Delhi – 110025, India

79 Anson Road, #06–04/06, Singapore 079906

Cambridge University Press is part of the University of Cambridge.

It furthers the University's mission by disseminating knowledge in the pursuit of education, learning, and research at the highest international levels of excellence.

www.cambridge.org
Information on this title: www.cambridge.org/9781108415194
DOI: 10.1017/9781108231596

First published 2018

A catalog record for this publication is available from the British Library.

Library of Congress Cataloging-in-Publication Data
Names: Vershynin, Roman, 1974– author.
Title: High-dimensional probability : an introduction with applications in data science / Roman Vershynin, University of Michigan.
Description: Cambridge : Cambridge University Press, 2018. |
Series: Cambridge series in statistical and probabilistic mathematics ; 47 |
Includes bibliographical references and index.
Identifiers: LCCN 2018016910 | ISBN 9781108415194
Subjects: LCSH: Probabilities. | Stochastic processes. | Random variables.
Classification: LCC QA273 .V4485 2018 | DDC 519.2–dc23
LC record available at https://lccn.loc.gov/2018016910

ISBN 978-1-108-41519-4 Hardback

Contents

Foreword

This book begins with an appetizer: the empirical method of B. Maurey for approximating points in the convex hull of a set by averages. It is a beautiful and fascinating example where probability theory can elegantly solve problems which at first sight have nothing to do with probabilities. It is a very gratifying experience to learn from this book that probability theory opens up a whole world of other mathematical areas (which may have been found very difficult to access before).

After presenting the necessary background material, the book goes straight into the heart of the matter in Chapter 3. Concentration in high dimensions is treated in an enlightening way. For example, the formula

$$\sqrt{n \pm O(\sqrt{n})} = \sqrt{n} \pm O(1)$$

in Remark 3.1.2 says it all in all its simplicity. Likewise for Figure 3.6, where a Gaussian point cloud is shown in high dimensions: it concentrates on a sphere with radius $\sqrt{n}$. This shape has hardly anything in common with the bell shape in dimension 2 or 3 – our low-dimensional intuition is useless! As another example where probability theory can make life easier, the book provides an insightful proof of Grothendieck's inequality. To understand any of the other proofs of Grothendieck's inequality (with "good" constants), I would probably need several years.

Let me mention another theme that is extremely well explained in the book: the isoperimetric inequality and how it leads to blow up. If a subset of the sphere covers at least 50 percent, then its coverage is exponentially close to 100 percent. The book also presents several extensions to other metric spaces, for example concentration on the Grassmannian. In this way it provides a first entrance into this area, and one would like to learn more. The supplied pointers allow one to do so.

The book is a joy to read. The author conveys the material as an exciting story and one keeps on reading. Participation in the development of the storyline is encouraged by the many exercises that are scattered throughout the text.

Other topics treated in this book are random matrices, empirical process theory, and sparse recovery, to name a few. The results are important for research in data science but also simply of beauty on their own. Many students and researchers may have heard the key words, and this is the book to find out what they are really about.

Sara van de Geer, ETH Zürich

Preface

Who is This Book For?

This is a textbook in probability in high dimensions with a view toward applications in data sciences. It is intended for doctoral and advanced masters students and beginning researchers in mathematics, statistics, electrical engineering, computational biology, and related areas who are looking to expand their knowledge of theoretical methods used in modern research in the data sciences.

Why This Book?

The data sciences are moving fast, and probabilistic methods often provide a foundation and inspiration for such advances. Today, a typical graduate probability course is no longer sufficient to acquire the level of mathematical sophistication that is expected from a beginning researcher in data sciences. The book is intended to partially cover this gap. It presents some key probabilistic methods and results that form an essential toolbox for a mathematical data scientist. It can be used as a textbook for a basic second course in probability with a view toward data science applications. It is also suitable for self-study.

What is This Book About?

High-dimensional probability is an area of probability theory that studies random objects in $\mathbb{R}^n$, where the dimension n can be very large. The book places particular emphasis on random vectors, random matrices, and random projections. It teaches basic theoretical skills for the analysis of these objects, which include concentration inequalities, covering and packing arguments, decoupling and symmetrization tricks, chaining and comparison techniques for stochastic processes, combinatorial reasoning based on the VC dimension, and a lot more.

The study of high-dimensional probability provides vital theoretical tools for applications in data science. The book integrates theory with applications for covariance estimation, semidefinite programming, networks, elements of statistical learning, error correcting codes, clustering, matrix completion, dimension reduction, sparse signal recovery, and sparse regression.

Prerequisites

The essential prerequisites for reading this book are a rigorous course in probability theory (of the Masters or Ph.D. level), an excellent command of undergraduate linear algebra, and general familiarity with basic notions about metrics, normed and Hilbert spaces, and linear operators. A knowledge of measure theory is not essential but would be helpful.

A Word on the Exercises

The exercises are integrated into the text. The reader can do them immediately to check his or her understanding of the material just presented, and to prepare better for later developments. The difficulty of the exercises is indicated by the number of coffee cups; it ranges from easy (☕) to hard (☕☕☕☕). A pointing hand (☞) means that a hint is available at the end of the book.

Related Reading

The book covers only a fraction of the theoretical apparatus of high-dimensional probability and illustrates it with only a sample of data science applications. Each chapter in this book concludes with a Notes section, which has pointers to other texts on the subject matter of the chapter. A few particularly useful sources are noted here. The now classical book [8] showcases the probabilistic method in applications to discrete mathematics and computer science. The forthcoming book [19] will present a panorama of mathematical data science, focusing on applications in computer science. Both these books will be accessible to graduate and advanced undergraduate students. The lecture notes [206] are pitched at graduate students and present more theoretical material in high-dimensional probability.

Acknowledgements

The feedback from my many colleagues was instrumental in preparing this book. My special thanks go to Florent Benaych-Georges, Jennifer Bryson, Lukas Grätz, Rémi Gribonval, Ping Hsu, Mike Izbicki, George Linderman Cong Ma, Galyna Livshyts, Jelani Nelson, Ekkehard Schnoor, Martin Spindler, Dominik Stöger, Tim Sullivan, Terence Tao, Joel Tropp, Katarzyna Wyczesany, Yifei Shen, and Haoshu Xu, for many valuable suggestions and corrections, and in particular to Sjoerd Dirksen, Larry Goldstein, Wu Han, Han Wu, and Mahdi Soltanolkotabi for detailed proofreading of the book. I am grateful to Can Le, Jennifer Bryson, and my son Ivan Vershynin for their help with many of the pictures.

Appetizer

Using Probability to Cover a Geometric Set

We begin our study of high-dimensional probability with an elegant argument that showcases the usefulness of probabilistic reasoning in geometry.

Recall that a *convex combination* of points $z_1, \dots, z_m \in \mathbb{R}^n$ is a linear combination with coefficients that are non-negative and sum to 1, i.e., it is a sum of the form

$$\sum_{i=1}^{m} \lambda_i z_i \quad \text{where} \quad \lambda_i \geq 0 \quad \text{and} \quad \sum_{i=1}^{m} \lambda_i = 1. \tag{0.1}$$

The *convex hull* of a set $T \subset \mathbb{R}^n$ is the set of all convex combinations of all finite collections of points in T:

$$\text{conv}(T) := \{\text{convex combinations of } z_1, \dots, z_m \in T \text{ for } m \in \mathbb{N}\};$$

see Figure 0.1 for illustration.

The number m of elements defining a convex combination in $\mathbb{R}^n$ is not restricted a priori. However, the classical theorem of Caratheodory states that one can always take $m \leq n + 1$.

Theorem 0.0.1 (Caratheodory's theorem) *Every point in the convex hull of a set $T \subset \mathbb{R}^n$ can be expressed as a convex combination of at most $n + 1$ points from T.*

The bound $n + 1$ cannot be improved, as it is clearly attained for a simplex T (a set of $n + 1$ points in general positions). Suppose, however, that we want only to *approximate* a point $x \in \text{conv}(T)$ rather than to represent it exactly as a convex combination. Can we do

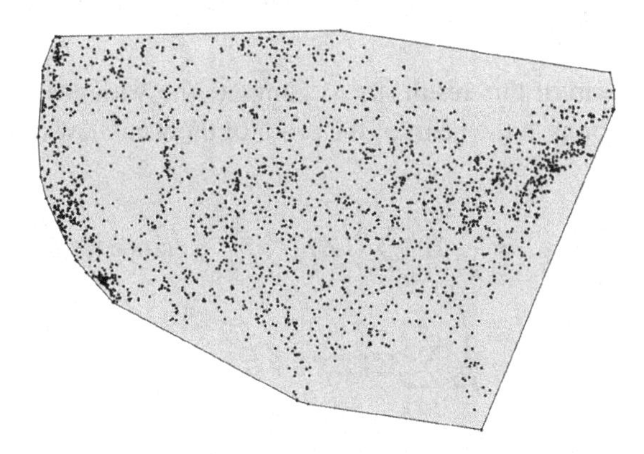

Figure 0.1 The convex hull of a set of points representing major US cities.

this with fewer than $n+1$ points? We now show that it is possible, and actually the number of required points does not need to depend on the dimension n at all!

Theorem 0.0.2 (Approximate form of Caratheodory's theorem) *Consider a set $T \subset \mathbb{R}^n$ whose diameter*[1] *is bounded by* 1. *Then, for every point $x \in \mathrm{conv}(T)$ and every integer k, one can find points $x_1, \dots, x_k \in T$ such that*

$$\left\| x - \frac{1}{k}\sum_{j=1}^{k} x_j \right\|_2 \le \frac{1}{\sqrt{k}}.$$

There are two reasons why this result is surprising. First, the number of points k in convex combinations does not depend on the dimension n. Second, the coefficients of convex combinations can be made all equal. (Note, however, that repetitions among the points x_i are allowed.)

Proof Our argument is known as the *empirical method* of B. Maurey.

Translating T if necessary, we may assume that not only the diameter but also the *radius* of T is bounded by 1, i.e.,

$$\|t\|_2 \le 1 \quad \text{for all } t \in T. \tag{0.2}$$

Fix a point $x \in \mathrm{conv}(T)$ and express it as a convex combination of some vectors $z_1, \dots, z_m \in T$ as in (0.1). Now, interpret the definition of the convex combination (0.1) probabilistically, with the λ_i taking the roles of probabilities. Specifically, we can define a random vector Z that takes the value z_i with probability λ_i:

$$\mathbb{P}\left\{Z = z_i\right\} = \lambda_i, \quad i = 1, \dots, m.$$

(This is possible by the fact that the weights λ_i are non-negative and sum to 1.) Then

$$\mathbb{E}\, Z = \sum_{i=1}^{m} \lambda_i z_i = x.$$

Consider independent copies $Z_1, Z_2, \dots$ of Z. By the strong law of large numbers,

$$\frac{1}{k}\sum_{j=1}^{k} Z_j \to x \quad \text{almost surely as } k \to \infty.$$

To get a quantitative form of this result, let us compute the variance of $\frac{1}{k}\sum_{j=1}^{k} Z_j$. (Incidentally, this computation is at the heart of the proof of the weak law of large numbers.) We obtain

$$\begin{aligned}
\mathbb{E}\left\| x - \frac{1}{k}\sum_{j=1}^{k} Z_j \right\|_2^2 &= \frac{1}{k^2}\,\mathbb{E}\left\| \sum_{j=1}^{k} (Z_j - x) \right\|_2^2 \quad (\text{since } \mathbb{E}(Z_i - x) = 0) \\
&= \frac{1}{k^2}\sum_{j=1}^{k} \mathbb{E}\,\|Z_j - x\|_2^2.
\end{aligned}$$

[1] The diameter of T is defined as $\mathrm{diam}(T) = \sup\{\|s - t\|_2 : s, t \in T\}$. We have assumed that $\mathrm{diam}(T) = 1$ for simplicity. For a general set T, the bound in the theorem changes to $\mathrm{diam}(T)/\sqrt{k}$. Check this!

The last identity is just a higher-dimensional version of the basic fact that the variance of a sum of independent random variables equals the sum of the variances; see Exercise 0.0.3 below.

It remains to bound the variances of the terms. We have

$$\begin{aligned} \mathbb{E}\,\|Z_j - x\|_2^2 &= \mathbb{E}\,\|Z - \mathbb{E}\,Z\|_2^2 \\ &= \mathbb{E}\,\|Z\|_2^2 - \|\mathbb{E}\,Z\|_2^2 \quad \text{(another variance identity; see Exercise 0.0.3)} \\ &\leq \mathbb{E}\,\|Z\|_2^2 \leq 1 \quad \text{(since } Z \in T \text{ and using (0.2)).} \end{aligned}$$

We have shown that

$$\mathbb{E}\left\| x - \frac{1}{k}\sum_{j=1}^{k} Z_j \right\|_2^2 \leq \frac{1}{k}.$$

Therefore, there exists a realization of the random variables $Z_1, \ldots, Z_k$ such that

$$\left\| x - \frac{1}{k}\sum_{j=1}^{k} Z_j \right\|_2^2 \leq \frac{1}{k}.$$

Since by construction each Z_j takes values in T, the proof is complete. ■

Exercise 0.0.3 ☕☕ Check the following variance identities, which we used in the proof of Theorem 0.0.2.

(a) Let $Z_1, \ldots, Z_k$ be independent mean-zero random vectors in $\mathbb{R}^n$. Show that

$$\mathbb{E}\left\| \sum_{j=1}^{k} Z_j \right\|_2^2 = \sum_{j=1}^{k} \mathbb{E}\,\|Z_j\|_2^2.$$

(b) Let Z be a random vector in $\mathbb{R}^n$. Show that

$$\mathbb{E}\,\|Z - \mathbb{E}\,Z\|_2^2 = \mathbb{E}\,\|Z\|_2^2 - \|\mathbb{E}\,Z\|_2^2.$$

Let us give one application of Theorem 0.0.2 in computational geometry. Suppose that we are given a subset $P \subset \mathbb{R}^n$ and asked to cover it by balls of a given radius ε; see Figure 0.2. What is the smallest number of balls needed, and how should we place them?

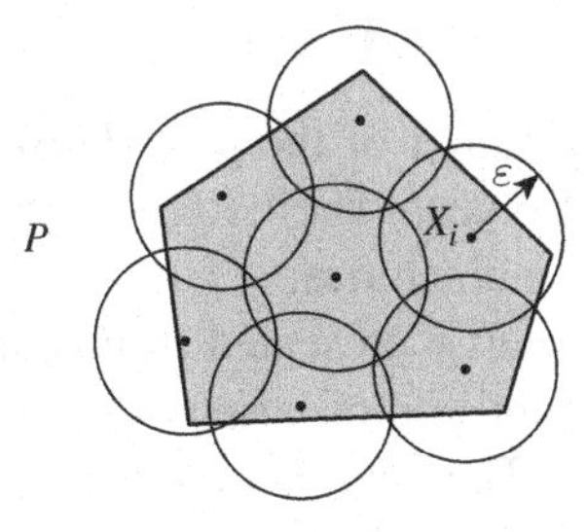

Figure 0.2 The covering problem asks how many balls of radius ε are needed to cover a given set P in $\mathbb{R}^n$ and where to place these balls.

Corollary 0.0.4 (Covering polytopes by balls) *Let P be a polytope in $\mathbb{R}^n$ with N vertices and whose diameter is bounded by 1. Then P can be covered by at most $N^{\lceil 1/\varepsilon^2 \rceil}$ Euclidean balls of radii $\varepsilon > 0$.*

Proof Let us define the centers of the balls as follows. Let $k := \lceil 1/\varepsilon^2 \rceil$ and consider the set

$$\mathcal{N} := \left\{ \frac{1}{k} \sum_{j=1}^{k} x_j : x_j \text{ are vertices of } P \right\}.$$

We claim that the family of ε-balls centered at $\mathcal{N}$ satisfies the conclusion of the corollary. To check this, note that the polytope P is the convex hull of the set of its vertices, which we denote by T. Thus we can apply Theorem 0.0.2 to any point $x \in P = \text{conv}(T)$ and deduce that x is within a distance $1/\sqrt{k} \leq \varepsilon$ from some point in $\mathcal{N}$. This shows that the ε-balls centered at $\mathcal{N}$ do indeed cover P.

To bound the cardinality of $\mathcal{N}$, note that there are N^k ways to choose k out of N vertices with repetition. Thus $|\mathcal{N}| \leq N^k = N^{\lceil 1/\varepsilon^2 \rceil}$. The proof is complete. ■

In this book we will learn several other approaches to the covering problem in relation to packing (Section 4.2), entropy and coding (Section 4.3), and random processes (Chapters 7 and 8).

To finish this section, let us show how to slightly improve Corollary 0.0.4.

Exercise 0.0.5 (The sum of binomial coefficients) ☕☕ Prove the inequalities

$$\left(\frac{n}{m}\right)^m \leq \binom{n}{m} \leq \sum_{k=0}^{m} \binom{n}{k} \leq \left(\frac{en}{m}\right)^m$$

for all integers $m \in [1, n]$. ☞

Exercise 0.0.6 (Improved covering) ☕☕ Check that, in Corollary 0.0.4,

$$(C + C\varepsilon^2 N)^{\lceil 1/\varepsilon^2 \rceil}$$

Euclidean balls suffice. Here C is a suitable absolute constant. (Note that this bound is slightly stronger than $N^{\lceil 1/\varepsilon^2 \rceil}$ for small ε.) ☞

0.1 Notes

In this appetizer we gave an illustration of the *probabilistic method*, where one employs randomness to construct a useful object. The book [8] presents many illustrations of the probabilistic method, mainly in combinatorics.

The empirical method of B. Maurey presented in this section was originally proposed in [162]. B. Carl used it to get bounds on covering numbers [48] including those stated in Corollary 0.0.4 and Exercise 0.0.6. The bound in Exercise 0.0.6 is sharp [48, 49].

1

Preliminaries on Random Variables

In this chapter we recall some basic concepts and results of probability theory. The reader should already be familiar with most of this material, which is routinely taught in introductory probability courses.

Expectation, variance, and moments of random variables are introduced in Section 1.1. Some classical inequalities can be found in Section 1.2. The two fundamental limit theorems of probability – the law of large numbers and the central limit theorem – are recalled in Section 1.3.

1.1 Basic Quantities Associated with Random Variables

In basic courses in probability theory, one learns about the two most important quantities associated with a random variable X, namely the *expectation*[1] (also called the *mean*) and *variance*. They will be denoted in this book by[2]

$$\mathbb{E}\, X \quad \text{and} \quad \mathrm{Var}(X) = \mathbb{E}(X - \mathbb{E}\, X)^2.$$

Let us recall some other classical quantities and functions that describe probability distributions. The *moment generating function of* X is defined as

$$M_X(t) = \mathbb{E}\, e^{tX}, \quad t \in \mathbb{R}.$$

For $p > 0$, the *pth moment* of X is defined as $\mathbb{E}\, X^p$, and the *pth absolute moment* is $\mathbb{E}\,|X|^p$.

It is useful to take the pth root of the moments, which leads to the notion of the L^p *norm* of a random variable:

$$\|X\|_{L^p} = (\mathbb{E}\,|X|^p)^{1/p}, \quad p \in (0, \infty).$$

This definition can be extended to $p = \infty$ by the essential supremum of $|X|$:

$$\|X\|_{L^\infty} = \operatorname{ess\,sup} |X|.$$

For fixed p and a given probability space $(\Omega, \Sigma, \mathbb{P})$, the classical vector space $L^p = L^p(\Omega, \Sigma, \mathbb{P})$ consists of all random variables X on Ω with finite L^p norm, that is,

[1] If you have studied measure theory, you will recall that the expectation $\mathbb{E}\, X$ of a random variable X on a probability space $(\Omega, \Sigma, \mathbb{P})$ is, by definition, the Lebesgue integral of the function $X : \Omega \to \mathbb{R}$. This makes all theorems on Lebesgue integration applicable in probability theory for expectations of random variables.

[2] Throughout this book, we omit brackets and simply write $\mathbb{E}\, f(X)$. Thus, nonlinear functions bind before an expectation.

$$L^p = \{X \colon \|X\|_{L^p} < \infty\}.$$

If $p \in [1, \infty]$, the quantity $\|X\|_{L^p}$ is a norm and L^p is a *Banach space*. This fact follows from Minkowski's inequality, which we recall in (1.4). For $p < 1$, the triangle inequality fails and $\|X\|_{L^p}$ is not a norm.

The exponent $p = 2$ is special in that L^2 is not only a Banach space but also a *Hilbert space*. The inner product and the corresponding norm on L^2 are given by

$$\langle X, Y \rangle_{L^2} = \mathbb{E}\, XY, \quad \|X\|_{L^2} = (\mathbb{E}\,|X|^2)^{1/2}. \tag{1.1}$$

Then the *standard deviation* of X can be expressed as

$$\|X - \mathbb{E}\, X\|_{L^2} = \sqrt{\operatorname{Var}(X)} = \sigma(X).$$

Similarly, we can express the *covariance* of random variables X and Y as

$$\operatorname{cov}(X, Y) = \mathbb{E}((X - \mathbb{E}\, X)(Y - \mathbb{E}\, Y)) = \langle X - \mathbb{E}\, X, Y - \mathbb{E}\, Y \rangle_{L^2}. \tag{1.2}$$

Remark 1.1.1 (Geometry of random variables) When we consider random variables as vectors in the Hilbert space L^2, the identity (1.2) gives a *geometric interpretation* of the notion of covariance: the more the vectors $X - \mathbb{E}\, X$ and $Y - \mathbb{E}\, Y$ are aligned with each other, the larger are their inner product and covariance.

1.2 Some Classical Inequalities

Jensen's inequality states that for any random variable X and a *convex*[3] function $\varphi \colon \mathbb{R} \to \mathbb{R}$, we have

$$\varphi(\mathbb{E}\, X) \leq \mathbb{E}\, \varphi(X).$$

As a simple consequence of Jensen's inequality, $\|X\|_{L^p}$ is an *increasing function in* p, that is

$$\|X\|_{L^p} \leq \|X\|_{L^q} \quad \text{for any } 0 \leq p \leq q = \infty. \tag{1.3}$$

This inequality follows since $\phi(x) = x^{q/p}$ is a convex function if $q/p \geq 1$.

Minkowski's inequality states that for any $p \in [1, \infty]$ and any random variables $X, Y \in L^p$, we have

$$\|X + Y\|_{L^p} \leq \|X\|_{L^p} + \|Y\|_{L^p}. \tag{1.4}$$

This can be viewed as the *triangle inequality*, which implies that $\|\cdot\|_{L^p}$ is a norm when $p \in [1, \infty]$.

The *Cauchy–Schwarz inequality* states that, for any random variables $X, Y \in L^2$, we have

$$|\mathbb{E}\, XY| \leq \|X\|_{L^2}\, \|Y\|_{L^2}.$$

The more general *Hölder's inequality* states that if $p, q \in (1, \infty)$ are conjugate exponents, that is, $1/p + 1/q = 1$, then the random variables $X \in L^p$ and $Y \in L^q$ satisfy

[3] By definition, a function φ is *convex* if $\varphi(\lambda x + (1 - \lambda)y) \leq \lambda\varphi(x) + (1 - \lambda)\varphi(y)$ for all $\lambda \in [0, 1]$ and all vectors x, y in the domain of φ.

$$|\mathbb{E} XY| \le \|X\|_{L^p} \|Y\|_{L^q}.$$

This inequality also holds for the pair $p = 1, q = \infty$.

As we recall from basic probability concepts, the *distribution* of a random variable X is, intuitively, the information about what values X takes with what probabilities. More rigorously, the distribution of X is determined by the *cumulative distribution function* (CDF) of X, defined as

$$F_X(t) = \mathbb{P}\left\{X \le t\right\}, \quad t \in \mathbb{R}.$$

It is often more convenient to work with the *tails* of random variables, namely with

$$\mathbb{P}\left\{X > t\right\} = 1 - F_X(t).$$

There is an important connection between the tails and the expectation (and more generally, the moments) of a random variable. The following identity is typically used to bound the expectation by the tails.

Lemma 1.2.1 (Integral identity) *Let X be a non-negative random variable. Then*

$$\mathbb{E} X = \int_0^\infty \mathbb{P}\left\{X > t\right\} dt.$$

The two sides of this identity are either finite or infinite simultaneously.

Proof We can represent any non-negative real number x via the identity[4]

$$x = \int_0^x 1\, dt = \int_0^\infty \mathbf{1}_{\{t<x\}}\, dt.$$

Substitute the random variable X for x and take expectation of both sides. This gives

$$\mathbb{E} X = \mathbb{E} \int_0^\infty \mathbf{1}_{\{t<X\}}\, dt = \int_0^\infty \mathbb{E}\, \mathbf{1}_{\{t<X\}}\, dt = \int_0^\infty \mathbb{P}\left\{t < X\right\} dt.$$

To change the order of expectation and integration in the second equality, we used the Fubini–Tonelli theorem. The proof is complete. ■

Exercise 1.2.2 (Generalization of integral identity)☕ Prove the following extension of Lemma 1.2.1, which is valid for any random variable X (not necessarily non-negative):

$$\mathbb{E} X = \int_0^\infty \mathbb{P}\left\{X > t\right\} dt - \int_{-\infty}^0 \mathbb{P}\left\{X < t\right\} dt.$$

Exercise 1.2.3 (pth moment via the tail)☕ Let X be a random variable and $p \in (0, \infty)$. Show that

$$\mathbb{E}\,|X|^p = \int_0^\infty p t^{p-1}\, \mathbb{P}\left\{|X| > t\right\} dt$$

whenever the right-hand side is finite. ☞

[4] Here and later in this book, $\mathbf{1}_E$ denotes the *indicator* of the event E; it is the function that takes the value 1 if E occurs and 0 otherwise.

Another classical tool, Markov's inequality, can be used to bound the tail in terms of the expectation.

Proposition 1.2.4 (Markov's inequality) *For any non-negative random variable X and $t > 0$, we have*

$$\mathbb{P}\left\{X \geq t\right\} \leq \frac{\mathbb{E}\, X}{t}.$$

Proof Fix $t > 0$. We can represent any real number x via the identity

$$x = x\mathbf{1}_{\{x \geq t\}} + x\mathbf{1}_{\{x < t\}}.$$

Substitute the random variable X for x and take the expectation of both sides. This gives

$$\begin{aligned}\mathbb{E}\, X &= \mathbb{E}\, X\mathbf{1}_{\{X \geq t\}} + \mathbb{E}\, X\mathbf{1}_{\{X < t\}} \\ &\geq \mathbb{E}\, t\mathbf{1}_{\{X \geq t\}} + 0 = t\,\mathbb{P}\left\{X \geq t\right\}.\end{aligned}$$

Dividing both sides by t, we complete the proof. ∎

A well-known consequence of Markov's inequality is Chebyshev's inequality. It offers a better, quadratic, dependence on t and, instead of controlling a one-side tail, it quantifies the *concentration* of X about its mean.

Corollary 1.2.5 (Chebyshev's inequality) *Let X be a random variable with mean μ and variance σ^2. Then, for any $t > 0$, we have*

$$\mathbb{P}\left\{|X - \mu| \geq t\right\} \leq \frac{\sigma^2}{t^2}.$$

Exercise 1.2.6 Deduce Chebyshev's inequality by squaring both sides of the bound $|X - \mu| \geq t$ and applying Markov's inequality.

Remark 1.2.7 In Proposition 2.5.2 we will establish relations among the three basic quantities associated with random variables – the moment generating functions, the L^p norms, and the tails.

1.3 Limit Theorems

The study of *sums of independent random variables* is a core part of classical probability theory. Recall that the identity

$$\operatorname{Var}(X_1 + \cdots + X_N) = \operatorname{Var}(X_1) + \cdots + \operatorname{Var}(X_N)$$

holds for any independent random variables $X_1, \ldots, X_N$. If, furthermore, the X_i each have the same distribution, with mean μ and variance σ^2, then dividing both sides by N we see that

$$\operatorname{Var}\left(\frac{1}{N}\sum_{i=1}^{N} X_i\right) = \frac{\sigma^2}{N}. \tag{1.5}$$

Thus, the variance of the *sample mean* $\frac{1}{N}\sum_{i=1}^{N} X_i$ of the sample $\{X_1, \dots, X_N\}$ shrinks to zero as $N \to \infty$. This indicates that, for large N, we should expect that the sample mean concentrates tightly about its expectation μ. One of the most important results in probability theory – the law of large numbers – states precisely this.

Theorem 1.3.1 (Strong law of large numbers) *Let $X_1, X_2, \dots$ be a sequence of i.i.d. random variables with mean μ. Consider the sum*

$$S_N = X_1 + \cdots X_N.$$

Then, as $N \to \infty$,

$$\frac{S_N}{N} \to \mu \quad \textit{almost surely}.$$

The next result, the central limit theorem, goes one step further. It identifies the limiting distribution of the (properly scaled) sum of the X_i as the *normal* distribution, also called the *Gaussian* distribution. Recall that the *standard normal* distribution, denoted $N(0,1)$, has density

$$f(x) = \frac{1}{\sqrt{2\pi}} e^{-x^2/2}, \quad x \in \mathbb{R}. \tag{1.6}$$

Theorem 1.3.2 (Lindeberg–Lévy central limit theorem) *Let $X_1, X_2, \dots$ be a sequence of i.i.d. random variables with mean μ and variance σ^2. Consider the sum*

$$S_N = X_1 + \cdots + X_N$$

and normalize it to obtain a random variable with zero mean and unit variance as follows:

$$Z_N := \frac{S_N - \mathbb{E}\, S_N}{\sqrt{\operatorname{Var}(S_N)}} = \frac{1}{\sigma\sqrt{N}} \sum_{i=1}^{N} (X_i - \mu).$$

Then, as $N \to \infty$,

$$Z_N \to N(0,1) \quad \textit{in distribution}.$$

Convergence in distribution means that the CDF of the normalized sum converges pointwise to the CDF of the standard normal distribution. We can express this in terms of tails as follows. Thus, for every $t \in \mathbb{R}$ we have

$$\mathbb{P}\left\{Z_N \ge t\right\} \to \mathbb{P}\left\{g \ge t\right\} = \frac{1}{\sqrt{2\pi}} \int_t^{\infty} e^{-x^2/2}\, dx$$

as $N \to \infty$, where $g \sim N(0,1)$ is a standard normal random variable.

Exercise 1.3.3☕ Let $X_1, X_2, \dots$ be a sequence of i.i.d. random variables with mean μ and finite variance. Show that

$$\mathbb{E}\left|\frac{1}{N}\sum_{i=1}^{N} X_i - \mu\right| = O\Big(\frac{1}{\sqrt{N}}\Big) \quad \text{as } N \to \infty.$$

One remarkable special case of the central limit theorem occurs when the X_i are Bernoulli random variables with some fixed parameter $p \in (0, 1)$, denoted

$$X_i \sim \mathrm{Ber}(p).$$

Recall that this means that the X_i take the values 1 and 0 with probabilities p and $1 - p$ respectively; also recall that $\mathbb{E} X_i = p$ and $\mathrm{Var}(X_i) = p(1 - p)$. The sum

$$S_N := X_1 + \cdots + X_N$$

is said to have the *binomial distribution* $\mathrm{Binom}(N, p)$. The central limit theorem (Theorem 1.3.2) yields that, as $N \to \infty$,

$$\frac{S_N - Np}{\sqrt{Np(1-p)}} \to N(0, 1) \quad \text{in distribution.} \tag{1.7}$$

This special case of the central limit theorem is called the *de Moivre–Laplace theorem.*

Now suppose that $X_i \sim \mathrm{Ber}(p_i)$, with parameters p_i that *decay to zero* as $N \to \infty$ so fast that the sum S_N has mean $O(1)$ instead of being proportional to N. The central limit theorem fails in this regime. A different result, which we are about to state, says that S_N still converges but to the *Poisson* instead of the normal distribution.

Recall that a random variable Z has a *Poisson distribution* with parameter λ, denoted

$$Z \sim \mathrm{Pois}(\lambda),$$

if it takes values in $\{0, 1, 2, \dots\}$ with probabilities

$$\mathbb{P}\left\{Z = k\right\} = e^{-\lambda}\frac{\lambda^k}{k!}, \quad k = 0, 1, 2, \dots \tag{1.8}$$

Theorem 1.3.4 (Poisson limit theorem) *Let $X_{N,i}$, $1 \le i \le N$, be independent random variables $X_{N,i} \sim \mathrm{Ber}(p_{N,i})$, and let $S_N = \sum_{i=1}^N X_{N,i}$. Assume that, as $N \to \infty$,*

$$\max_{i \le N} p_{N,i} \to 0 \quad \text{and} \quad \mathbb{E} S_N = \sum_{i=1}^N p_{N,i} \to \lambda < \infty.$$

Then, as $N \to \infty$,

$$S_N \to \mathrm{Pois}(\lambda) \quad \text{in distribution.}$$

1.4 Notes

The material presented in this chapter is included in most graduate probability textbooks. In particular, proofs of the strong law of large numbers (Theorem 1.3.1) and the Lindeberg–Lévy central limit theorem (Theorem 1.3.2) can be found e.g. in [70, Sections 1.7 and 2.4] and [22, Sections 6 and 27].

2

Concentration of Sums of Independent Random Variables

This chapter introduces the reader to the rich topic of concentration inequalities. After motivating the subject in Section 2.1, we prove some basic concentration inequalities: Hoeffding's in Sections 2.2 and 2.6, Chernoff's in Section 2.3, and Bernstein's in Section 2.8. Another goal of this chapter is to introduce two important classes of distributions: sub-gaussian in Section 2.5 and sub-exponential in Section 2.7. These classes form a natural "habitat" in which many results of high-dimensional probability and its applications will be developed. We give two quick applications of concentration inequalities for randomized algorithms in Section 2.2 and random graphs in Section 2.4. Many more applications are given later in the book.

2.1 Why Concentration Inequalities?

Concentration inequalities quantify how a random variable X deviates around its mean μ. They usually take the form of two-sided bounds for the tails of $X - \mu$, such as

$$\mathbb{P}\left\{|X - \mu| > t\right\} \leq \text{something small.}$$

The simplest concentration inequality is Chebyshev's inequality (Corollary 1.2.5). It is very general but often too weak. Let us illustrate this with the example of the binomial distribution.

Question 2.1.1 *Toss a fair coin N times. What is the probability that we get at least $3N/4$ heads?*

Let S_N denote the number of heads. Then

$$\mathbb{E}\, S_N = \frac{N}{2}, \quad \text{Var}(S_N) = \frac{N}{4}.$$

Chebyshev's inequality bounds the probability of getting at least $3N/4$ heads as follows:

$$\mathbb{P}\left\{S_N \geq \frac{3}{4}N\right\} \leq \mathbb{P}\left\{\left|S_N - \frac{N}{2}\right| \geq \frac{N}{4}\right\} \leq \frac{4}{N}. \tag{2.1}$$

So the probability converges to zero at least *linearly* in N.

Is this the right rate of decay, or we should expect something faster? Let us approach the same question using the central limit theorem. To do this, we represent S_N as a sum of independent random variables:

$$S_N = \sum_{i=1}^{N} X_i$$

where the X_i are independent Bernoulli random variables with parameter 1/2, i.e. $\mathbb{P}\{X_i = 0\} = \mathbb{P}\{X_i = 1\} = 1/2$. (These X_i are the indicators of heads.) The De Moivre–Laplace central limit theorem (1.7) states that the distribution of the normalized number of heads,

$$Z_N = \frac{S_N - N/2}{\sqrt{N/4}},$$

converges to the standard normal distribution $N(0, 1)$. Thus we should anticipate that for large N, we have

$$\mathbb{P}\{S_N \geq 3N/4\} = \mathbb{P}\left\{Z_N \geq \sqrt{N/4}\right\} \approx \mathbb{P}\left\{g \geq \sqrt{N/4}\right\} \tag{2.2}$$

where $g \sim N(0, 1)$. To understand how this quantity decays in N, we will now obtain a good bound on the tails of the normal distribution.

Proposition 2.1.2 (Tails of the normal distribution) *Let $g \sim N(0, 1)$. Then, for all $t > 0$, we have*

$$\left(\frac{1}{t} - \frac{1}{t^3}\right) \frac{1}{\sqrt{2\pi}} e^{-t^2/2} \leq \mathbb{P}\{g \geq t\} \leq \frac{1}{t} \frac{1}{\sqrt{2\pi}} e^{-t^2/2}.$$

In particular, for $t \geq 1$ the tail is bounded by the density:

$$\mathbb{P}\{g \geq t\} \leq \frac{1}{\sqrt{2\pi}} e^{-t^2/2}. \tag{2.3}$$

Proof To obtain an upper bound on the tail

$$\mathbb{P}\{g \geq t\} = \frac{1}{\sqrt{2\pi}} \int_t^\infty e^{-x^2/2}\, dx,$$

let us make the change of variables $x = t + y$. This gives

$$\mathbb{P}\{g \geq t\} = \frac{1}{\sqrt{2\pi}} \int_0^\infty e^{-t^2/2}\, e^{-ty}\, e^{-y^2/2}\, dy \leq \frac{1}{\sqrt{2\pi}} e^{-t^2/2} \int_0^\infty e^{-ty} dy,$$

where we have used that $e^{-y^2/2} \leq 1$. Since the last integral equals $1/t$, the desired upper bound on the tail follows.

The lower bound follows from the identity

$$\int_t^\infty (1 - 3x^{-4}) e^{-x^2/2}\, dx = \left(\frac{1}{t} - \frac{1}{t^3}\right) e^{-t^2/2}.$$

This completes the proof. ■

Returning to (2.2), we see that we should expect the probability of having at least $3N/4$ heads to be smaller than

$$\frac{1}{\sqrt{2\pi}} e^{-N/8}. \tag{2.4}$$

This quantity decays to zero *exponentially* fast in N, which is much better than the linear decay in (2.1) that follows from Chebyshev's inequality.

Unfortunately, (2.4) does not follow rigorously from the central limit theorem. Although the approximation by the normal density in (2.2) is valid, the error of approximation cannot be ignored. And, unfortunately, *the error decays too slowly* – even more slowly than linearly in N. This can be seen from the following sharp quantitative version of the central limit theorem.

Theorem 2.1.3 (Berry–Esseen central limit theorem) *In the setting of Theorem 1.3.2, for every N and every $t \in \mathbb{R}$ we have*

$$\left|\mathbb{P}\left\{Z_N \geq t\right\} - \mathbb{P}\left\{g \geq t\right\}\right| \leq \frac{\rho}{\sqrt{N}}.$$

Here $\rho = \mathbb{E}\,|X_1 - \mu|^3/\sigma^3$ and $g \asymp N(0, 1)$.

Thus the approximation error in (2.2) is of order $1/\sqrt{N}$, which ruins the desired exponential decay (2.4).

Can we improve the approximation error involved in using the central limit theorem? In general, no. If N is even then the probability of getting exactly $N/2$ heads is

$$\mathbb{P}\left\{S_N = N/2\right\} = 2^{-N}\binom{N}{N/2} \asymp \frac{1}{\sqrt{N}};$$

the last estimate can be obtained using Stirling's approximation. (Do it!) Hence, $\mathbb{P}\left\{Z_N = 0\right\} \asymp 1/\sqrt{N}$. On the other hand, since the normal distribution is continuous, we have $\mathbb{P}\left\{g = 0\right\} = 0$. Thus the approximation error here has to be of order $1/\sqrt{N}$.

Let us summarize our situation. The central limit theorem offers an approximation of a sum of independent random variables $S_N = X_1 + \cdots + X_N$ by the normal distribution. The normal distribution is especially nice owing to its very light, exponentially decaying, tails. At the same time, the error of approximation in the central limit theorem decays too slowly, even more slowly than linear. This large error is a roadblock toward proving concentration properties for random variables S_N with light, exponentially decaying, tails.

In order to resolve this issue, we will develop alternative, direct, approaches to concentration which bypass the central limit theorem.

Exercise 2.1.4 (Truncated normal distribution)☕ Let $g \sim N(0, 1)$. Show that, for all $t \geq 1$, we have

$$\mathbb{E}\, g^2 \mathbf{1}_{\{g>t\}} = t\, \frac{1}{\sqrt{2\pi}} e^{-t^2/2} + \mathbb{P}\left\{g > t\right\} \leq \left(t + \frac{1}{t}\right) \frac{1}{\sqrt{2\pi}} e^{-t^2/2}.$$ ☞

2.2 Hoeffding's Inequality

We start with a particularly simple concentration inequality, which holds for sums of i.i.d. *symmetric Bernoulli* random variables.

Definition 2.2.1 (Symmetric Bernoulli distribution) A random variable X has a *symmetric Bernoulli* distribution (also called a *Rademacher* distribution) if it takes values -1 and 1 with probabilities $1/2$ each, i.e.,

$$\mathbb{P}\{X=-1\}=\mathbb{P}\{X=1\}=\frac{1}{2}.$$

Clearly, a random variable X has the (usual) Bernoulli distribution with parameter $1/2$ if and only if $Z = 2X - 1$ has a symmetric Bernoulli distribution.

Theorem 2.2.2 (Hoeffding's inequality) *Let $X_1, \ldots, X_N$ be independent symmetric Bernoulli random variables, and let $a = (a_1, \ldots, a_N) \in \mathbb{R}^N$. Then, for any $t \geq 0$, we have*

$$\mathbb{P}\left\{\sum_{i=1}^{N} a_i X_i \geq t\right\} \leq \exp\left(-\frac{t^2}{2\|a\|_2^2}\right).$$

Proof We can assume without loss of generality that $\|a\|_2 = 1$. (Why?)

Let us recall how we deduced Chebyshev's inequality (Corollary 1.2.5): we squared both sides and applied Markov's inequality. Let us do something similar here. But instead of squaring both sides, let us multiply by a fixed parameter $\lambda > 0$ (to be chosen later) and exponentiate. This gives

$$\begin{aligned}\mathbb{P}\left\{\sum_{i=1}^{N} a_i X_i \geq t\right\} &= \mathbb{P}\left\{\exp\left(\lambda\sum_{i=1}^{N} a_i X_i\right) \geq \exp(\lambda t)\right\} \\ &\leq e^{-\lambda t}\, \mathbb{E}\exp\left(\lambda\sum_{i=1}^{N} a_i X_i\right). \end{aligned} \tag{2.5}$$

In the last step we applied Markov's inequality (Proposition 1.2.4).

We have thus reduced the problem to bounding the *moment generating function* (MGF) of the sum $\sum_{i=1}^N a_i X_i$. Recall that the MGF of the sum is the product of the MGFs of the terms; this follows immediately from the independence of the X_i. Thus

$$\mathbb{E}\exp\left(\lambda\sum_{i=1}^{N} a_i X_i\right) = \prod_{i=1}^{N}\mathbb{E}\exp(\lambda a_i X_i). \tag{2.6}$$

Let us fix i. Since X_i takes values -1 and 1 with probabilities $1/2$ each, we have

$$\mathbb{E}\exp(\lambda a_i X_i) = \frac{\exp(\lambda a_i) + \exp(-\lambda a_i)}{2} = \cosh(\lambda a_i).$$

Exercise 2.2.3 (Bounding the hyperbolic cosine)☕ Show that

$$\cosh(x) \leq \exp(x^2/2) \quad \text{for all } x \in \mathbb{R}.$$

☞

This bound shows that

$$\mathbb{E}\exp(\lambda a_i X_i) \leq \exp(\lambda^2 a_i^2/2).$$

Substituting into (2.6) and then into (2.5), we obtain

$$\mathbb{P}\left\{\sum_{i=1}^{N} a_i X_i \geq t\right\} \leq e^{-\lambda t} \prod_{i=1}^{N} \exp(\lambda^2 a_i^2/2) = \exp\Big(-\lambda t + \frac{\lambda^2}{2}\sum_{i=1}^{N} a_i^2\Big)$$
$$= \exp\Big(-\lambda t + \frac{\lambda^2}{2}\Big).$$

In the last identity, we used the assumption that $\|a\|_2 = 1$.

This bound holds for arbitrary $\lambda > 0$. It remains to optimize in λ; the minimum is clearly attained for $\lambda = t$. With this choice, we obtain

$$\mathbb{P}\left\{\sum_{i=1}^{N} a_i X_i \geq t\right\} \leq \exp(-t^2/2).$$

This completes the proof of Hoeffding's inequality. ■

We can view Hoeffding's inequality as a concentration version of the central limit theorem. Indeed, the most that we may expect from a concentration inequality is that the tail of $\sum a_i X_i$ behaves similarly to the tail of the normal distribution. And, for all practical purposes, Hoeffding's tail bound does that. With the normalization $\|a\|_2 = 1$, Hoeffding's inequality provides the tail $e^{-t^2/2}$, which is exactly the same as the bound for the standard normal tail in (2.3). This is good news. We have been able to obtain the same *exponentially light* tails for sums as for the normal distribution, even though the difference of these two distributions is not exponentially small.

Armed with Hoeffding's inequality, we can now return to Question 2.1.1 regarding the bounding of the probability of at least $3N/4$ heads in N tosses of a fair coin. After rescaling from Bernoulli to symmetric Bernoulli, we obtain that this probability is *exponentially small* in N, namely

$$\mathbb{P}\left\{\text{at least } 3N/4 \text{ heads}\right\} \leq \exp(-N/8).$$

(Check this.)

Remark 2.2.4 (Non-asymptotic results) It should be stressed that, unlike the classical limit theorems of probability theory, Hoeffding's inequality is *non-asymptotic* in that it holds for all fixed N as opposed to $N \to \infty$. The larger the value of N, the stronger the inequality becomes. As we will see later, the non-asymptotic nature of concentration inequalities like that of Hoeffding makes them attractive in applications in the data sciences, where N often corresponds to the *sample size*.

We can easily derive a version of Hoeffding's inequality for *two-sided tails* $\mathbb{P}\left\{|S| \geq t\right\}$ where $S = \sum_{i=1}^{N} a_i X_i$. Indeed, applying Hoeffding's inequality for $-X_i$ instead of X_i, we obtain a bound on $\mathbb{P}\left\{-S \geq t\right\}$. Combining the two bounds, we obtain the bound

$$\mathbb{P}\left\{|S| \geq t\right\} = \mathbb{P}\left\{S \geq t\right\} + \mathbb{P}\left\{-S \geq t\right\}.$$

Thus the bound doubles, and we obtain:

Theorem 2.2.5 (Hoeffding's inequality, two-sided) *Let $X_1, \dots, X_N$ be independent symmetric Bernoulli random variables, and let $a = (a_1, \dots, a_N) \in \mathbb{R}^N$. Then, for any $t > 0$, we have*

$$\mathbb{P}\left\{\left|\sum_{i=1}^N a_i X_i\right| \ge t\right\} \le 2\exp\left(-\frac{t^2}{2\|a\|_2^2}\right).$$

Our proof above of Hoeffding's inequality, which is based on bounding the moment generating function, is quite flexible. It applies far beyond the canonical example of the symmetric Bernoulli distribution. For example, the following extension of Hoeffding's inequality is valid for general bounded random variables.

Theorem 2.2.6 (Hoeffding's inequality for general bounded random variables) *Let $X_1, \dots, X_N$ be independent random variables. Assume that $X_i \in [m_i, M_i]$ for every i. Then, for any $t > 0$, we have*

$$\mathbb{P}\left\{\sum_{i=1}^N (X_i - \mathbb{E} X_i) \ge t\right\} \le \exp\left(-\frac{2t^2}{\sum_{i=1}^N (M_i - m_i)^2}\right).$$

Exercise 2.2.7 ☕☕ Prove Theorem 2.2.6, possibly with some absolute constant instead of 2 in the tail.

Exercise 2.2.8 (Boosting randomized algorithms) ☕☕ Imagine we have an algorithm for solving some decision problem (e.g., is a given number p a prime?). Suppose that the algorithm makes a decision at random and returns the correct answer with probability $1/2 + \delta$, for some $\delta > 0$, which is just a bit better than a random guess. To improve the performance, we run the algorithm N times and take the majority vote. Show that, for any $\varepsilon \in (0, 1)$, the answer is correct with probability at least $1 - \varepsilon$, as long as $N \ge (1/2)\delta^{-2}\ln(\varepsilon^{-1})$. ☞

Exercise 2.2.9 (Robust estimation of the mean) ☕☕☕ Suppose that we want to estimate the mean μ of a random variable X from a sample $X_1, \dots, X_N$ drawn independently from the distribution of X. We want an ε-accurate estimate, i.e. one that falls in the interval $(\mu - \varepsilon, \mu + \varepsilon)$.

(a) Show that a sample of size $N = O(\sigma^2/\varepsilon^2)$ is sufficient to compute an ε-accurate estimate with probability at least $3/4$, where $\sigma^2 = \operatorname{Var} X$.[1] ☞
(b) Show that a sample of size $N = O(\log(\delta^{-1})\sigma^2/\varepsilon^2)$ is sufficient to compute an ε-accurate estimate with probability at least $1 - \delta$. ☞

Exercise 2.2.10 (Small ball probabilities) ☕☕ Let $X_1, \dots, X_N$ be *non-negative* independent random variables with continuous distributions. Assume that the densities of X_i are uniformly bounded by 1.

[1] More accurately, this claim means that there exists an absolute constant C such that if $N \ge C\sigma^2/\varepsilon^2$ then $\mathbb{P}\left\{|\hat{\mu} - \mu| \le \varepsilon\right\} \ge 3/4$. Here $\hat{\mu}$ is the sample mean; see the hint.

(a) Show that the MGF of X_i satisfies

$$\mathbb{E}\exp(-tX_i) \leq \frac{1}{t} \quad \text{for all } t > 0.$$

(b) Deduce that, for any $\varepsilon > 0$, we have

$$\mathbb{P}\left\{\sum_{i=1}^{N} X_i \leq \varepsilon N\right\} \leq (e\varepsilon)^N.$$

☞

2.3 Chernoff's Inequality

As we have noted, Hoeffding's inequality is quite sharp for symmetric Bernoulli random variables. But the general form of Hoeffding's inequality (Theorem 2.2.6) is sometimes too conservative and does not give sharp results. This happens, for example, when the X_i are Bernoulli random variables with parameters p_i so small that we expect S_N to have an approximately Poisson distribution according to Theorem 1.3.4. However, Hoeffding's inequality is not sensitive to the magnitudes of the p_i, and the Gaussian tail bound that it gives is very far from the true, Poisson, tail. In this section we study Chernoff's inequality, which is sensitive to the magnitudes of the p_i.

Theorem 2.3.1 (Chernoff's inequality) *Let X_i be independent Bernoulli random variables with parameters p_i. Consider their sum $S_N = \sum_{i=1}^{N} X_i$ and denote its mean by $\mu = \mathbb{E}\, S_N$. Then, for any $t > \mu$, we have*

$$\mathbb{P}\left\{S_N \geq t\right\} \leq e^{-\mu}\left(\frac{e\mu}{t}\right)^t.$$

Proof We will use the same method – based on the moment generating function – as we did in the proof of Hoeffding's inequality, Theorem 2.2.2. We repeat the first steps of that argument, leading to (2.5) and (2.6): multiply both sides of the inequality $S_N \geq t$ by a parameter λ, exponentiate, and then use Markov's inequality and independence. This gives

$$\mathbb{P}\left\{S_N \geq t\right\} \leq e^{-\lambda t} \prod_{i=1}^{N} \mathbb{E}\exp(\lambda X_i). \tag{2.7}$$

It remains to bound the MGF of each Bernoulli random variable X_i separately. Since X_i takes the value 1 with probability p_i and the value 0 with probability $1 - p_i$, we have

$$\mathbb{E}\exp(\lambda X_i) = e^{\lambda} p_i + (1 - p_i) = 1 + (e^{\lambda} - 1)p_i \leq \exp\left((e^{\lambda} - 1)p_i\right).$$

In the last step, we used the numeric inequality $1 + x \leq e^x$. Consequently,

$$\prod_{i=1}^{N} \mathbb{E}\exp(\lambda X_i) \leq \exp\left((e^{\lambda} - 1)\sum_{i=1}^{N} p_i\right) = \exp\left((e^{\lambda} - 1)\mu\right).$$

Substituting this into (2.7), we obtain

$$\mathbb{P}\left\{S_N \geq t\right\} \leq e^{-\lambda t}\exp\left((e^{\lambda} - 1)\mu\right).$$

This bound holds for any $\lambda > 0$. Substituting the value $\lambda = \ln(t/\mu)$, which is positive by the assumption $t > \mu$, and simplifying the expression, we complete the proof. ∎

Exercise 2.3.2 (Chernoff's inequality: lower tails) Modify the proof of Theorem 2.3.1 to obtain the following bound on the lower tail. For any $t < \mu$, we have

$$\mathbb{P}\left\{S_N \le t\right\} \le e^{-\mu}\left(\frac{e\mu}{t}\right)^t.$$

Exercise 2.3.3 (Poisson tails) Let $X \sim \text{Pois}(\lambda)$. Show that, for any $t > \lambda$, we have

$$\mathbb{P}\left\{X \ge t\right\} \le e^{-\lambda}\left(\frac{e\lambda}{t}\right)^t. \tag{2.8}$$

☞

Remark 2.3.4 (Poisson tails) Note that the Poisson tail bound (2.8) is quite sharp. Indeed, the probability mass function (1.8) of $X \sim \text{Pois}(\lambda)$ can be approximated via Stirling's formula $k! \approx \sqrt{2\pi k}(k/e)^k$ as follows:

$$\mathbb{P}\left\{X = k\right\} \approx \frac{1}{\sqrt{2\pi k}}\, e^{-\lambda}\left(\frac{e\lambda}{k}\right)^k. \tag{2.9}$$

So our bound (2.8) on the *entire tail* of X has essentially the same form as the probability of hitting *one value* k (the smallest value) in that tail. The difference between these two quantities is the factor $\sqrt{2\pi k}$, which is negligible since both these quantities are exponentially small in k.

Exercise 2.3.5 (Chernoff's inequality: small deviations) Show that, in the setting of Theorem 2.3.1, for $\delta \in (0, 1]$ we have

$$\mathbb{P}\left\{|S_N - \mu| \ge \delta\mu\right\} \le 2e^{-c\mu\delta^2},$$

where $c > 0$ is an absolute constant. ☞

Exercise 2.3.6 (Poisson distribution near the mean) Let $X \sim \text{Pois}(\lambda)$. Show that for $t \in (0, \lambda]$, we have

$$\mathbb{P}\left\{|X - \lambda| \ge t\right\} \le 2\exp\left(-\frac{ct^2}{\lambda}\right).$$

☞

Remark 2.3.7 (Large and small deviations) Exercises 2.3.3 and 2.3.6 indicate two different behaviors of the tail of the Poisson distribution $\text{Pois}(\lambda)$. In the small-deviation regime, near the mean λ, the tail of $\text{Pois}(\lambda)$ is like that for the normal distribution $N(\lambda, \lambda)$. In the large-deviation regime, far to the right from the mean, the tail is heavier and decays like $(\lambda/t)^t$; see Figure 2.1.

Exercise 2.3.8 (Normal approximation to Poisson) Let $X \sim \text{Pois}(\lambda)$. Show that, as $\lambda \to \infty$, we have

$$\frac{X - \lambda}{\sqrt{\lambda}} \to N(0, 1) \quad \text{in distribution.}$$

☞

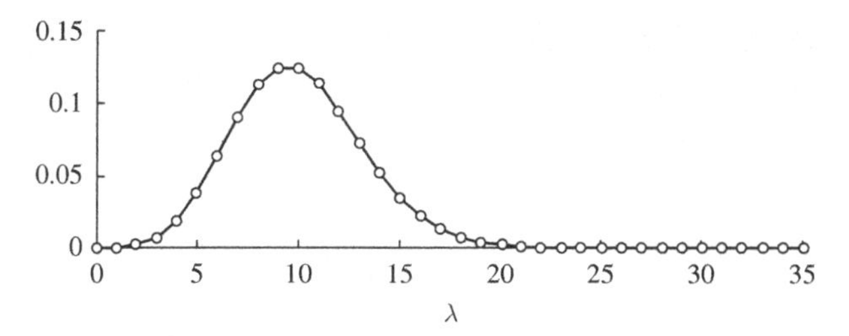

Figure 2.1 The probability mass function of the Poisson distribution Pois(λ) with $\lambda = 10$. The distribution is approximately normal near the mean λ, but to the right of the mean the tail is heavier.

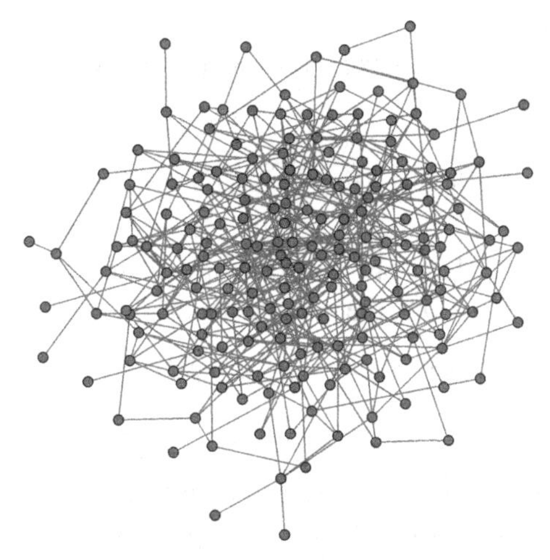

Figure 2.2 A random graph from the Erdös–Rényi model $G(n, p)$ with $n = 200$ and $p = 1/40$.

2.4 Application: Degrees of Random Graphs

We now give an application of Chernoff's inequality to a classical object in probability: *random graphs*.

The most thoroughly studied model of random graphs is the classical *Erdös–Rényi model* $G(n, p)$, which is constructed on a set of n vertices by connecting every pair of distinct vertices independently with probability p. Figure 2.2 shows an example of a random graph $G \sim G(n, p)$. In applications, the Erdös–Rényi model often appears as the simplest stochastic model for large, real-world, *networks*.

The *degree* of a vertex in the graph is the number of edges incident to that vertex. The expected degree of every vertex in $G(n, p)$ clearly equals

$$(n-1)p =: d.$$

(Check!) We will show that relatively *dense graphs*, those where $d \gtrsim \log n$, are almost *regular*, with high probability, which means that the degrees of all vertices approximately equal d.

Proposition 2.4.1 (Dense graphs are almost regular) *There is an absolute constant C such that the following holds. Consider a random graph $G \sim G(n, p)$ with expected degree satisfying $d \geq C \log n$. Then, with high probability (for example, 0.9), the following occurs: all vertices of G have degrees between $0.9d$ and $1.1d$.*

Proof The argument uses a combination of Chernoff's inequality and a *union bound.* Let us fix a vertex i of the graph. The degree of i, which we denote d_i, is a sum of $n - 1$ independent $\text{Ber}(p)$ random variables (the indicators of the edges incident to i). Thus we can apply Chernoff's inequality, which yields

$$\mathbb{P}\left\{|d_i - d| \geq 0.1d\right\} \leq 2e^{-cd}.$$

(Here, we have used the version of Chernoff's inequality given in Exercise 2.3.5.)

This bound holds for each fixed vertex i. Next, we can "unfix" i by taking the union bound over all n vertices. We obtain

$$\mathbb{P}\left\{\exists i \leq n \colon |d_i - d| \geq 0.1d\right\} \leq \sum_{i=1}^{n} \mathbb{P}\left\{|d_i - d| \geq 0.1d\right\} \leq n\, 2e^{-cd}.$$

If $d \geq C \log n$ for a sufficiently large absolute constant C, the probability is bounded by 0.1. This means that, with probability 0.9, the complementary event occurs and we have

$$\mathbb{P}\left\{\forall i \leq n \colon |d_i - d| < 0.1d\right\} \geq 0.9.$$

This completes the proof. ∎

Sparser graphs, those for which $d = o(\log n)$, are no longer almost regular, but there are still useful bounds on their degrees. The following series of exercises makes these claims clear. In all of them we shall assume that the graph size n grows to infinity but we will not assume the connection probability p to be constant in n.

Exercise 2.4.2 (Bounding the degrees of sparse graphs)☕ Consider a random graph $G \sim G(n, p)$ with expected degrees $d = O(\log n)$. Show that with high probability (say, 0.9), all the vertices of G have degree $O(\log n)$. ☞

Exercise 2.4.3 (Bounding the degrees of very sparse graphs)☕☕ Consider a random graph $G \sim G(n, p)$ with expected degrees $d = O(1)$. Show that with high probability (say, 0.9), all the vertices of G have degree

$$O\left(\frac{\log n}{\log \log n}\right).$$

Now we pass to the lower bounds. The next exercise shows that Proposition 2.4.1 does not hold for sparse graphs.

Exercise 2.4.4 (Sparse graphs are not almost regular)☕☕☕ Consider a random graph $G \sim G(n, p)$ with expected degrees $d = o(\log n)$. Show that, with high probability (say, 0.9), G has a vertex with degree $10d$.[2] ☞

[2] We assume here that $10d$ is an integer. There is nothing particular about the factor 10; it could be replaced by any other constant.

Moreover, very sparse graphs, those for which $d = O(1)$, are even farther from regular. The next exercise gives a lower bound on the degrees that matches the upper bound we gave in Exercise 2.4.3.

Exercise 2.4.5 (Very sparse graphs are far from being regular) Consider a random graph $G \sim G(n, p)$ with expected degrees $d = O(1)$. Show that, with high probability (say, 0.9), G has a vertex whose degree is at least of order

$$\frac{\log n}{\log \log n}.$$

2.5 Sub-Gaussian Distributions

So far, we have studied concentration inequalities that apply only for Bernoulli random variables X_i. It would be useful to extend these results to a wider class of distributions. At the very least we may expect that the normal distribution belongs to this class, since we think of concentration results as quantitative versions of the central limit theorem.

So let us ask which random variables X_i must obey a concentration inequality like Hoeffding's in Theorem 2.2.5, namely

$$\mathbb{P}\left\{\left|\sum_{i=1}^{N} a_i X_i\right| \geq t\right\} \leq 2\exp\left(-\frac{ct^2}{\|a\|_2^2}\right).$$

If the sum $\sum_{i=1}^{N} a_i X_i$ consists of a single term X_i, this inequality reads as

$$\mathbb{P}\left\{|X_i| > t\right\} \leq 2e^{-ct^2}.$$

This gives us an automatic restriction: if we want Hoeffding's inequality to hold, we must assume that the random variables X_i have sub-gaussian tails.

This class of distributions, which we call *sub-gaussian*, deserves special attention. It is sufficiently wide as it contains Gaussian, Bernoulli, and all bounded distributions. And, as we will see shortly, concentration results like Hoeffding's inequality can indeed be proved for all sub-gaussian distributions. This makes the family of sub-gaussian distributions a natural, and in many cases canonical, class where one can develop various results in high-dimensional probability theory and its applications.

We now explore several equivalent approaches to sub-gaussian distributions, examining the behavior of their tails, moments, and moment generating functions. To pave our way, let us recall how these quantities behave for the standard normal distribution.

Let $X \sim N(0, 1)$. Then using (2.3) and symmetry, we obtain the following tail bound:

$$\mathbb{P}\left\{|X| \geq t\right\} \leq 2e^{-t^2/2} \quad \text{for all } t \geq 0. \tag{2.10}$$

(Deduce this formally!) In the next exercise we obtain a bound on the absolute moments and L^p norms of the normal distribution.

Exercise 2.5.1 (Moments of the normal distribution) Show that, for each $p \geq 1$, the random variable $X \sim N(0, 1)$ satisfies

$$\|X\|_{L^p} = (\mathbb{E}\,|X|^p)^{1/p} = \sqrt{2}\Big(\frac{\Gamma((1+p)/2)}{\Gamma(1/2)}\Big)^{1/p}.$$

Deduce that

$$\|X\|_{L^p} = O(\sqrt{p}) \quad \text{as } p \to \infty. \tag{2.11}$$

Finally, a classical formula gives the moment generating function of $X \sim N(0, 1)$:

$$\mathbb{E}\exp(\lambda X) = e^{\lambda^2/2} \quad \text{for all } \lambda \in \mathbb{R}. \tag{2.12}$$

2.5.1 Sub-Gaussian Properties

Now let X be a general random variable. The following proposition states that the properties we just considered are equivalent – a sub-gaussian tail decay as in (2.10), the growth of moments as in (2.11), and the growth of the moment generating function as in (2.12). The proof of this result is quite useful; it shows how to transform one type of information about random variables into another.

Proposition 2.5.2 (Sub-gaussian properties) *Let X be a random variable. Then the following properties are equivalent; the parameters $K_i > 0$ appearing in these properties differ from each other by at most an absolute constant factor.*[3]

(i) The tails of X satisfy

$$\mathbb{P}\{|X| \geq t\} \leq 2\exp(-t^2/K_1^2) \quad \text{for all } t \geq 0.$$

(ii) The moments of X satisfy

$$\|X\|_{L^p} = (\mathbb{E}|X|^p)^{1/p} \leq K_2\sqrt{p} \quad \text{for all } p \geq 1.$$

(iii) The MGF of X^2 satisfies

$$\mathbb{E}\exp(\lambda^2 X^2) \leq \exp(K_3^2\lambda^2) \quad \text{for all } \lambda \text{ such that } |\lambda| \leq \frac{1}{K_3}.$$

(iv) The MGF of X^2 is bounded at some point, namely

$$\mathbb{E}\exp(X^2/K_4^2) \leq 2.$$

Moreover, if $\mathbb{E}X = 0$ then properties (i)–(iv) *are also equivalent to the following property.*

(v) The MGF of X satisfies

$$\mathbb{E}\exp(\lambda X) \leq \exp(K_5^2\lambda^2) \quad \text{for all } \lambda \in \mathbb{R}.$$

Proof (i) $\Rightarrow$ (ii) Assume that property (i) holds. By homogeneity and rescaling X to X/K_1 we can assume that $K_1 = 1$. Applying the integral identity (Lemma 1.2.1) for $|X|^p$, we obtain

$$\begin{aligned}
\mathbb{E}|X|^p &= \int_0^\infty \mathbb{P}\{|X|^p \geq u\}\, du \\
&= \int_0^\infty \mathbb{P}\{|X| \geq t\}\, pt^{p-1}\, dt \quad \text{(by change of variables } u = t^p)
\end{aligned}$$

[3] The precise meaning of this equivalence is the following. There exists an absolute constant C such that property i implies property j with parameter $K_j \leq CK_i$ for any two properties $i, j = 1, \ldots, 5$.

$$\leq \int_0^\infty 2e^{-t^2} p t^{p-1}\, dt \quad \text{(by property (i))}$$
$$= p\Gamma(p/2) \quad \text{(set } t^2 = s \text{ and use definition of Gamma function)}$$
$$\leq 3p(p/2)^{p/2} \quad \text{(since } \Gamma(x) \leq 3x^x \text{ for all } x \geq 1/2\text{)}.$$

Taking the pth root yields property (ii) with $K_2 \leq 3$.

(ii) $\Rightarrow$ (iii) Assume that property (ii) holds. As before, by homogeneity we may assume that $K_2 = 1$. Recalling the Taylor series expansion of the exponential function, we obtain

$$\mathbb{E}\exp(\lambda^2 X^2) = \mathbb{E}\left(1 + \sum_{p=1}^{\infty} \frac{(\lambda^2 X^2)^p}{p!}\right) = 1 + \sum_{p=1}^{\infty} \frac{\lambda^{2p}\, \mathbb{E}(X^{2p})}{p!}.$$

Property (ii) guarantees that $\mathbb{E}[X^{2p}] \leq (2p)^p$, while Stirling's approximation yields $p! \geq (p/e)^p$. Substituting these two bounds, we get

$$\mathbb{E}\exp(\lambda^2 X^2) \leq 1 + \sum_{p=1}^{\infty} \frac{(2\lambda^2 p)^p}{(p/e)^p} = \sum_{p=0}^{\infty} (2e\lambda^2)^p = \frac{1}{1 - 2e\lambda^2},$$

provided that $2e\lambda^2 < 1$, in which case the geometric series above converges. To bound this quantity further, we can use the numeric inequality $1/(1-x) \leq e^{2x}$, which is valid for $x \in [0, 1/2]$. It follows that

$$\mathbb{E}\exp(\lambda^2 X^2) \leq \exp(4e\lambda^2) \quad \text{for all } \lambda \text{ satisfying } |\lambda| \leq \frac{1}{2\sqrt{e}}.$$

This yields property (iii) with $K_3 = 2\sqrt{e}$.

(iii) $\Rightarrow$ (iv) is trivial.

(iv) $\Rightarrow$ (i) Assume that property (iv) holds. As before, we may assume that $K_4 = 1$. Then

$$\mathbb{P}\{|X| \geq t\} = \mathbb{P}\{e^{X^2} \geq e^{t^2}\}$$
$$\leq e^{-t^2}\, \mathbb{E}\, e^{X^2} \quad \text{(by Markov's inequality, Proposition 1.2.4)}$$
$$\leq 2e^{-t^2} \quad \text{(by property (iv))}.$$

This proves property (i) with $K_1 = 1$.

To prove the second part of the proposition, we show that (iii) $\Rightarrow$ (v) and (v) $\Rightarrow$ (i).

(iii) $\Rightarrow$ (v) Assume that property (iii) holds; as before we can assume that $K_3 = 1$. Let us use the numeric inequality $e^x \leq x + e^{x^2}$, which is valid for all $x \in \mathbb{R}$. Then

$$\mathbb{E}\, e^{\lambda X} \leq \mathbb{E}\left(\lambda X + e^{\lambda^2 X^2}\right)$$
$$= \mathbb{E}\, e^{\lambda^2 X^2} \quad \text{(since } \mathbb{E}\, X = 0 \text{ by assumption)}$$
$$\leq e^{\lambda^2} \quad \text{if } |\lambda| \leq 1,$$

where in the last line we used property (iii). Thus we have proved property (v) in the range $|\lambda| \leq 1$. Now assume that $|\lambda| \geq 1$. Here we can use the numeric inequality $2\lambda x \leq \lambda^2 + x^2$, which is valid for all λ and x. It follows that

$$\mathbb{E}\, e^{\lambda X} \le e^{\lambda^2/2}\, \mathbb{E}\, e^{X^2/2} \le e^{\lambda^2/2} \exp(1/2) \quad \text{(by property (iii))}$$
$$\le e^{\lambda^2} \quad \text{(since } |\lambda| \ge 1).$$

This proves property (v) with $K_5 = 1$.

(v) $\Rightarrow$ (i) Assume that property (v) holds; we can assume that $K_5 = 1$. We will use some ideas from the proof of Hoeffding's inequality (Theorem 2.2.2). Let $\lambda > 0$ be a parameter to be chosen later. Then

$$\mathbb{P}\{X \ge t\} = \mathbb{P}\{e^{\lambda X} \ge e^{\lambda t}\}$$
$$\le e^{-\lambda t}\, \mathbb{E}\, e^{\lambda X} \quad \text{(by Markov's inequality)}$$
$$\le e^{-\lambda t} e^{\lambda^2} \quad \text{(by property (v))}$$
$$= e^{-\lambda t + \lambda^2}.$$

Optimizing in λ and thus choosing $\lambda = t/2$, we conclude that

$$\mathbb{P}\{X \ge t\} \le e^{-t^2/4}.$$

Repeating this argument for $-X$, we also obtain $\mathbb{P}\{X \le -t\} \le e^{-t^2/4}$. Combining these two bounds we conclude that

$$\mathbb{P}\{|X| \ge t\} \le 2e^{-t^2/4}.$$

Thus property (i) holds with $K_1 = 2$. The proposition is proved. ■

Remark 2.5.3 The constant 2 that appears in some properties in Proposition 2.5.2 does not have any special meaning; it can be replaced by any other absolute constant that is larger than 1. (Check!)

Exercise 2.5.4☕☕ Show that the condition $\mathbb{E}\, X = 0$ is necessary for property (v) to hold.

Exercise 2.5.5 (On property (iii) in Proposition 2.5.2)☕☕

(a) Show that if $X \sim N(0,1)$, the function $\lambda \mapsto \mathbb{E} \exp(\lambda^2 X^2)$ of X^2 is finite only in some bounded neighborhood of zero.
(b) Suppose that some random variable X satisfies $\mathbb{E} \exp(\lambda^2 X^2) \le \exp(K\lambda^2)$ for all $\lambda \in \mathbb{R}$ and some constant K. Show that X is a bounded random variable, i.e. $\|X\|_\infty < \infty$.

2.5.2 Definition and Examples of Sub-Gaussian Distributions

Definition 2.5.6 (Sub-gaussian random variables) A random variable X that satisfies one of the equivalent properties (i)–(iv) in Proposition 2.5.2 is called a *sub-gaussian random variable*. The *sub-gaussian norm* of X, denoted $\|X\|_{\psi_2}$, is defined to be the smallest K_4 in property (iv). In other words, we define

$$\|X\|_{\psi_2} = \inf\left\{t > 0 \colon\ \mathbb{E} \exp(X^2/t^2) \le 2\right\}. \tag{2.13}$$

Exercise 2.5.7☕☕ Check that $\|\cdot\|_{\psi_2}$ is indeed a norm on the space of sub-gaussian random variables.

Let us restate Proposition 2.5.2 in terms of the sub-gaussian norm. It states that every sub-gaussian random variable X satisfies the following bounds:

$$\mathbb{P}\{|X| \geq t\} \leq 2\exp(-ct^2/\|X\|_{\psi_2}^2) \quad \text{for all } t \geq 0; \tag{2.14}$$

$$\|X\|_{L^p} \leq C\|X\|_{\psi_2}\sqrt{p} \quad \text{for all } p \geq 1; \tag{2.15}$$

$$\mathbb{E}\exp(X^2/\|X\|_{\psi_2}^2) \leq 2;$$

$$\text{if } \mathbb{E}X = 0 \text{ then } \mathbb{E}\exp(\lambda X) \leq \exp(C\lambda^2\|X\|_{\psi_2}^2) \quad \text{for all } \lambda \in \mathbb{R}. \tag{2.16}$$

Here $C, c > 0$ are absolute constants. Moreover, up to absolute constant factors, $\|X\|_{\psi_2}$ is the smallest possible number that makes each of these inequalities valid.

Example 2.5.8 Here are some classical examples of sub-gaussian distributions.

(i) **(Gaussian)** As we have already noted, $X \sim N(0,1)$ is a sub-gaussian random variable with $\|X\|_{\psi_2} \leq C$, where C is an absolute constant. More generally, if $X \sim N(0,\sigma^2)$ then X is sub-gaussian with

$$\|X\|_{\psi_2} \leq C\sigma.$$

(Why?)

(ii) **(Bernoulli)** Let X be a random variable with symmetric Bernoulli distribution (recall Definition 2.2.1). Since $|X| = 1$, it follows that X is a sub-gaussian random variable with

$$\|X\|_{\psi_2} = \frac{1}{\sqrt{\ln 2}}.$$

(iii) **(Bounded)** More generally, any bounded random variable X is sub-gaussian with

$$\|X\|_{\psi_2} \leq C\|X\|_{\infty}, \tag{2.17}$$

where $C = 1/\sqrt{\ln 2}$.

Exercise 2.5.9☕ Check that the Poisson, exponential, Pareto, and Cauchy distributions are not sub-gaussian.

Exercise 2.5.10 (Maximum of sub-gaussian)☕☕☕ Let $X_1, X_2, \ldots,$ be an infinite sequence of sub-gaussian random variables which are not necessarily independent. Show that

$$\mathbb{E}\max_i \frac{|X_i|}{\sqrt{1+\log i}} \leq CK,$$

where $K = \max_i \|X_i\|_{\psi_2}$. Deduce that for every $N \geq 2$ we have

$$\mathbb{E}\max_{i\leq N} |X_i| \leq CK\sqrt{\log N}.$$

Exercise 2.5.11 (Lower bound)☕☕ Show that the bound in Exercise 2.5.10 is sharp. Let $X_1, X_2, \ldots, X_N$ be independent $N(0,1)$ random variables. Prove that

$$\mathbb{E}\max_{i\leq N} X_i \geq c\sqrt{\log N}.$$

2.6 General Hoeffding and Khintchine Inequalities

After all the work we did in characterizing sub-gaussian distributions in the previous section, we can now easily extend Hoeffding's inequality (Theorem 2.2.2) to general sub-gaussian distributions. But before we do this, let us deduce an important *rotation invariance* property of sums of independent sub-gaussians.

Recall that a sum of independent normal random variables X_i is normal. Indeed, if the $X_i \sim N(0, \sigma_i^2)$ are independent then

$$\sum_{i=1}^{N} X_i \sim N\Big(0, \sum_{i=1}^{N} \sigma_i^2\Big). \tag{2.18}$$

This fact is a form of the *rotation invariance* property of the normal distribution, which we will recall in Section 3.3.2 in more detail.

The rotation invariance property extends to general sub-gaussian distributions, albeit up to an absolute constant.

Proposition 2.6.1 (Sums of independent sub-gaussians) *Let $X_1, \dots, X_N$ be independent mean-zero sub-gaussian random variables. Then $\sum_{i=1}^N X_i$ is also a sub-gaussian random variable, and*

$$\Big\| \sum_{i=1}^{N} X_i \Big\|_{\psi_2}^2 \le C \sum_{i=1}^{N} \|X_i\|_{\psi_2}^2,$$

where C is an absolute constant.

Proof Let us analyze the moment generating function of the sum. For any $\lambda \in \mathbb{R}$, we have

$$\begin{aligned}
\mathbb{E} \exp\Big(\lambda \sum_{i=1}^{N} X_i\Big) &= \prod_{i=1}^{N} \mathbb{E} \exp(\lambda X_i) \quad \text{(by independence)} \\
&\le \prod_{i=1}^{N} \exp(C\lambda^2 \|X_i\|_{\psi_2}^2) \quad \text{(by sub-gaussian property (2.16))} \\
&= \exp(\lambda^2 K^2) \quad \text{where } K^2 := C \sum_{i=1}^{N} \|X_i\|_{\psi_2}^2.
\end{aligned}$$

To complete the proof, we just need to recall that the bound on the MGF that we have just proved characterizes sub-gaussian distributions. Indeed, the equivalence of properties (v) and (iv) in Proposition 2.5.2 and Definition 2.5.6 implies that the sum $\sum_{i=1}^N X_i$ is sub-gaussian and

$$\Big\| \sum_{i=1}^{N} X_i \Big\|_{\psi_2} \le C_1 K$$

where C_1 is an absolute constant. The proposition is proved. ■

The approximate rotation invariance property can be restated as a concentration inequality via (2.14):

Theorem 2.6.2 (General Hoeffding inequality) *Let $X_1, \dots, X_N$ be independent mean-zero sub-gaussian random variables. Then, for every $t \geq 0$, we have*

$$\mathbb{P}\left\{\left|\sum_{i=1}^{N} X_i\right| \geq t\right\} \leq 2\exp\left(-\frac{ct^2}{\sum_{i=1}^{N}\|X_i\|_{\psi_2}^2}\right).$$

To compare this general result with the specific case for Bernoulli distributions (Theorem 2.2.2), let us apply this result for $a_i X_i$ instead of X_i. We obtain a general form of Theorem 2.2.2 for sub-gaussian random variables.

Theorem 2.6.3 (General Hoeffding inequality) *Let $X_1, \dots, X_N$ be independent mean-zero sub-gaussian random variables, and let $a = (a_1, \dots, a_N) \in \mathbb{R}^N$. Then, for every $t \geq 0$, we have*

$$\mathbb{P}\left\{\left|\sum_{i=1}^{N} a_i X_i\right| \geq t\right\} \leq 2\exp\left(-\frac{ct^2}{K^2\|a\|_2^2}\right)$$

where $K = \max_i \|X_i\|_{\psi_2}$.

Exercise 2.6.4☕ Deduce Hoeffding's inequality for bounded random variables (Theorem 2.2.6) from Theorem 2.6.3, possibly with some other absolute constant instead of 2 in the exponent.

As an application of the general Hoeffding inequality, we can quickly derive the classical Khintchine inequality for the L^p norms of sums of independent random variables.

Exercise 2.6.5 (Khintchine's inequality)☕☕ Let $X_1, \dots, X_N$ be independent sub-gaussian random variables with zero means and unit variances, and let $a = (a_1, \dots, a_N) \in \mathbb{R}^N$. Prove that for every $p \in [2, \infty)$ we have

$$\left(\sum_{i=1}^{N} a_i^2\right)^{1/2} \leq \left\|\sum_{i=1}^{N} a_i X_i\right\|_{L^p} \leq CK\sqrt{p}\left(\sum_{i=1}^{N} a_i^2\right)^{1/2}$$

where $K = \max_i \|X_i\|_{\psi_2}$ and C is an absolute constant.

Exercise 2.6.6 (Khintchine's inequality for $p = 1$)☕☕☕ Show that, in the setting of Exercise 2.6.5, we have

$$c(K)\left(\sum_{i=1}^{N} a_i^2\right)^{1/2} \leq \left\|\sum_{i=1}^{N} a_i X_i\right\|_{L^1} \leq \left(\sum_{i=1}^{N} a_i^2\right)^{1/2}.$$

Here $K = \max_i \|X_i\|_{\psi_2}$ and $c(K) > 0$ is a quantity which may depend only on K. ☞

Exercise 2.6.7 (Khintchine's inequality for $p \in (0, 2)$)☕☕ State and prove a version of Khintchine's inequality for $p \in (0, 2)$. ☞

2.6.1 Centering

In results like Hoeffding's inequality, and in many other results that we will encounter later, we typically assume that the random variables X_i have zero means. If this is not the case, we can always center a variable X_i by subtracting the mean. Let us check that centering does not harm the sub-gaussian property.

First note the following simple centering inequality for the L^2 norm:

$$\|X - \mathbb{E} X\|_{L^2} \leq \|X\|_{L^2}. \tag{2.19}$$

(Check this!) Now let us prove a similar centering inequality for the sub-gaussian norm.

Lemma 2.6.8 (Centering) *If X is a sub-gaussian random variable then $X - \mathbb{E} X$ is sub-gaussian too and*

$$\|X - \mathbb{E} X\|_{\psi_2} \leq C\|X\|_{\psi_2},$$

where C is an absolute constant.

Proof Recall from Exercise 2.5.7 that $\| \cdot \|_{\psi_2}$ is a norm. Thus we can use the triangle inequality and get

$$\|X - \mathbb{E} X\|_{\psi_2} \leq \|X\|_{\psi_2} + \| \mathbb{E} X\|_{\psi_2}. \tag{2.20}$$

We only have to bound the second term. Note that, for any constant random variable a, we trivially have $\|a\|_{\psi_2} \lesssim |a|$ (recall 2.17).[4] Using this for $a = \mathbb{E} X$, we get

$$\begin{aligned}
\| \mathbb{E} X\|_{\psi_2} &\lesssim |\mathbb{E} X| \\
&\leq \mathbb{E} |X| \quad \text{(by Jensen's inequality)} \\
&= \|X\|_1 \\
&\lesssim \|X\|_{\psi_2} \quad \text{(using (2.15) with } p = 1).
\end{aligned}$$

Substituting this into (2.20), we complete the proof. ■

Exercise 2.6.9☕☕☕ Show that, unlike (2.19), the centering inequality in Lemma 2.6.8 does not hold with $C = 1$.

2.7 Sub-Exponential Distributions

The class of sub-gaussian distributions is natural and quite large. Nevertheless, it leaves out some important distributions whose tails are heavier than Gaussian. Here is one example. Consider a standard normal random vector $g = (g_1, \dots, g_N)$ in $\mathbb{R}^N$, whose coordinates g_i are independent $N(0, 1)$ random variables. It is useful in many applications to have a concentration inequality for the Euclidean norm of g, which is

$$\|g\|_2 = \left(\sum_{i=1}^{N} g_i^2 \right)^{1/2}.$$

[4] In this proof and later, the notation $a \lesssim b$ means that $a \leq Cb$ where C is some absolute constant.

Here we find ourselves in a strange situation. On the one hand, $\|g\|_2^2$ is a sum of independent random variables g_i^2, so we should expect some concentration to hold. On the other hand, although the g_i are sub-gaussian random variables, the g_i^2 are not. Indeed, recalling the behavior of Gaussian tails (Proposition 2.1.2) we have[5]

$$\mathbb{P}\left\{g_i^2 > t\right\} = \mathbb{P}\left\{|g| > \sqrt{t}\right\} \sim \exp\left(-(\sqrt{t})^2/2\right) = \exp(-t/2).$$

The tails of g_i^2 are like those for the exponential distribution and are strictly heavier than sub-gaussian. This prevents us from using Hoeffding's inequality (Theorem 2.6.2) if we want to study the concentration of $\|g\|_2$.

In this section we focus on the class of distributions that have at least an exponential tail decay, and in Section 2.8 we prove an analog of Hoeffding's inequality for them.

Our analysis here will be quite similar to what we did for sub-gaussian distributions in Section 2.5. The following is a version of Proposition 2.5.2 for sub-exponential distributions.

Proposition 2.7.1 (Sub-exponential properties) *Let X be a random variable. Then the following properties are equivalent; the parameters $K_i > 0$ appearing in these properties differ from each other by at most an absolute constant factor.*[6]

(i) The tails of X satisfy

$$\mathbb{P}\{|X| \geq t\} \leq 2\exp(-t/K_1) \quad \text{for all } t \geq 0.$$

(ii) The moments of X satisfy

$$\|X\|_{L^p} = (\mathbb{E}\,|X|^p)^{1/p} \leq K_2 p \quad \text{for all } p \geq 1.$$

(iii) The MGF of $|X|$ satisfies

$$\mathbb{E}\exp(\lambda|X|) \leq \exp(K_3\lambda) \quad \text{for all } \lambda \text{ such that } 0 \leq \lambda \leq 1/K_3.$$

(iv) The MGF of $|X|$ is bounded at some point, namely

$$\mathbb{E}\exp(|X|/K_4) \leq 2.$$

Moreover, if $\mathbb{E}\,X = 0$ then properties (i)–(iv) are also equivalent to the following property.

(v) The MGF of X satisfies

$$\mathbb{E}\exp(\lambda X) \leq \exp(K_5^2\lambda^2) \quad \text{for all } \lambda \text{ such that } |\lambda| \leq 1/K_5.$$

Proof We will prove the equivalence of properties (ii) and (v) only; the reader can check the other implications in Exercise 2.7.2.

[5] Here we have ignored the pre-factor $1/t$, which does not have much effect on the exponent.

[6] The precise meaning of this equivalence is the following. There exists an absolute constant C such that property i implies property j with parameter $K_j \leq CK_i$ for any two properties $i, j = 1, 2, 3, 4$.

(ii) $\Rightarrow$ (v) Without loss of generality we may assume that $K_2 = 1$. (Why?) Expanding the exponential function in a Taylor series, we obtain

$$\mathbb{E}\exp(\lambda X) = \mathbb{E}\left(1 + \lambda X + \sum_{p=2}^{\infty} \frac{(\lambda X)^p}{p!}\right) = 1 + \sum_{p=2}^{\infty} \frac{\lambda^p \,\mathbb{E}\, X^p}{p!},$$

where we have used the assumption that $\mathbb{E}\, X = 0$. Property (ii) guarantees that $\mathbb{E}\, X^p \le p^p$, while Stirling's approximation yields $p! \ge (p/e)^p$. Substituting these two bounds, we obtain

$$\mathbb{E}\exp(\lambda X) \le 1 + \sum_{p=2}^{\infty} \frac{(\lambda p)^p}{(p/e)^p} = 1 + \sum_{p=2}^{\infty} (e\lambda)^p = 1 + \frac{(e\lambda)^2}{1 - e\lambda},$$

provided that $|e\lambda| < 1$, in which case the geometric series above converges. Moreover, if $|e\lambda| \le 1/2$ then we can further bound the above quantity by

$$1 + 2e^2\lambda^2 \le \exp(2e^2\lambda^2).$$

Summarizing, we have shown that

$$\mathbb{E}\exp(\lambda X) \le \exp(2e^2\lambda^2) \quad \text{for all } \lambda \text{ satisfying } |\lambda| \le \frac{1}{2e}.$$

This yields property (v) with $K_5 = 2e$.

(v) $\Rightarrow$ (ii) Without loss of generality, we can assume that $K_5 = 1$. We will use the numeric inequality

$$|x|^p \le p^p(e^x + e^{-x}),$$

which is valid for all $x \in \mathbb{R}$ and $p > 0$. (Check it by dividing both sides by p^p and taking pth roots.) Substituting $x = X$ and taking expectations, we get

$$\mathbb{E}\,|X|^p \le p^p\left(\mathbb{E}\, e^X + \mathbb{E}\, e^{-X}\right).$$

Property (v) gives $\mathbb{E}\, e^X \le 1$ and $\mathbb{E}\, e^{-X} \le 1$. Thus

$$\mathbb{E}\,|X|^p \le 2p^p.$$

This yields property (ii) with $K_2 = 2$. ■

Exercise 2.7.2☕☕ Prove the equivalence of properties (i)–(iv) in Proposition 2.7.1 by modifying the proof of Proposition 2.5.2.

Exercise 2.7.3☕☕☕ More generally, consider the class of distributions whose tail decay is of the type $\exp(-ct^\alpha)$ or faster. Here $\alpha = 2$ corresponds to sub-gaussian distributions and $\alpha = 1$ to sub-exponential distributions. State and prove a version of Proposition 2.7.1 for such distributions.

Exercise 2.7.4☕ Argue that the bound in property (iii) in Proposition 2.7.1 cannot be extended to all λ such that $|\lambda| \le 1/K_3$.

Definition 2.7.5 (Sub-exponential random variables) A random variable X that satisfies one of the equivalent properties (i)–(iv) in Proposition 2.7.1 is called a *sub-exponential random variable*. The *sub-exponential norm* of X, denoted $\|X\|_{\psi_1}$, is defined to be the smallest K_3 in property (iii). In other words,

$$\|X\|_{\psi_1} = \inf\{t > 0 : \mathbb{E}\exp(|X|/t) \leq 2\}. \tag{2.21}$$

Sub-gaussian and sub-exponential distributions are closely related. First, any sub-gaussian distribution is clearly sub-exponential. (Why?) Second, the square of a sub-gaussian random variable is sub-exponential:

Lemma 2.7.6 (Sub-exponential is sub-gaussian squared) *A random variable X is sub-gaussian if and only if X^2 is sub-exponential. Moreover,*

$$\|X^2\|_{\psi_1} = \|X\|_{\psi_2}^2.$$

Proof This follows easily from the definition. Indeed, $\|X^2\|_{\psi_1}$ is the infimum of the numbers $K > 0$ satisfying $\mathbb{E}\exp(X^2/K) \leq 2$, while $\|X\|_{\psi_2}$ is the infimum of the numbers $L > 0$ satisfying $\mathbb{E}\exp(X^2/L^2) \leq 2$. So these two become the same definition with $K = L^2$. ■

More generally, the product of two sub-gaussian random variables is sub-exponential:

Lemma 2.7.7 (The product of sub-gaussians is sub-exponential) *Let X and Y be sub-gaussian random variables. Then XY is sub-exponential. Moreover,*

$$\|XY\|_{\psi_1} \leq \|X\|_{\psi_2} \|Y\|_{\psi_2}.$$

Proof Without loss of generality we may assume that $\|X\|_{\psi_2} = \|Y\|_{\psi_2} = 1$. (Why?) The lemma claims that if

$$\mathbb{E}\exp(X^2) \leq 2 \quad \text{and} \quad \mathbb{E}\exp(Y^2) \leq 2 \tag{2.22}$$

then $\mathbb{E}\exp(|XY|) \leq 2$. To prove this, let us use the elementary form of Young's inequality, which states that

$$ab \leq \frac{a^2}{2} + \frac{b^2}{2} \quad \text{for } a, b \in \mathbb{R}.$$

It yields

$$\begin{aligned}
\mathbb{E}\exp(|XY|) &\leq \mathbb{E}\exp\left(\frac{X^2}{2} + \frac{Y^2}{2}\right) \quad \text{(by Young's inequality)} \\
&= \mathbb{E}\left(\exp\left(\frac{X^2}{2}\right)\exp\left(\frac{Y^2}{2}\right)\right) \\
&\leq \frac{1}{2}\mathbb{E}\left(\exp(X^2) + \exp(Y^2)\right) \quad \text{(by Young's inequality)} \\
&= \frac{1}{2}(2 + 2) = 2 \quad \text{(by assumption (2.22))}.
\end{aligned}$$

The proof is complete. ■

Example 2.7.8 Let us mention a few examples of sub-exponential random variables. As we have just learned, all sub-gaussian random variables and their squares are sub-exponential; for example, g^2 for $g \sim N(\mu, \sigma)$. Apart from that, sub-exponential distributions include the exponential and Poisson distributions. Recall that X has an *exponential distribution* with rate $\lambda > 0$, denoted $X \sim \operatorname{Exp}(\lambda)$, if X is a non-negative random variable with tails

$$\mathbb{P}\left\{X \geq t\right\} = e^{-\lambda t} \quad \text{for } t \geq 0.$$

The mean, standard deviation, and sub-exponential norm of X are all of order $1/\lambda$:

$$\mathbb{E}\, X = \frac{1}{\lambda}, \quad \operatorname{Var}(X) = \frac{1}{\lambda^2}, \quad \|X\|_{\psi_1} = \frac{C}{\lambda}.$$

(Check this!)

Remark 2.7.9 (MGF near the origin) You may be surprised to see the same bound on the MGF near the origin for sub-gaussian and sub-exponential distributions. (Compare property (v) in Propositions 2.5.2 and 2.7.1.) This should not be very surprising, though: this kind of local bound is expected from a *general* random variable X with mean zero and unit variance. To see this, assume for simplicity that X is bounded. The MGF of X can be approximated using the first two terms of a Taylor expansion:

$$\mathbb{E} \exp(\lambda X) \approx \mathbb{E}\left(1 + \lambda X + \frac{\lambda^2 X^2}{2} + o(\lambda^2 X^2)\right) = 1 + \frac{\lambda^2}{2} \approx e^{\lambda^2/2}$$

as $\lambda \to 0$. For the standard *normal* distribution $N(0, 1)$, this approximation becomes an equality; see (2.12). For *sub-gaussian* distributions, Proposition 2.5.2 says that a bound like this holds for all λ and that this characterizes sub-gaussian distributions. And, for *sub-exponential* distributions, Proposition 2.7.1 says that this bound holds for small λ and that this characterizes sub-exponential distributions. For larger λ, no general bound can exist for sub-exponential distributions: indeed, for the *exponential* random variable $X \sim \operatorname{Exp}(1)$, the MGF is infinite for $\lambda \geq 1$. (Check this!)

Exercise 2.7.10 (Centering) Prove an analog of the centering lemma 2.6.8 for sub-exponential random variables X:

$$\|X - \mathbb{E}\, X\|_{\psi_1} \leq C\|X\|_{\psi_1}.$$

2.7.1 A More General View: Orlicz Spaces

Sub-gaussian distributions can be introduced within the more general framework of *Orlicz spaces*. A function $\psi\colon [0, \infty) \to [0, \infty)$ is called an *Orlicz function* if ψ is convex, increasing, and satisfies

$$\psi(0) = 0, \quad \psi(x) \to \infty \text{ as } x \to \infty.$$

For a given Orlicz function ψ, the Orlicz norm of a random variable X is defined as

$$\|X\|_{\psi} := \inf\{t > 0\colon\ \mathbb{E}\, \psi(|X|/t) \leq 1\}.$$

The *Orlicz space* $L_\psi = L_\psi(\Omega, \Sigma, \mathbb{P})$ consists of all random variables X on the probability space $(\Omega, \Sigma, \mathbb{P})$ with finite Orlicz norm, i.e.

$$L_\psi := \{X : \|X\|_\psi < \infty\}.$$

Exercise 2.7.11 Show that $\|X\|_\psi$ is indeed a norm on the space L_ψ.

It can also be shown that L_ψ is complete and thus a Banach space.

Example 2.7.12 (L^p space) Consider the function

$$\psi(x) = x^p,$$

which is obviously an Orlicz function for $p \geq 1$. The resulting Orlicz space L_ψ is the classical space L^p.

Example 2.7.13 (L_{ψ_2} space) Consider the function

$$\psi_2(x) := e^{x^2} - 1,$$

which is obviously an Orlicz function. The resulting Orlicz norm is exactly the sub-gaussian norm $\|\cdot\|_{\psi_2}$ that we defined in (2.13). The corresponding Orlicz space L_{ψ_2} consists of all sub-gaussian random variables.

Remark 2.7.14 We can easily locate L_{ψ_2} in the hierarchy of classical L^p spaces:

$$L^\infty \subset L_{\psi_2} \subset L^p \quad \text{for every } p \in [1, \infty).$$

The first inclusion follows from property (ii) of Proposition 2.5.2, and the second inclusion from the bound (2.17). Thus the space of sub-gaussian random variables L_{ψ_2} is smaller than all the L^p spaces, but it is still larger than the space of bounded random variables L^∞.

2.8 Bernstein's Inequality

We are ready to state and prove a concentration inequality for sums of independent sub-exponential random variables.

Theorem 2.8.1 (Bernstein's inequality) *Let $X_1, \ldots, X_N$ be independent mean-zero sub-exponential random variables. Then, for every $t \geq 0$, we have*

$$\mathbb{P}\left\{\left|\sum_{i=1}^N X_i\right| \geq t\right\} \leq 2\exp\left(-c\min\left(\frac{t^2}{\sum_{i=1}^N \|X_i\|_{\psi_1}^2}, \frac{t}{\max_i \|X_i\|_{\psi_1}}\right)\right),$$

where $c > 0$ is an absolute constant.

Proof We begin the proof in the same way as we argued about other concentration inequalities for $S = \sum_{i=1}^N X_i$, e.g. Theorems 2.2.2 and 2.3.1. Multiply both sides of the inequality $S \geq t$ by a parameter λ, exponentiate, and then use Markov's inequality and independence. This leads to the bound (2.7), which is

$$\mathbb{P}\left\{S \geq t\right\} \leq e^{-\lambda t} \prod_{i=1}^{N} \mathbb{E} \exp(\lambda X_i). \tag{2.23}$$

To bound the MGF of each term X_i, we use property (v) in Proposition 2.7.1. It says that if λ is small enough that

$$|\lambda| \leq \frac{c}{\max_i \|X_i\|_{\psi_1}} \tag{2.24}$$

then $\mathbb{E} \exp(\lambda X_i) \leq \exp\left(C\lambda^2 \|X_i\|_{\psi_1}^2\right)$.[7] Substituting this into (2.23), we obtain

$$\mathbb{P}\{S \geq t\} \leq \exp\left(-\lambda t + C\lambda^2\sigma^2\right), \quad \text{where } \sigma^2 = \sum_{i=1}^{N} \|X_i\|_{\psi_1}^2.$$

Now we minimize this expression in λ subject to the constraint (2.24). The optimal choice is

$$\lambda = \min\left(\frac{t}{2C\sigma^2}, \frac{c}{\max_i \|X_i\|_{\psi_1}}\right),$$

for which we obtain

$$\mathbb{P}\{S \geq t\} \leq \exp\left(-\min\left(\frac{t^2}{4C\sigma^2}, \frac{ct}{2\max_i \|X_i\|_{\psi_1}}\right)\right).$$

Repeating this argument for $-X_i$ instead of X_i, we obtain the same bound for $\mathbb{P}\{-S \geq t\}$. A combination of these two bounds completes the proof. ■

To put Theorem 2.8.1 in a more convenient form, let us apply it for $a_i X_i$ instead of X_i.

Theorem 2.8.2 (Bernstein's inequality) *Let $X_1, \ldots, X_N$ be independent, mean-zero sub-exponential random variables, and let $a = (a_1, \ldots, a_N) \in \mathbb{R}^N$. Then, for every $t \geq 0$, we have*

$$\mathbb{P}\left\{\left|\sum_{i=1}^{N} a_i X_i\right| \geq t\right\} \leq 2\exp\left(-c\min\left(\frac{t^2}{K^2\|a\|_2^2}, \frac{t}{K\|a\|_\infty}\right)\right)$$

where $K = \max_i \|X_i\|_{\psi_1}$.

In the special case where $a_i = 1/N$, we obtain a form of Bernstein's inequality for averages:

Corollary 2.8.3 (Bernstein's inequality) *Let $X_1, \ldots, X_N$ be independent mean-zero, sub-exponential random variables. Then, for every $t \geq 0$, we have*

$$\mathbb{P}\left\{\left|\frac{1}{N}\sum_{i=1}^{N} X_i\right| \geq t\right\} \leq 2\exp\left(-c\min\left(\frac{t^2}{K^2}, \frac{t}{K}\right)N\right)$$

where $K = \max_i \|X_i\|_{\psi_1}$.

[7] Recall that by Proposition 2.7.1 and the definition of the sub-exponential norm, property (v) holds for a value of K_5 that is within an absolute constant factor of $\|X\|_{\psi_1}$.

This result can be considered as a quantitative form of the *law of large numbers* for the averages $\frac{1}{N}\sum_{i=1}^N X_i$.

Let us compare Bernstein's inequality (Theorem 2.8.1) with Hoeffding's inequality (Theorem 2.6.2). The obvious difference is that Bernstein's bound has *two tails*, as it would if the sum $S_N = \sum X_i$ were a mixture of sub-gaussian and sub-exponential distributions. The sub-gaussian tail is of course expected from the central limit theorem. But the sub-exponential tails of the terms X_i are too heavy to be able to produce a sub-gaussian tail everywhere, so a sub-exponential tail should be expected, too. In fact, the sub-exponential tail in Theorem 2.8.1 is produced by a *single term* X_i in the sum, the one with the maximal sub-exponential norm. Indeed, this term alone has a tail of magnitude $\exp(-ct/\|X_i\|_{\psi_1})$.

We have already seen a similar mixture of two tails, one for small deviations and the other for large deviations, in our analysis of Chernoff's inequality; see Remark 2.3.7. To put Bernstein's inequality in the same perspective, let us normalize the sum as in the central limit theorem and apply Theorem 2.8.2. We obtain[8]

$$\mathbb{P}\left\{\left|\frac{1}{\sqrt{N}}\sum_{i=1}^N X_i\right| \ge t\right\} \le \begin{cases} 2\exp(-ct^2), & t \le C\sqrt{N}, \\ 2\exp(-t\sqrt{N}), & t \ge C\sqrt{N}. \end{cases}$$

Thus, in the *small-deviation* regime, where $t \le C\sqrt{N}$, we have the same sub-gaussian tail bound as if the sum had a *normal distribution* with constant variance. Note that this domain widens as N increases and the central limit theorem becomes more powerful. For *large deviations*, where $t \ge C\sqrt{N}$, the sum has a heavier, *sub-exponential*, tail bound, which can be due to the contribution of a single term X_i. We illustrate this in Figure 2.3.

Let us mention the following strengthening of Bernstein's inequality under the stronger assumption that the random variables X_i are bounded.

Theorem 2.8.4 (Bernstein's inequality for bounded distributions) *Let $X_1, \dots, X_N$ be independent mean-zero random variables such that $|X_i| \le K$ all i. Then, for every $t \ge 0$, we have*

$$\mathbb{P}\left\{\left|\sum_{i=1}^N X_i\right| \ge t\right\} \le 2\exp\left(-\frac{t^2/2}{\sigma^2 + Kt/3}\right).$$

Here $\sigma^2 = \sum_{i=1}^N \mathbb{E} X_i^2$ is the variance of the sum.

We leave the proof of this theorem to the next two exercises.

Exercise 2.8.5 (A bound on MGF) Let X be a mean-zero random variable such that $|X| \le K$. Prove the following bound on the MGF of X:

$$\mathbb{E}\exp(\lambda X) \le \exp(g(\lambda)\,\mathbb{E} X^2) \quad \text{where } g(\lambda) = \frac{\lambda^2/2}{1 - |\lambda|K/3},$$

provided that $|\lambda| < 3/K$. ☞

[8] For simplicity, we have suppressed the dependence on K here by allowing the constants c, C to depend on K.

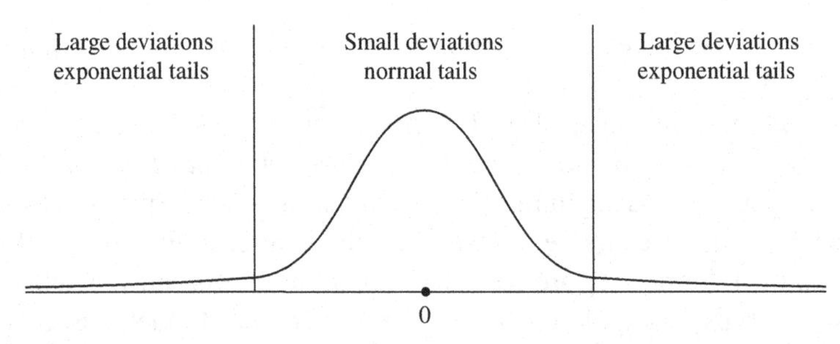

Figure 2.3 Bernstein's inequality for a sum of sub-exponential random variables gives a mixture of two tails: sub-gaussian for small deviations and sub-exponential for large deviations.

Exercise 2.8.6 Deduce Theorem 2.8.4 from the bound in Exercise 2.8.5.

2.9 Notes

The topic of concentration inequalities is very wide, and we will continue to examine it in Chapter 5. We refer the reader to [8, Appendix A], [148, Chapter 4], [126], [29], [76, Chapter 7], [11, Section 3.5.4], [170, Chapter 1], and [14, Chapter 4] for various versions of Hoeffding's, Chernoff's, and Bernstein's inequalities and related results.

Proposition 2.1.2 on the tails of the normal distribution is taken from [70, Theorem 1.4]. The proof of the Berry–Esseen central limit theorem (Theorem 2.1.3) with an extra factor 3 on the right-hand side can be found e.g. in [70, Section 2.4.d]; the best currently known factor is ≈ 0.47 [116].

It is worthwhile mentioning two important concentration inequalities that were omitted in this chapter. One is the *bounded differences inequality*, also called *McDiarmid's inequality*, which works not only for sums but for general functions of independent random variables. It is a generalization of Hoeffding's inequality (Theorem 2.2.6).

Theorem 2.9.1 (Bounded differences inequality) *Let $X_1, \dots, X_N$ be independent random variables.*[9] *Let $f \colon \mathbb{R}^n \to \mathbb{R}$ be a measurable function. Assume that the value of $f(x)$ can change by at most $c_i > 0$ under an arbitrary change*[10] *of a single coordinate of $x \in \mathbb{R}^n$. Then, for any $t > 0$, we have*

$$\mathbb{P}\left\{f(X) - \mathbb{E}\, f(X) \geq t\right\} \leq \exp\left(-\frac{2t^2}{\sum_{i=1}^N c_i^2}\right)$$

where $X = (X_1, \dots, X_n)$.

Another result worth mentioning is *Bennett's inequality*, which can be regarded as a generalization of Chernoff's inequality.

[9] The theorem remains valid if the random variables X_i take values in an abstract set $\mathcal{X}$ and $f \colon \mathcal{X} \to \mathbb{R}$.

[10] This means that for any index i and any $x_1, \dots, x_n, x_i'$, we have
$|f(x_1, \dots, x_{i-1}, x_i, x_{i+1}, \dots, x_n) - f(x_1, \dots, x_{i-1}, x_i', x_{i+1}, \dots, x_n)| \leq c_i$.

Theorem 2.9.2 (Bennett's inequality) *Let $X_1, \dots, X_N$ be independent random variables. Assume that $|X_i - \mathbb{E} X_i| \le K$ almost surely for every i. Then, for any $t > 0$, we have*

$$\mathbb{P}\left\{\sum_{i=1}^{N}(X_i - \mathbb{E} X_i) \ge t\right\} \le \exp\Big(-\frac{\sigma^2}{K^2} h\Big(\frac{Kt}{\sigma^2}\Big)\Big)$$

where $\sigma^2 = \sum_{i=1}^N \mathrm{Var}(X_i)$ is the variance of the sum, and $h(u) = (1+u)\log(1+u) - u$.

In the small-deviation regime, where $u := Kt/\sigma^2 \ll 1$, we have asymptotically $h(u) \approx u^2$ and Bennett's inequality gives approximately the Gaussian tail bound $\approx \exp(-t^2/\sigma^2)$. In the large-deviation regime, say where $u \gg Kt/\sigma^2 \ge 2$, we have $h(u) \ge \frac{1}{2} u \log u$, and Bennett's inequality gives a Poisson-like tail $(\sigma^2/Kt)^{t/2K}$.

Both the bounded differences inequality and Bennett's inequality can be proved by the same general method as Hoeffding's inequality (Theorem 2.2.2) and Chernoff's inequality (Theorem 2.3.1), namely by bounding the moment generating function of the sum. This method was pioneered by Sergei Bernstein in the 1920s and 1930s. Our presentation of Chernoff's inequality in Section 2.3 mostly follows [148, Chapter 4].

Section 2.4 scratches the surface of the rich theory of *random graphs*. The books [25, 105] offer a comprehensive introduction to the random graph theory.

The presentation in Sections 2.5–2.8 mostly follows [216]; see [76, Chapter 7] for some more elaborate results. For sharp versions of Khintchine's inequalities in Exercises 2.6.5–2.6.7 and related results, see e.g. [189, 93, 114, 151].

3

Random Vectors in High Dimensions

In this chapter we study the distributions of random vectors $X = (X_1, \dots, X_n) \in \mathbb{R}^n$, where the dimension n is typically very large. Examples of high-dimensional distributions abound in data science. For instance, computational biologists study the expressions of $n \sim 10^4$ genes in the human genome, which can be modeled as a random vector $X = (X_1, \dots, X_n)$ that encodes the gene expressions of a person randomly drawn from a given population.

Life in high dimensions presents new challenges, which stem from the fact that there is *exponentially more room* in higher dimensions than in lower dimensions. For example, in $\mathbb{R}^n$ the volume of a cube of side 2 is 2^n times larger than the volume of a unit cube, even though the sides of the cubes are just a factor 2 apart (see Figure 3.1). The abundance of room in higher dimensions makes many algorithmic tasks exponentially more difficult, a phenomenon known as the *curse of dimensionality*.

Probability in high dimensions offers an array of tools to circumvent these difficulties; some examples will be given in this chapter. We start by examining the Euclidean norm $\|X\|_2$ of a random vector X with independent coordinates, and we show in Section 3.1 that the norm concentrates tightly about its mean. Further basic results and examples of high-dimensional distributions (multivariate normal, spherical, Bernoulli, frames, etc.) are covered in Section 3.2, in which we also discuss principal component analysis, a powerful data exploratory procedure.

In Section 3.5 we give a probabilistic proof of the classical inequality of Grothendieck and an application to semidefinite optimization. We show that one can sometimes relax hard optimization problems to tractable, semidefinite, programs and use Grothendieck's inequality to analyze the quality of such relaxations. In Section 3.6 we give a remarkable example of a semidefinite relaxation of a hard optimization problem – finding the maximum cut of a given graph. We present there the classical Goemans–Williamson randomized

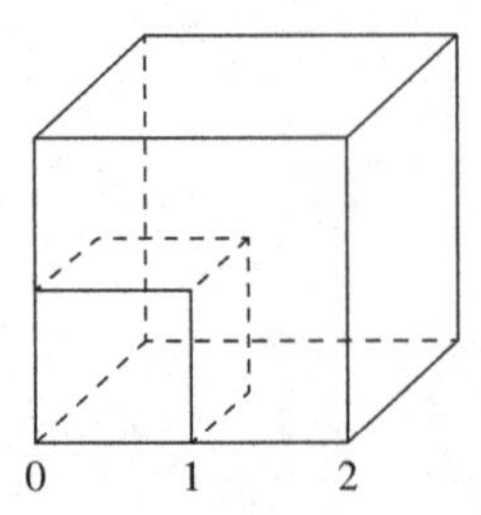

Figure 3.1 The abundance of room in high dimensions: the larger cube has volume exponentially larger than the smaller cube.

approximation algorithm for the maximum-cut problem. In Section 3.7 we give an alternative proof of Grothendieck's inequality (that leads to almost the best constant) by introducing the kernel trick, a method that has significant applications in machine learning.

3.1 Concentration of the Norm

Where in the space $\mathbb{R}^n$ is a random vector $X = (X_1, \dots, X_n)$ likely to be located? Assume that the coordinates X_i are independent random variables with zero means and unit variances. What length do we expect X to have? We have

$$\mathbb{E}\,\|X\|_2^2 = \mathbb{E}\sum_{i=1}^n X_i^2 = \sum_{i=1}^n \mathbb{E}\,X_i^2 = n,$$

so we should expect the length of X to be

$$\|X\|_2 \approx \sqrt{n}.$$

We will see now that X is indeed very close to $\sqrt{n}$ with high probability.

Theorem 3.1.1 (Concentration of the norm) *Let $X = (X_1, \dots, X_n) \in \mathbb{R}^n$ be a random vector with independent sub-gaussian coordinates X_i that satisfy $\mathbb{E}\,X_i^2 = 1$. Then*

$$\Big\| \|X\|_2 - \sqrt{n} \Big\|_{\psi_2} \le CK^2,$$

where $K = \max_i \|X_i\|_{\psi_2}$ and C is an absolute constant.[1]

Proof For simplicity, we assume that $K \ge 1$. (Argue that one can make this assumption.) We shall apply Bernstein's deviation inequality for the normalized sum of independent mean-zero random variables

$$\frac{1}{n}\|X\|_2^2 - 1 = \frac{1}{n}\sum_{i=1}^n (X_i^2 - 1).$$

Since the random variable X_i is sub-gaussian, $X_i^2 - 1$ is sub-exponential, and, more precisely,

$$\begin{aligned} \|X_i^2 - 1\|_{\psi_1} &\le C\|X_i^2\|_{\psi_1} \quad \text{(by centering, see Exercise 2.7.10)} \\ &= C\|X_i\|_{\psi_2}^2 \quad \text{(by Lemma 2.7.6)} \\ &\le CK^2. \end{aligned}$$

Applying Bernstein's inequality (Corollary 2.8.3), we obtain for any $u \ge 0$ that

$$\mathbb{P}\left\{ \left| \frac{1}{n}\|X\|_2^2 - 1 \right| \ge u \right\} \le 2\exp\Big(-\frac{cn}{K^4}\min(u^2, u)\Big). \tag{3.1}$$

(Here we have used $K^4 \ge K^2$ since we assumed that $K \ge 1$.)

[1] From now on, we will always denote various positive absolute constants by C, c, C_1, c_1 without mentioning this explicitly.

This is a good concentration inequality for $\|X\|_2^2$, from which we are going to deduce a concentration inequality for $\|X\|_2$. To make the link, we can use the following elementary observation, which is valid for all numbers $z \geq 0$:

$$|z-1| \geq \delta \quad \text{implies} \quad |z^2-1| \geq \max(\delta, \delta^2). \tag{3.2}$$

(Check it!) We obtain for any $\delta \geq 0$ that

$$\begin{aligned}\mathbb{P}\left\{\left|\frac{1}{\sqrt{n}}\|X\|_2 - 1\right| \geq \delta\right\} &\leq \mathbb{P}\left\{\left|\frac{1}{n}\|X\|_2^2 - 1\right| \geq \max(\delta, \delta^2)\right\} \quad \text{(by (3.2))} \\ &\leq 2\exp\left(-\frac{cn}{K^4}\delta^2\right) \quad \text{(by (3.1) for } u = \max(\delta, \delta^2)).\end{aligned}$$

Changing variables to $t = \delta\sqrt{n}$, we obtain the desired sub-gaussian tail

$$\mathbb{P}\left\{\left|\|X\|_2 - \sqrt{n}\right| \geq t\right\} \leq 2\exp\left(-\frac{ct^2}{K^4}\right) \quad \text{for all } t \geq 0. \tag{3.3}$$

From Section 2.5.2, this is equivalent to the conclusion of the theorem. ■

Remark 3.1.2 (Deviation) Theorem 3.1.1 states that, with high probability, X takes values very close to the sphere of radius $\sqrt{n}$. In particular, with high probability (say, 0.99), X even stays within a *constant distance* from that sphere. Such small and constant deviations could be surprising at first sight, so let us explain this intuitively. The square of the norm, $S_n := \|X\|_2^2$ has mean n and standard deviation $O(\sqrt{n})$. (Why?) Thus $\|X\|_2 = \sqrt{S_n}$ ought to deviate by $O(1)$ around $\sqrt{n}$. This is so because

$$\sqrt{n \pm O(\sqrt{n})} = \sqrt{n} \pm O(1);$$

see Figure 3.2 for an illustration.

Remark 3.1.3 (Anisotropic distributions) After we develop more tools, we will prove a generalization of Theorem 3.1.1 for *anisotropic* random vectors X; see Theorem 6.3.2.

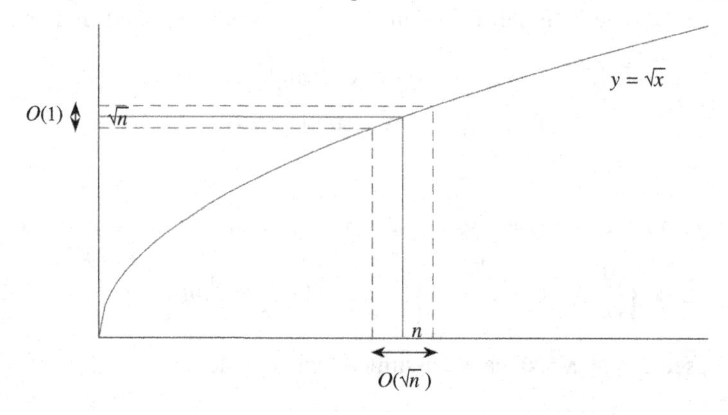

Figure 3.2 Concentration of the norm of a random vector X in $\mathbb{R}^n$. While $\|X\|_2^2$ deviates by $O(\sqrt{n})$ around n, $\|X\|_2$ deviates by $O(1)$ around $\sqrt{n}$.

Exercise 3.1.4 (Expectation of the norm)☕☕☕

(a) Deduce from Theorem 3.1.1 that

$$\sqrt{n} - CK^2 \le \mathbb{E}\,\|X\|_2 \le \sqrt{n} + CK^2.$$

(b) Can CK^2 be replaced by $o(1)$, a quantity that vanishes as $n \to \infty$?

Exercise 3.1.5 (Variance of the norm)☕☕☕ Deduce from Theorem 3.1.1 that

$$\mathrm{Var}(\|X\|_2) \le CK^4.$$

☞

The result of the last exercise actually holds not only for sub-gaussian distributions but for all distributions with bounded fourth moment.

Exercise 3.1.6 (Variance of the norm under finite-moment assumptions)☕☕☕ Let $X = (X_1, \dots, X_n) \in \mathbb{R}^n$ be a random vector with independent coordinates X_i that satisfy $\mathbb{E}\,X_i^2 = 1$ and $\mathbb{E}\,X_i^4 \le K^4$. Show that

$$\mathrm{Var}(\|X\|_2) \le CK^4.$$

☞

Exercise 3.1.7 (Small ball probabilities)☕☕ Let $X = (X_1, \dots, X_n) \in \mathbb{R}^n$ be a random vector with independent coordinates X_i having continuous distributions. Assume that the densities of X_i are uniformly bounded by 1. Show that, for any $\varepsilon > 0$, we have

$$\mathbb{P}\left\{\|X\|_2 \le \varepsilon\sqrt{n}\right\} \le (C\varepsilon)^n.$$

☞

3.2 Covariance Matrices and Principal Component Analysis

In the last section we considered a special class of random variables, those with independent coordinates. Before we study more general situations, let us recall a few basic notions about high-dimensional distributions which the reader may have already seen in basic courses.

The concept of the *mean* of a random variable generalizes in a straightforward way for a random vector X taking values in $\mathbb{R}^n$. The notion of variance is replaced in high dimensions by the *covariance matrix* of a random vector $X \in \mathbb{R}^n$, defined as follows:

$$\mathrm{cov}(X) = \mathbb{E}\left((X - \mu)(X - \mu)^\mathsf{T}\right) = \mathbb{E}\,XX^\mathsf{T} - \mu\mu^\mathsf{T} \quad \text{where } \mu = \mathbb{E}\,X.$$

Thus $\mathrm{cov}(X)$ is an $n \times n$ symmetric positive-semidefinite matrix. The formula for covariance is a direct high-dimensional generalization of the definition of the variance of a random variable Z, which is

$$\mathrm{Var}(Z) = \mathbb{E}(Z - \mu)^2 = \mathbb{E}\,Z^2 - \mu^2 \quad \text{where } \mu = \mathbb{E}\,Z.$$

The entries of $\mathrm{cov}(X)$ are the *covariances* of the pairs of coordinates of $X = (X_1, \dots, X_n)$:

$$\mathrm{cov}(X)_{ij} = \mathbb{E}\left((X_i - \mathbb{E}\,X_i)(X_j - \mathbb{E}\,X_j)\right).$$

It is sometimes useful to consider the *second-moment matrix* of a random vector X, defined as

$$\Sigma = \Sigma(X) = \mathbb{E}\,XX^\mathsf{T}.$$

The second moment matrix is a higher-dimensional generalization of the second moment $\mathbb{E} Z^2$ of a random variable Z. By translation (replacing X with $X - \mu$), we can assume in many problems that X has zero mean and thus that the covariance and second moment matrices are equal:

$$\mathrm{cov}(X) = \Sigma(X).$$

This observation allows us in future to focus mostly on the second-moment matrix $\Sigma = \Sigma(X)$ rather than on the covariance $\mathrm{cov}(X)$.

Like the covariance matrix, the second-moment matrix Σ is also an $n \times n$ symmetric and positive-semidefinite matrix. The spectral theorem for such matrices says that all eigenvalues s_i of Σ are real and non-negative. Moreover, Σ can be expressed via spectral decomposition as

$$\Sigma = \sum_{i=1}^{n} s_i u_i u_i^{\mathsf{T}},$$

where $u_i \in \mathbb{R}^n$ are the eigenvectors of Σ. We usually arrange the terms in this sum so that the eigenvalues s_i are decreasing.

3.2.1 Principal Component Analysis

The spectral decomposition of Σ is of utmost importance in applications where the distribution of a random vector X in $\mathbb{R}^n$ represents data, for example the genetic data we mentioned at the start of the chapter. The eigenvector u_1 corresponding to the largest eigenvalue s_1 defines the first *principal direction*. This is the direction in which the distribution is most extended, and it explains most of the variability in the data. The next eigenvector u_2 (corresponding to the next largest eigenvalue s_2) defines the next principal direction; it best explains the remaining variations in the data, and so on. This is illustrated in Figure 3.3.

It often happens with real data that only a few eigenvalues s_i are large and can be considered as informative; the remaining eigenvalues are small and considered as noise. In such situations, a few principal directions can explain most variability in the data. Even though the data are presented in a high-dimensional space $\mathbb{R}^n$, such data are essentially *low dimensional* and cluster near the low-dimensional subspace E spanned by the first few principal components.

The most basic data analysis algorithm, called *principal component analysis* (PCA), computes the first few principal components and then projects the data in $\mathbb{R}^n$ onto the subspace

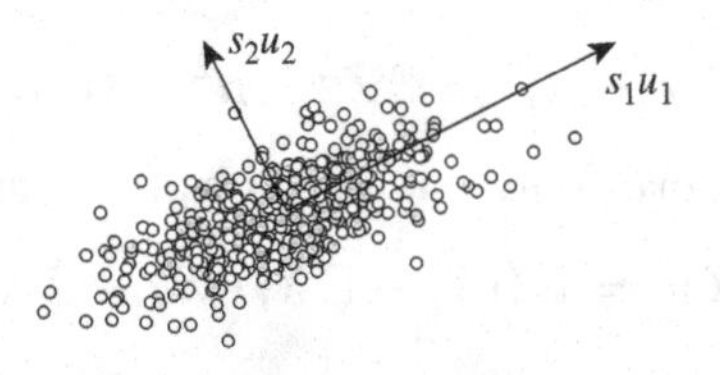

Figure 3.3 Illustration of principal component analysis. Two hundred sample points are shown from a distribution in $\mathbb{R}^2$. The covariance matrix Σ has eigenvalues s_i and eigenvectors u_i.

E spanned by them. This considerably reduces the dimension of the data and simplifies data analysis. For example, if E is two- or three-dimensional, PCA allows one to visualize the data.

3.2.2 Isotropy

We might remember from a basic probability course how it is often convenient to assume that random variables have zero means and unit variances. This is also true in higher dimensions, where the notion of isotropy generalizes the assumption of unit variance.

Definition 3.2.1 (Isotropic random vectors) A random vector X in $\mathbb{R}^n$ is called *isotropic* if

$$\Sigma(X) = \mathbb{E}\, XX^\mathsf{T} = I_n$$

where I_n denotes the identity matrix in $\mathbb{R}^n$.

Recall that any random variable X with positive variance can be reduced by translation and dilation to the *standard score* – a random variable Z with zero mean and unit variance, namely

$$Z = \frac{X - \mu}{\sqrt{\operatorname{Var}(X)}}.$$

The following exercise gives a high-dimensional version of the standard score.

Exercise 3.2.2 (Reduction to isotropy)☕

(a) Let Z be an isotropic mean-zero random vector in $\mathbb{R}^n$. Let $\mu \in \mathbb{R}^n$ be a fixed vector and Σ be a fixed $n \times n$ positive-semidefinite matrix. Check that the random vector

$$X := \mu + \Sigma^{1/2} Z$$

has mean μ and covariance matrix $\operatorname{cov}(X) = \Sigma$.

(b) Let X be a random vector with mean μ and invertible covariance matrix $\Sigma = \operatorname{cov}(X)$. Check that the random vector

$$Z := \Sigma^{-1/2}(X - \mu)$$

is an isotropic mean-zero random vector.

These observations will allow us, in many future results about random vectors, to assume without loss of generality that they have zero means and are isotropic.

3.2.3 Properties of Isotropic Distributions

Lemma 3.2.3 (Characterization of isotropy) *A random vector X in $\mathbb{R}^n$ is isotropic if and only if*

$$\mathbb{E}\, \langle X, x\rangle^2 = \|x\|_2^2 \quad \textit{for all } x \in \mathbb{R}^n.$$

Proof Recall that two symmetric $n \times n$ matrices A and B are equal if and only if $x^\mathsf{T} A x = x^\mathsf{T} B x$ for all $x \in \mathbb{R}^n$. (Check this!) Thus X is isotropic if and only if

$$x^\mathsf{T} \left(\mathbb{E}\, X X^\mathsf{T} \right) x = x^\mathsf{T} I_n x \quad \text{for all } x \in \mathbb{R}^n.$$

The left-hand side of this identity equals $\mathbb{E}\, \langle X, x \rangle^2$ and the right-hand side is $\|x\|_2^2$. This completes the proof. ■

If x is a unit vector in Lemma 3.2.3, we can view $\langle X, x \rangle$ as a one-dimensional marginal of the distribution of X obtained by projecting X onto the direction of x. Then X is isotropic if and only if *all one-dimensional marginals of X have unit variance*. Informally, this means that an isotropic distribution is extended evenly in all directions.

Lemma 3.2.4 *Let X be an isotropic random vector in $\mathbb{R}^n$. Then*

$$\mathbb{E}\, \|X\|_2^2 = n.$$

Moreover, if X and Y are two independent isotropic random vectors in $\mathbb{R}^n$, then

$$\mathbb{E}\, \langle X, Y \rangle^2 = n.$$

Proof To prove the first part, we have

$$\begin{aligned}
\mathbb{E}\, \|X\|_2^2 &= \mathbb{E}\, X^\mathsf{T} X = \mathbb{E}\, \mathrm{tr}(X^\mathsf{T} X) \quad \text{(viewing } X^\mathsf{T} X \text{ as a } 1 \times 1 \text{ matrix)} \\
&= \mathbb{E}\, \mathrm{tr}(X X^\mathsf{T}) \quad \text{(by the cyclic property of the trace)} \\
&= \mathrm{tr}(\mathbb{E}\, X X^\mathsf{T}) \quad \text{(by linearity)} \\
&= \mathrm{tr}(I_n) \quad \text{(by isotropy)} \\
&= n.
\end{aligned}$$

To prove the second part, we use a conditioning argument. Fix a realization of Y and take the conditional expectation (with respect to X), which we denote $\mathbb{E}_X$. The law of total expectation says that

$$\mathbb{E}\, \langle X, Y \rangle^2 = \mathbb{E}_Y\, \mathbb{E}_X \left(\langle X, Y \rangle^2 \,|\, Y \right),$$

where by $\mathbb{E}_Y$ we denote the expectation with respect to Y. To compute the inner expectation, we apply Lemma 3.2.3 with $x = Y$ and conclude that the inner expectation equals $\|Y\|_2^2$. Thus

$$\begin{aligned}
\mathbb{E}\, \langle X, Y \rangle^2 &= \mathbb{E}_Y\, \|Y\|_2^2 \\
&= n \quad \text{(by the first part of the lemma).}
\end{aligned}$$

The proof is complete. ■

Remark 3.2.5 (Almost-orthogonality of independent vectors) Let us normalize the random vectors X and Y in Lemma 3.2.4, setting

$$\overline{X} := \frac{X}{\|X\|_2} \quad \text{and} \quad \overline{Y} := \frac{Y}{\|Y\|_2}.$$

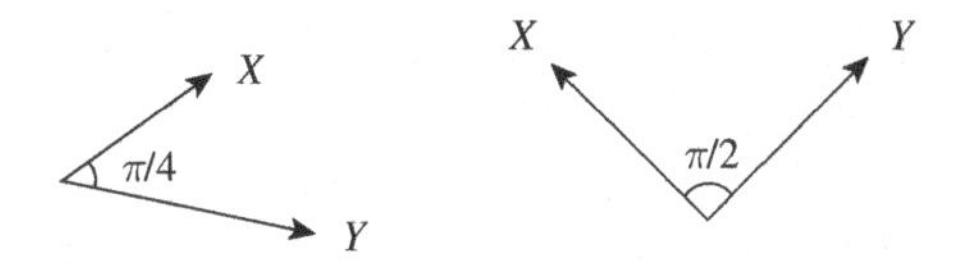

Figure 3.4 Independent isotropic random vectors tend to be almost orthogonal in high dimensions but not in low dimensions. On the plane, the average angle is $\pi/4$, while in high dimensions it is close to $\pi/2$.

Lemma 3.2.4 basically tells us that[2] $\|X\|_2 \asymp \sqrt{n}$, $\|Y\|_2 \asymp \sqrt{n}$, and $\langle X, Y \rangle \asymp \sqrt{n}$, with high probability, which implies that

$$\left|\langle \overline{X}, \overline{Y} \rangle\right| \asymp \frac{1}{\sqrt{n}}.$$

Thus, in high-dimensional spaces independent and isotropic random vectors tend to be *almost orthogonal*; see Figure 3.4.

This may sound surprising since this is not the case in low dimensions. For example, the angle between two random independent and uniformly distributed directions on the plane has mean $\pi/4$. (Check!) But in higher dimensions, there is much more room as we mentioned in the beginning of this chapter. This is an intuitive reason why random directions in high-dimensional spaces tend to be very far from each other, i.e. almost orthogonal.

Exercise 3.2.6 (Distance between independent isotropic vectors) Let X and Y be independent mean-zero isotropic random vectors in $\mathbb{R}^n$. Check that

$$\mathbb{E}\,\|X - Y\|_2^2 = 2n.$$

3.3 Examples of High-Dimensional Distributions

In this section we give several basic examples of isotropic high-dimensional distributions.

3.3.1 Spherical and Bernoulli Distributions

The coordinates of an isotropic random vector are always uncorrelated (why?), but they are not necessarily independent. An example of this situation is the *spherical distribution*, where a random vector X is uniformly distributed[3] on the Euclidean sphere in $\mathbb{R}^n$ with center at the origin and radius $\sqrt{n}$:

$$X \sim \mathrm{Unif}\big(\sqrt{n}\, S^{n-1}\big).$$

Exercise 3.3.1 Show that the spherically distributed random vector X is isotropic. Argue that the coordinates of X are not independent.

[2] This argument is not entirely rigorous, since Lemma 3.2.4 is about expectation and not high probability. To make it more rigorous, one can use Theorem 3.1.1 about concentration of the norm.

[3] More rigorously, we say that X is uniformly distributed on $\sqrt{n}\, S^{n-1}$ if, for every (Borel) subset $E \subset S^{n-1}$, the probability $\mathbb{P}\{X \in E\}$ equals the ratio of the $(n-1)$-dimensional areas of E and S^{n-1}.

A good example of a discrete isotropic distribution in $\mathbb{R}^n$ is the *symmetric Bernoulli* distribution. We say that a random vector $X = (X_1, \dots, X_n)$ is symmetric Bernoulli if the coordinates X_i are independent, symmetric, Bernoulli random variables. Equivalently, we may say that X is uniformly distributed on the unit discrete cube in $\mathbb{R}^n$:

$$X \sim \text{Unif}\big(\{-1, 1\}^n\big).$$

The symmetric Bernoulli distribution is isotropic. (Check!)

More generally, we may consider any random vector $X = (X_1, \dots, X_n)$ whose coordinates X_i are independent random variables with zero mean and unit variance. Then X is an isotropic vector in $\mathbb{R}^n$. (Why?)

3.3.2 Multivariate Normal

One of the most important high-dimensional distributions is Gaussian, or multivariate normal. Recall that a random vector $g = (g_1, \dots, g_n)$ has the *standard normal distribution* in $\mathbb{R}^n$, denoted

$$g \sim N(0, I_n),$$

if the coordinates g_i are independent standard normal random variables $N(0, 1)$. The density of Z is then the product of the n standard normal densities (1.6):

$$f(x) = \prod_{i=1}^{n} \frac{1}{\sqrt{2\pi}} e^{-x_i^2/2} = \frac{1}{(2\pi)^{n/2}} e^{-\|x\|_2^2/2}, \quad x \in \mathbb{R}^n. \tag{3.4}$$

The standard normal distribution is isotropic. (Why?)

Note that the standard normal density (3.4) is *rotation invariant*, since $f(x)$ depends only on the length, not the direction of x. We can equivalently express this observation as follows.

Proposition 3.3.2 (Rotation invariance) *Consider a random vector $g \sim N(0, I_n)$ and a fixed orthogonal matrix U. Then*

$$Ug \sim N(0, I_n).$$

Exercise 3.3.3 (Rotation invariance) Deduce the following properties from the rotation invariance of the normal distribution.

(a) Consider a random vector $g \sim N(0, I_n)$ and a fixed vector $u \in \mathbb{R}^n$. Then

$$\langle g, u \rangle \sim N(0, \|u\|_2^2).$$

(b) Consider independent random variables $X_i \sim N(0, \sigma_i^2)$. Then

$$\sum_{i=1}^{n} X_i \sim N(0, \sigma^2) \quad \text{where} \quad \sigma^2 = \sum_{i=1}^{n} \sigma_i^2.$$

(c) Let G be an $m \times n$ Gaussian random matrix, i.e., the entries of G are independent $N(0, 1)$ random variables. Let $u \in \mathbb{R}^n$ be a fixed unit vector. Then

$$Gu \sim N(0, I_m).$$

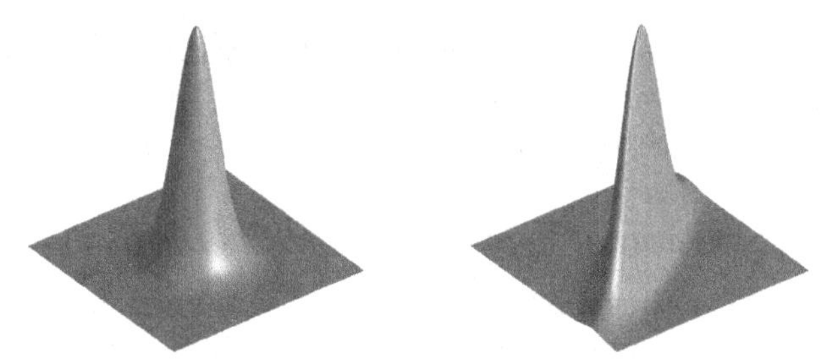

Figure 3.5 The densities of the isotropic distribution $N(0, I_2)$ and a non-isotropic distribution $N(0, \Sigma)$.

Let us also recall the notion of the *general* normal distribution $N(\mu, \Sigma)$. Consider a vector $\mu \in \mathbb{R}^n$ and an invertible $n \times n$ positive-semidefinite matrix Σ. According to Exercise 3.2.2, the random vector $X := \mu + \Sigma^{1/2} Z$ has mean μ and covariance matrix $\operatorname{cov}(X) = \Sigma$. Such an X is said to have a general normal distribution in $\mathbb{R}^n$, denoted

$$X \sim N(\mu, \Sigma).$$

Summarizing, we have $X \sim N(\mu, \Sigma)$ if and only if

$$Z := \Sigma^{-1/2}(X - \mu) \sim N(0, I_n).$$

The density of $X \sim N(\mu, \Sigma)$ can be computed by the change of variables formula, and it equals

$$f_X(x) = \frac{1}{(2\pi)^{n/2} \det(\Sigma)^{1/2}} \exp\left(-(x-\mu)^{\mathsf{T}} \Sigma^{-1} (x-\mu)/2\right), \quad x \in \mathbb{R}^n. \tag{3.5}$$

Figure 3.5 shows examples of two densities of multivariate normal distributions.

An important observation is that the coefficients of a random vector $X \sim N(\mu, \Sigma)$ are independent if and only if they are uncorrelated. (In this case $\Sigma = I_n$.)

Exercise 3.3.4 (Characterization of normal distribution)☕☕☕ Let X be a random vector in $\mathbb{R}^n$. Show that X has a multivariate normal distribution if and only if every one-dimensional marginal $\langle X, \theta \rangle$, $\theta \in \mathbb{R}^n$, has a (univariate) normal distribution. ☞

Exercise 3.3.5☕ Let $X \sim N(0, I_n)$.

(a) Show that, for any fixed vectors $u, v \in \mathbb{R}^n$, we have

$$\mathbb{E}(\langle X, u \rangle \langle X, v \rangle) = \langle u, v \rangle. \tag{3.6}$$

(b) Given a vector $u \in \mathbb{R}^n$, consider the random variable $X_u := \langle X, u \rangle$. From Exercise 3.3.3 we know that $X_u \sim N(0, \|u\|_2^2)$. Check that

$$\|X_u - X_v\|_{L^2} = \|u - v\|_2$$

for any fixed vectors $u, v \in \mathbb{R}^n$. (Here $\|\cdot\|_{L^2}$ denotes the norm in the Hilbert space L^2 of random variables, which we introduced in (1.1).)

Exercise 3.3.6 Let G be an $m \times n$ Gaussian random matrix, i.e., the entries of G are independent $N(0, 1)$ random variables. Let $u, v \in \mathbb{R}^n$ be unit orthogonal vectors. Prove that Gu and Gv are independent $N(0, I_m)$ random vectors. ☞

3.3.3 Similarity of Normal and Spherical Distributions

Contradicting our low-dimensional intuition, the standard normal distribution $N(0, I_n)$ in high dimensions is *not* concentrated close to the origin, where the density is maximal. Instead, it is concentrated *in a thin spherical shell* around the sphere of radius $\sqrt{n}$, a shell of width $O(1)$. Indeed, the concentration inequality (3.3) for the norm of $g \sim N(0, I_n)$ states that

$$\mathbb{P}\left\{\left|\|g\|_2 - \sqrt{n}\right| \geq t\right\} \leq 2\exp(-ct^2) \quad \text{for all } t \geq 0. \tag{3.7}$$

This observation suggests that the normal distribution should be quite similar to the uniform distribution on the sphere. Let us clarify the relation.

Exercise 3.3.7 (Normal and spherical distributions) Let us represent $g \sim N(0, I_n)$ in polar form as

$$g = r\theta,$$

where $r = \|g\|_2$ is the length and $\theta = g/\|g\|_2$ is the direction of g. Prove the following:

(a) The length r and direction θ are independent random variables.
(b) The direction θ is uniformly distributed on the unit sphere S^{n-1}.

The concentration inequality (3.7) says that $r = \|g\|_2 \approx \sqrt{n}$ with high probability, so

$$g \approx \sqrt{n}\,\theta \sim \text{Unif}\big(\sqrt{n}S^{n-1}\big).$$

In other words, the standard normal distribution in high dimensions is close to the uniform distribution on the sphere of radius $\sqrt{n}$, i.e.

$$N(0, I_n) \approx \text{Unif}\big(\sqrt{n}S^{n-1}\big). \tag{3.8}$$

Figure 3.6 illustrates this fact, which goes against our intuition, trained in low dimensions.

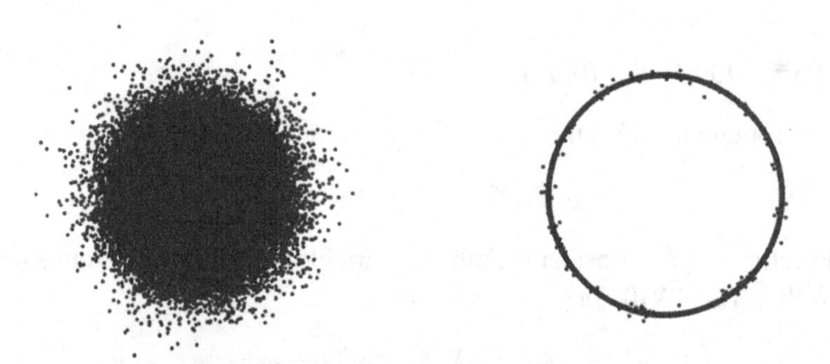

Figure 3.6 A Gaussian point cloud in two dimensions (left) and its intuitive visualization in high dimensions (right). In high dimensions, the standard normal distribution is very close to the uniform distribution on the sphere of radius $\sqrt{n}$.

3.3.4 Frames

For an example of an extremely discrete distribution, consider a *coordinate random vector* X uniformly distributed in the set $\{\sqrt{n}\, e_i\}_{i=1}^n$, where $\{e_i\}_{i=1}^n$ is the canonical basis of $\mathbb{R}^n$:

$$X \sim \operatorname{Unif}\left\{\sqrt{n}\, e_i : i = 1, \ldots, n\right\}.$$

Then X is an isotropic random vector in $\mathbb{R}^n$. (Check!)

Of all high-dimensional distributions, the Gaussian is often the most convenient for obtaining results, so we may think of it as the "best" distribution. The coordinate distribution, the most discrete of all distributions, is the "worst".

A general class of discrete, isotropic, distributions arises in the area of signal processing under the name of *frames*.

Definition 3.3.8 A *frame* is a set of vectors $\{u_i\}_{i=1}^N$ in $\mathbb{R}^n$ which obeys an approximate Parseval's identity, i.e. there exist numbers $A, B > 0$, called *frame bounds*, such that

$$A\|x\|_2^2 \le \sum_{i=1}^N \langle u_i, x\rangle^2 \le B\|x\|_2^2 \quad \text{for all } x \in \mathbb{R}^n.$$

If $A = B$ then the set $\{u_i\}_{i=1}^N$ is called a *tight frame*.

Exercise 3.3.9 ☕☕ Show that $\{u_i\}_{i=1}^N$ is a tight frame in $\mathbb{R}^n$ with bound A if and only if

$$\sum_{i=1}^N u_i u_i^\mathsf{T} = A I_n. \tag{3.9}$$

☞

Multiplying both sides of (3.9) by a vector x, we see that

$$\sum_{i=1}^N \langle u_i, x\rangle\, u_i = Ax \quad \text{for any } x \in \mathbb{R}^n. \tag{3.10}$$

This is a *frame expansion* of a vector x, and it should look familiar. Indeed, if $\{u_i\}$ is an orthonormal basis then (3.10) is just a classical basis expansion of x, and it holds with $A = 1$.

We can think of tight frames as generalizations of orthogonal bases *without the linear independence* requirement. Any orthonormal basis in $\mathbb{R}^n$ is clearly a tight frame. But so is the "Mercedez–Benz frame", a set of three equidistant points on a circle in $\mathbb{R}^2$, shown in Figure 3.7.

Now we are ready to connect the concept of frames to probability. We will show that tight frames correspond to isotropic distributions, and vice versa.

Lemma 3.3.10 (Tight frames and isotropic distributions)

(i) *Consider a tight frame $\{u_i\}_{i=1}^N$ in $\mathbb{R}^n$ with frame bounds $A = B$. Let X be a random vector that is uniformly distributed in the set of frame elements, i.e.,*

$$X \sim \operatorname{Unif}\{u_i : i = 1, \ldots, N\}.$$

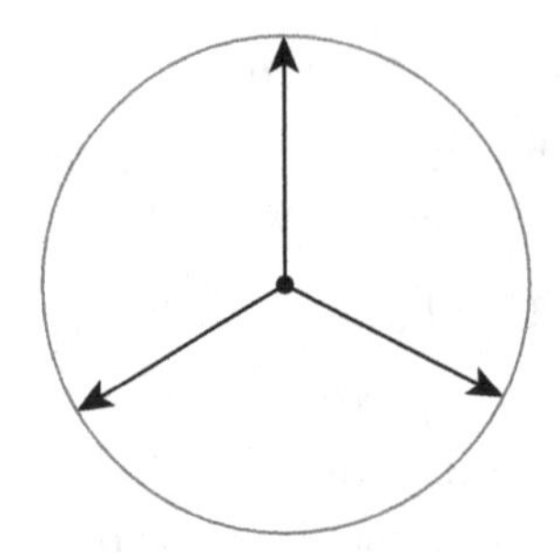

Figure 3.7 The Mercedes–Benz frame. A set of equidistant points on the circle forms a tight frame in $\mathbb{R}^2$.

Then $(N/A)^{1/2}X$ is an isotropic random vector in $\mathbb{R}^n$.

(ii) *Consider an isotropic random vector X in $\mathbb{R}^n$ that takes a finite set of values x_i with probabilities p_i each, $i = 1, \dots, N$. Then the vectors*

$$u_i := \sqrt{p_i}\, x_i, \quad i = 1, \dots, N,$$

form a tight frame in $\mathbb{R}^N$ with bounds $A = B = 1$.

Proof (i) Without loss of generality, we can assume that $A = N$. (Why?) The assumptions and (3.9) imply that

$$\sum_{i=1}^{N} u_i u_i^\mathsf{T} = N I_n.$$

Dividing both sides by N and interpreting $N^{-1}\sum_{i=1}^{N}$ as an expectation, we conclude that X is isotropic.

(ii) The isotropy of X means that

$$\mathbb{E}\, X X^\mathsf{T} = \sum_{i=1}^{N} p_i x_i x_i^\mathsf{T} = I_n.$$

Denoting $u_i := \sqrt{p_i}\, x_i$, we obtain (3.9) with $A = 1$. ■

3.3.5 Isotropic Convex Sets

Our last example of a high-dimensional distribution comes from convex geometry. Consider a bounded convex set K in $\mathbb{R}^n$ with non-empty interior; such sets are called *convex bodies*. Let X be a random vector uniformly distributed in K according to the probability measure given by the normalized volume in K:

$$X \sim \mathrm{Unif}(K).$$

Assume that $\mathbb{E}\, X = 0$ (translate K appropriately to achieve this) and denote the covariance matrix of X by Σ. Then, by Exercise 3.2.2, the random vector $Z := \Sigma^{-1/2}X$ is isotropic. Note that Z is uniformly distributed in the linearly transformed copy of K:

$$Z \sim \mathrm{Unif}(\Sigma^{-1/2}K).$$

Figure 3.8 The convex body K on the left is transformed into the isotropic convex body TK on the right. The pre-conditioner T is computed from the covariance matrix Σ of K as $T = \Sigma^{-1/2}$.

(Why?) Summarizing, we have found a linear transformation $T := \Sigma^{-1/2}$ which makes the uniform distribution on TK isotropic. The body TK is sometimes called isotropic itself.

In algorithmic convex geometry, one can think of the isotropic convex body TK as a *well-conditioned* version of K, with T playing the role of a pre-conditioner; see Figure 3.8. Algorithms related to convex bodies K (such as an algorithm for computing the volume of K) tend to work better for well-conditioned bodies K.

3.4 Sub-Gaussian Distributions in Higher Dimensions

The concept of sub-gaussian distributions, which we introduced in Section 2.5, can be extended to higher dimensions. To see how this is done, recall from Exercise 3.3.4 that the multivariate normal distribution can be characterized through its *one-dimensional marginals*, or projections onto lines: a random vector X has a normal distribution in $\mathbb{R}^n$ if and only if the one-dimensional marginals $\langle X, x\rangle$ are normal for all $x \in \mathbb{R}^n$. Guided by this characterization, it is natural to define multivariate sub-gaussian distributions as follows.

Definition 3.4.1 (Sub-gaussian random vectors) A random vector X in $\mathbb{R}^n$ is called *sub-gaussian* if the one-dimensional marginals $\langle X, x\rangle$ are sub-gaussian random variables for all $x \in \mathbb{R}^n$. The *sub-gaussian norm* of X is defined as

$$\|X\|_{\psi_2} = \sup_{x \in S^{n-1}} \| \langle X, x\rangle \|_{\psi_2}.$$

A good example of a sub-gaussian random vector is a random vector with independent sub-gaussian coordinates:

Lemma 3.4.2 (Sub-gaussian distributions with independent coordinates) *Let $X = (X_1, \dots, X_n) \in \mathbb{R}^n$ be a random vector with independent mean-zero sub-gaussian coordinates X_i. Then X is a sub-gaussian random vector, and*

$$\|X\|_{\psi_2} \le C \max_{i \le n} \|X_i\|_{\psi_2}.$$

Proof This is an easy consequence of the fact that a sum of independent sub-gaussian random variables is sub-gaussian, which we proved in Proposition 2.6.1. Indeed, for a fixed unit vector $x = (x_1, \dots, x_n) \in S^{n-1}$ we have

$$\| \langle X, x\rangle \|_{\psi_2}^2 = \Big\| \sum_{i=1}^n x_i X_i \Big\|_{\psi_2}^2 \le C \sum_{i=1}^n x_i^2 \|X_i\|_{\psi_2}^2 \quad \text{(by Proposition 2.6.1)}$$

$$\leq C \max_{i \leq n} \|X_i\|_{\psi_2}^2 \quad (\text{using that } \sum_{i=1}^n x_i^2 = 1).$$

This completes the proof. ■

Exercise 3.4.3 This exercise clarifies the role of the independence of coordinates in Lemma 3.4.2.

(a) Let $X = (X_1, \ldots, X_n) \in \mathbb{R}^n$ be a random vector with sub-gaussian coordinates X_i. Show that X is a sub-gaussian random vector.
(b) Nevertheless, find an example of a random vector X with

$$\|X\|_{\psi_2} \gg \max_{i \leq n} \|X_i\|_{\psi_2}.$$

Many important high-dimensional distributions are sub-gaussian, but some are not. We now explore some basic distributions.

3.4.1 Gaussian and Bernoulli Distributions

As we have already noted, the *multivariate normal distribution* $N(\mu, \Sigma)$ is sub-gaussian. Moreover, the standard normal random vector $X \sim N(0, I_n)$ has sub-gaussian norm of order $O(1)$:

$$\|X\|_{\psi_2} \leq C.$$

(Indeed, all one-dimensional marginals of X are $N(0, 1)$.)

Next, consider the multivariate *symmetric Bernoulli* distribution, which we introduced in Section 3.3.1. A random vector X with this distribution has independent, symmetric Bernoulli coordinates, so Lemma 3.4.2 yields that

$$\|X\|_{\psi_2} \leq C.$$

3.4.2 Discrete Distributions

Let us now pass to discrete distributions. The extreme example, mentioned in Section 3.3.4, is the *coordinate distribution*. Recall that a random vector X with the coordinate distribution is uniformly distributed in the set $\{\sqrt{n}e_i : i = 1, \ldots, n\}$, where e_i denotes the n-element set of the canonical basis vectors in $\mathbb{R}^n$.

Is X sub-gaussian? Formally, yes. In fact, every distribution supported in a finite set is sub-gaussian. (Why?) But, unlike the Gaussian and Bernoulli distributions, the coordinate distribution has a very large sub-gaussian norm.

Exercise 3.4.4 Show that

$$\|X\|_{\psi_2} \asymp \sqrt{\frac{n}{\log n}}.$$

Such a large norm makes it useless to think of X as a sub-gaussian random vector.

More generally, discrete distributions do not make nice sub-gaussian distributions unless they are supported on exponentially large sets:

Exercise 3.4.5 ☕☕☕☕ Let X be an isotropic random vector supported in a finite set $T \subset \mathbb{R}^n$. Show that in order for X to be sub-gaussian with $\|X\|_{\psi_2} = O(1)$, the cardinality of the set must be exponentially large in n:

$$|T| \geq e^{cn}.$$

In particular, this observation rules out *frames* (see Section 3.3.4) as good sub-gaussian distributions unless they have exponentially many terms (in which case they are mostly useless in practice).

3.4.3 Uniform Distribution on the Sphere

In all our previous examples, sub-gaussian random vectors that were useful had independent coordinates. This is not necessary, however. A good example is the uniform distribution on the sphere of radius $\sqrt{n}$, which we discussed in Section 3.3.1. We will show that it is sub-gaussian by reducing it to the Gaussian distribution $N(0, I_n)$.

Theorem 3.4.6 (Uniform distribution on the sphere is sub-gaussian) *Let X be a random vector uniformly distributed on the Euclidean sphere in $\mathbb{R}^n$ with center at the origin and radius $\sqrt{n}$:*

$$X \sim \mathrm{Unif}\big(\sqrt{n}\, S^{n-1}\big).$$

Then X is sub-gaussian, and

$$\|X\|_{\psi_2} \leq C.$$

Proof Consider a standard normal random vector $g \sim N(0, I_n)$. As we noted in Exercise 3.3.7, the direction $g/\|g\|_2$ is uniformly distributed on the unit sphere S^{n-1}. Thus, by rescaling we can represent a random vector $X \sim \mathrm{Unif}\big(\sqrt{n}\, S^{n-1}\big)$ as

$$X = \sqrt{n}\,\frac{g}{\|g\|_2}.$$

We need to show that all one-dimensional marginals $\langle X, x\rangle$ are sub-gaussian. By rotation invariance we may assume that $x = (1, 0, \ldots, 0)$, in which case $\langle X, x\rangle = X_1$, the first coordinate of X. We want to bound the tail probability

$$p(t) := \mathbb{P}\left\{|X_1| \geq t\right\} = \mathbb{P}\left\{\frac{|g_1|}{\|g\|_2} \geq \frac{t}{\sqrt{n}}\right\}.$$

The concentration of the norm (Theorem 3.1.1) implies that

$$\|g\|_2 \approx \sqrt{n} \quad \text{with high probability.}$$

This reduces the problem to bounding $\mathbb{P}\left\{|g_1| \geq t\right\}$ but, as we know from (2.3), this tail is sub-gaussian.

Let us do this argument more carefully. Theorem 3.1.1 implies that

$$\big\| \|g\|_2 - \sqrt{n} \big\|_{\psi_2} \leq C.$$

Thus the event

$$\mathcal{E} := \left\{ \|g\|_2 \geq \frac{\sqrt{n}}{2} \right\}$$

is likely: by (2.14) its complement $\mathcal{E}^c$ has probability

$$\mathbb{P}(\mathcal{E}^c) \leq 2\exp(-cn). \tag{3.11}$$

Then the tail probability can be bounded as follows:

$$\begin{aligned} p(t) &\leq \mathbb{P}\left\{ \frac{|g_1|}{\|g\|_2} \geq \frac{t}{\sqrt{n}} \text{ and } \mathcal{E} \right\} + \mathbb{P}(\mathcal{E}^c) \\ &\leq \mathbb{P}\left\{ |g_1| \geq \frac{t}{2} \text{ and } \mathcal{E} \right\} + 2\exp(-cn) \quad \text{(by definition of } \mathcal{E} \text{ and (3.11))} \\ &\leq 2\exp(-t^2/8) + 2\exp(-cn) \quad \text{(drop } \mathcal{E} \text{ and use (2.3)).} \end{aligned}$$

Consider two cases. If $t \leq \sqrt{n}$ then $2\exp(-cn) \leq 2\exp(-ct^2/8)$, and we conclude that

$$p(t) \leq 4\exp(-c't^2)$$

as desired. In the opposite case where $t > \sqrt{n}$, the tail probability $p(t) = \mathbb{P}\{|X_1| \geq t\}$ trivially equals zero, since we always have $|X_1| \leq \|X\|_2 = \sqrt{n}$. This completes the proof. ■

Exercise 3.4.7 (Uniform distribution on the Euclidean ball) Extend Theorem 3.4.6 for the uniform distribution on the Euclidean ball $B(0, \sqrt{n})$ in $\mathbb{R}^n$ centered at the origin and with radius $\sqrt{n}$. Namely, show that a random vector

$$X \sim \text{Unif}\big(B(0, \sqrt{n})\big)$$

is sub-gaussian, and

$$\|X\|_{\psi_2} \leq C.$$

Remark 3.4.8 (Projective limit theorem) Theorem 3.4.6 should be compared to the so-called projective central limit theorem. It states that the marginals of the uniform distribution on the sphere become asymptotically normal as n increases, see Figure 3.9. Precisely, if $X \sim \text{Unif}\left(\sqrt{n}\, S^{n-1}\right)$ then for any fixed unit vector x we have

$$\langle X, x \rangle \to N(0,1) \quad \text{in distribution as } n \to \infty.$$

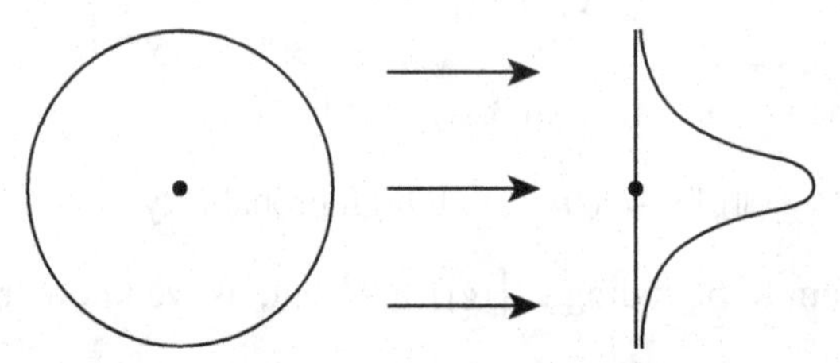

Figure 3.9 The projective central limit theorem: the projection of a uniform distribution on the sphere of radius $\sqrt{n}$ onto a line converges to the normal distribution $N(0, 1)$ as $n \to \infty$.

Thus we can view Theorem 3.4.6 as a concentration version of the projective limit theorem, in the same sense as Hoeffding's inequality in Section 2.2 is a concentration version of the classical central limit theorem.

3.4.4 Uniform Distribution on Convex Sets

To conclude this section, let us return to the class of uniform distributions on *convex sets*, which we discussed in Section 3.3.5. Let K be a convex body and

$$X \sim \mathrm{Unif}(K)$$

be an isotropic random vector. Is X always sub-gaussian?

For some bodies K, this is the case. Examples include the Euclidean ball of radius $\sqrt{n}$ (by Exercise 3.4.7) and the unit cube $[-1, 1]^n$ (according to Lemma 3.4.2). For some other bodies, this is not true:

Exercise 3.4.9 ☕☕☕ Consider a ball of the ℓ_1 norm in $\mathbb{R}^n$:

$$K := \left\{x \in \mathbb{R}^n : \ \|x\|_1 \le r\right\}.$$

(a) Show that the uniform distribution on K is isotropic for some $r \asymp n$.
(b) Show that the subgaussian norm of this distribution is *not* bounded by an absolute constant as the dimension n grows.

Nevertheless, it is possible to prove a weaker result for a general isotropic convex body K. The random vector $X \sim \mathrm{Unif}(K)$ has all *sub-exponential* marginals, and

$$\| \langle X, x \rangle \|_{\psi_1} \le C$$

for all unit vectors x. This result follows from C. Borell's lemma, which itself is a consequence of the Brunn–Minkowski inequality; see [79, Section 2.2.b$_3$].

Exercise 3.4.10 ☕☕ Show that the concentration inequality in Theorem 3.1.1 may not hold for a general isotropic sub-gaussian random vector X. Thus, independence of the coordinates of X is an essential requirement in that result.

3.5 Application: Grothendieck's Inequality and Semidefinite Programming

In this and the next section, we use high-dimensional Gaussian distributions to pursue some problems that have seemingly nothing to do with probability. Here we give a probabilistic proof of Grothendieck's inequality, a remarkable result which we will use later in the analysis of some computationally hard problems.

Theorem 3.5.1 (Grothendieck's inequality) *Consider an $m \times n$ matrix (a_{ij}) of real numbers. Assume that, for any numbers $x_i, y_j \in \{-1, 1\}$, we have*

$$\left| \sum_{i,j} a_{ij} x_i y_j \right| \le 1.$$

Then, for any Hilbert space H and any vectors $u_i, v_j \in H$ satisfying $\|u_i\| = \|v_j\| = 1$, we have

$$\left| \sum_{i,j} a_{ij} \langle u_i, v_j \rangle \right| \le K,$$

where $K \le 1.783$ is an absolute constant.

There is apparently nothing random in the statement of this theorem, but our proof will be probabilistic. We will actually give two proofs. The proof given in this section will yield a much worse bound on the constant K, namely $K \le 288$. In Section 3.7 we present an alternative argument that yields the bound $K \le 1.783$, as stated in Theorem 3.5.1.

Before we pass to the argument, let us make one simple observation.

Exercise 3.5.2

(a) Check that the assumption of Grothendieck's inequality can be equivalently stated as follows:

$$\left| \sum_{i,j} a_{ij} x_i y_j \right| \le \max_i |x_i| \max_j |y_j| \tag{3.12}$$

for any real numbers x_i and y_j.

(b) Show that the conclusion of Grothendieck's inequality can be equivalently stated as follows:

$$\left| \sum_{i,j} a_{ij} \langle u_i, v_j \rangle \right| \le K \max_i \|u_i\| \max_j \|v_j\| \tag{3.13}$$

for any Hilbert space H and any vectors $u_i, v_j \in H$.

Proof of Theorem 3.5.1 with $K \le 288$ Step 1: Reductions. Note that Grothendieck's inequality becomes trivial if we allow the value of K to depend on the matrix $A = (a_{ij})$. (For example, $K = \sum_{ij} |a_{ij}|$ would work – check!) Let us choose $K = K(A)$ to be the *smallest* number that makes the conclusion (3.13) valid for a given matrix A, any Hilbert space H, and any vectors $u_i, v_j \in H$. Our goal is to show that K does *not* depend on the matrix A or the dimensions m and n.

Without loss of generality,[4] we may do this for a specific Hilbert space H, namely for $\mathbb{R}^N$ equipped with the Euclidean norm $\|\cdot\|_2$. Let us fix vectors $u_i, v_j \in \mathbb{R}^N$ which realize the smallest K, that is

[4] To see this, we can first trivially replace H with the subspace of H spanned by the vectors u_i and v_j (and with the norm inherited from H). This subspace has dimension at most $N := m + n$. Next, we recall the basic fact that all N-dimensional Hilbert spaces are isometric with each other, and in particular they are isometric to $\mathbb{R}^N$ with norm $\|\cdot\|_2$. The isometry can be constructed by identifying orthogonal bases of those spaces.

$$\sum_{i,j} a_{ij} \langle u_i, v_j \rangle = K, \quad \|u_i\|_2 = \|v_j\|_2 = 1.$$

Step 2: Introducing randomness. The main idea of the proof is to realize the vectors u_i, v_j via Gaussian random variables

$$U_i := \langle g, u_i \rangle \quad \text{and} \quad V_j := \langle g, v_j \rangle, \quad \text{where } g \sim N(0, I_N).$$

As we noted in Exercise 3.3.5, U_i and V_j are standard normal random variables whose correlations follow exactly the inner products of the vectors u_i and v_j:

$$\mathbb{E}\, U_i V_j = \langle u_i, v_j \rangle.$$

Thus

$$K = \sum_{i,j} a_{ij} \langle u_i, v_j \rangle = \mathbb{E} \sum_{i,j} a_{ij} U_i V_j. \tag{3.14}$$

Assume for a moment that the random variables U_i and V_j are bounded almost surely by some constant – say, by R. Then the assumption (3.12) of Grothendieck's inequality would yield $\left| \sum_{i,j} a_{ij} U_i V_j \right| \le R^2$ almost surely, and (3.14) would then give $K \le R^2$.

Step 3: Truncation. Of course, this reasoning is flawed: the random variables $U_i, V_j \sim N(0,1)$ are not bounded almost surely. To fix this argument, we can employ a useful *truncation* trick. Let us fix some truncation level $R \ge 1$ and decompose the random variables as follows:

$$U_i = U_i^- + U_i^+ \quad \text{where} \quad U_i^- = U_i \, \mathbf{1}_{\{|U_i| \le R\}} \quad \text{and} \quad U_i^+ = U_i \, \mathbf{1}_{\{|U_i| > R\}}.$$

We then decompose $V_j = V_j^- + V_j^+$ similarly. Now U_i^- and V_j^- are bounded by R almost surely, as we desired. The remainder terms U_i^+ and V_j^+ are small in the L^2 norm: indeed, the bound in Exercise 2.1.4 gives

$$\|U_i^+\|_{L^2}^2 \le 2\left(R + \frac{1}{R}\right) \frac{1}{\sqrt{2\pi}} e^{-R^2/2} < \frac{4}{R^2}, \tag{3.15}$$

and similarly for V_j^+.

Step 4: Breaking up the sum. The sum in (3.14) becomes

$$K = \mathbb{E} \sum_{i,j} a_{ij} (U_i^- + U_i^+)(V_j^- + V_j^+).$$

When we expand the product in each term we obtain four sums, which we proceed to bound individually. The first sum,

$$S_1 := \mathbb{E} \sum_{i,j} a_{ij} U_i^- V_j^-,$$

is the easiest to control. By construction, the random variables U_i^- and V_j^- are bounded almost surely by R. Thus, just as we explained above, we can use the assumption (3.12) of Grothendieck's inequality to get $S_1 \le R^2$.

We are not able to use the same reasoning for the second sum,

$$S_2 := \mathbb{E} \sum_{i,j} a_{ij} U_i^+ V_j^-,$$

since the random variable U_i^+ is unbounded. Instead, we will view the random variables U_i^+ and V_j^- as elements of the Hilbert space L^2 with inner product $\langle X, Y \rangle_{L^2} = \mathbb{E}\, XY$. The second sum becomes

$$S_2 = \sum_{i,j} a_{ij} \left\langle U_i^+, V_j^- \right\rangle_{L^2}. \tag{3.16}$$

Recall from (3.15) that $\|U_i^+\|_{L^2} < 2/R$ and $\|V_j^-\|_{L^2} \le \|V_j\|_{L^2} = 1$ by construction. Then, applying the conclusion (3.13) of Grothendieck's inequality for the Hilbert space $H = L^2$, we find that[5]

$$S_2 \le K \frac{2}{R}.$$

The third and fourth sums, $S_3 := \mathbb{E} \sum_{i,j} a_{ij} U_i^- V_j^+$ and $S_4 := \mathbb{E} \sum_{i,j} a_{ij} U_i^+ V_j^+$, can be both bounded in the same way as S_2. (Check!)

Step 5: Putting everything together. Putting the four sums together, we conclude from (3.14) that

$$K \le R^2 + \frac{6K}{R}.$$

Choosing $R = 12$ (for example) and solve the resulting inequality, we obtain $K \le 288$. The theorem is proved. ■

Exercise 3.5.3 (Symmetric matrices, $x_i = y_i$) ☕☕☕ Deduce the following version of Grothendieck's inequality for symmetric $n \times n$ matrices $A = (a_{ij})$ with real entries. Assume that, for any numbers $x_i \in \{-1, 1\}$, we have

$$\left| \sum_{i,j} a_{ij} x_i x_j \right| \le 1.$$

Then, for any Hilbert space H and any vectors $u_i, v_j \in H$ satisfying $\|u_i\| = \|v_j\| = 1$, we have

$$\left| \sum_{i,j} a_{ij} \left\langle u_i, v_j \right\rangle \right| \le 2K, \tag{3.17}$$

where K is the absolute constant from Grothendieck's inequality. ☞

[5] It might seem weird that we are able to apply the inequality that we are trying to prove. Remember, however, that we chose K at the beginning of the proof as the best number that makes Grothendieck's inequality valid. This is the K we are using here.

3.5.1 Semidefinite Programming

One application area where Grothendieck's inequality can be particularly helpful is the analysis of certain computationally hard problems. A powerful approach to such problems is to try and *relax* them to computationally simpler and more tractable problems. This is often done using semidefinite programming, with Grothendieck's inequality guaranteeing the quality of such relaxations.

Definition 3.5.4 A *semidefinite program* is an optimization problem of the following type:

$$\text{maximize } \langle A, X\rangle: \quad X \succeq 0, \quad \langle B_i, X\rangle = b_i \text{ for } i = 1, \ldots, m. \tag{3.18}$$

Here A and the B_i are given $n \times n$ matrices and the b_i are given real numbers. The running "variable" X is an $n \times n$ positive-semidefinite matrix, indicated by the notation $X \succeq 0$. The inner product

$$\langle A, X\rangle = \mathrm{tr}(A^{\mathsf{T}} X) = \sum_{i,j=1}^{n} A_{ij} X_{ij} \tag{3.19}$$

is the canonical inner product on the space of $n \times n$ matrices.

Note in passing that if we *minimize* instead of maximize in (3.18), we still get a semidefinite program. (To see this, replace A with $-A$.)

Every semidefinite program is a *convex program*, which maximizes a linear function $\langle A, X\rangle$ over a convex set of matrices. Indeed, the set of positive-semidefinite matrices is convex (why?), and so is its intersection with the linear subspace defined by the constraints $\langle B_i, X\rangle = b_i$.

This is good news since convex programs are generally algorithmically tractable. A variety of computationally efficient solvers, for example, interior point methods, are available for general convex programs and for semidefinite programs (3.18) in particular.

Semidefinite Relaxations

Semidefinite programs can be designed to provide computationally efficient relaxations of computationally hard problems, such as this one:

$$\text{maximize } \sum_{i,j=1}^{n} A_{ij} x_i x_j: \quad x_i = \pm 1 \text{ for } i = 1, \ldots, n, \tag{3.20}$$

where A is a given $n \times n$ symmetric matrix. This is an *integer optimization problem*. The feasible set consists of 2^n vectors $x = (x_i) \in \{-1, 1\}^n$, so finding the maximum by exhaustive search would take exponential time. Is there a smarter way to solve the problem? This is not likely: the problem (3.20) is known to be computationally hard in general (NP-hard).

Nonetheless, we can "relax" the problem (3.20) to a semidefinite program that can compute the maximum *approximately*, up to a constant factor. To formulate such a relaxation, let us replace in (3.20) the numbers $x_i = \pm 1$ by their higher-dimensional analogs – unit vectors

X_i in $\mathbb{R}^n$. Thus we consider the following optimization problem:

$$\text{maximize} \sum_{i,j=1}^{n} A_{ij} \langle X_i, X_j \rangle : \quad \|X_i\|_2 = 1 \text{ for } i = 1, \ldots, n. \tag{3.21}$$

Exercise 3.5.5 Show that the optimization (3.21) is equivalent to the following semidefinite program:

$$\text{maximize } \langle A, X \rangle : \quad X \succeq 0, \quad X_{ii} = 1 \text{ for } i = 1, \ldots, n. \tag{3.22}$$

☞

The Guarantee of Relaxation

We now see how Grothendieck's inequality guarantees the accuracy of semidefinite relaxations: the semidefinite program (3.21) approximates the maximum value in the integer optimization problem (3.20) up to an absolute constant factor.

Theorem 3.5.6 *Let* $\mathrm{INT}(A)$ *denote the maximum in the integer optimization problem* (3.20) *and* $\mathrm{SDP}(A)$ *denote the maximum in the semidefinite problem* (3.21). *Then*

$$\mathrm{INT}(A) \leq \mathrm{SDP}(A) \leq 2K\, \mathrm{INT}(A)$$

where $K \leq 1.783$ *is the constant in Grothendieck's inequality.*

Proof The first bound follows with $X_i = (x_i, 0, 0, \ldots, 0)^\mathsf{T}$. The second bound follows from Grothendieck's inequality for symmetric matrices in Exercise 3.5.3. (Argue that one can drop absolute values in this exercise.) ■

Although Theorem 3.5.6 allows us to approximate the maximum value in (3.20), it is not obvious how to compute an x_i that attains this approximate value. Can we translate the vectors (X_i) that give a solution of the semidefinite program (3.21) into labels $x_i = \pm 1$ that approximately solve (3.20)? In the next section, we illustrate this using the example of a remarkable NP-hard problem on graphs – the maximum cut problem.

Exercise 3.5.7 Let A be an $m \times n$ matrix. Consider the optimization problem

$$\text{maximize} \sum_{i,j} A_{ij} \langle X_i, Y_j \rangle : \quad \|X_i\|_2 = \|Y_j\|_2 = 1 \text{ for all } i, j$$

over $X_i, Y_j \in \mathbb{R}^k$. Formulate this problem as a semidefinite program. ☞

3.6 Application: Maximum Cut for Graphs

We now illustrate the utility of semidefinite relaxations for the problem of finding the *maximum cut* of a graph, which is a well-known NP-hard problem discussed in the computer science literature.

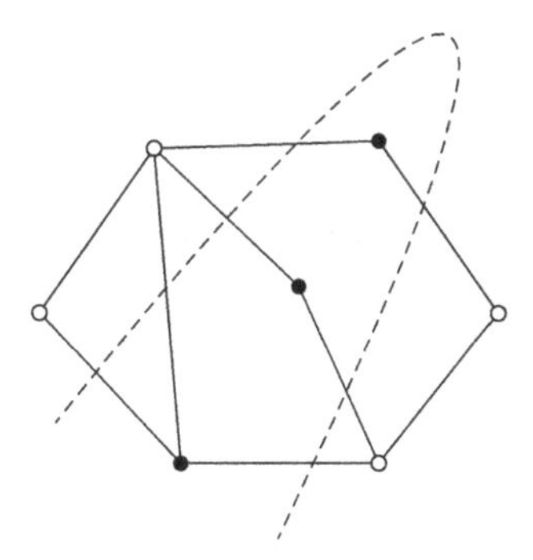

Figure 3.10 The dashed line illustrates the maximum cut of this graph, obtained by partitioning the vertices into black and white ones. Here MAX-CUT$(G) = 7$.

3.6.1 Graphs and Cuts

An undirected *graph* $G = (V, E)$ is defined as a set V of vertices together with a set E of edges; each edge is an unordered pair of vertices. Here we consider finite *simple* graphs – those with finitely many vertices and with no loops or multiple edges.

Definition 3.6.1 (Maximum cut) Suppose that we partition the set of vertices of a graph G into two disjoint sets. The *cut is the number of edges crossing between these two sets. The* maximum cut of G, denoted MAX-CUT(G), is obtained by maximizing the cut over all partitions of vertices; see Figure 3.10 for an illustration.

Computing the maximum cut of a given graph is known to be a computationally hard problem.

3.6.2 A Simple 0.5-Approximation Algorithm

We now try to relax the maximum-cut problem to a semidefinite program, following the method introduced in Section 3.5.1. To do this, we need to translate the problem into the language of linear algebra.

Definition 3.6.2 (Adjacency matrix) The *adjacency matrix* A of a graph G on n vertices is a symmetric $n \times n$ matrix whose entries are defined as $A_{ij} = 1$ if vertices i and j are connected by an edge and $A_{ij} = 0$ otherwise.

Let us label the vertices of G by the integers $1, \ldots, n$. A partition of the vertices into two sets can be described using a vector of labels

$$x = (x_i) \in \{-1, 1\}^n,$$

the sign of x_i indicating to which subset the vertex i belongs. For example, the three black vertices in Figure 3.10 could have labels $x_i = 1$ and the four white vertices labels $x_i = -1$. The cut of G corresponding to the partition given by x is simply the number of edges between the vertices with labels of opposite signs, i.e.,

$$\mathrm{CUT}(G,x) = \frac{1}{2} \sum_{i,j:\, x_i x_j = -1} A_{ij} = \frac{1}{4} \sum_{i,j=1}^{n} A_{ij}(1 - x_i x_j). \tag{3.23}$$

(The factor $1/2$ prevents the double counting of edges (i, j) and (j, i).) The maximum cut is then obtained by maximizing $\mathrm{CUT}(G, x)$ over all x, that is

$$\text{MAX-CUT}(G) = \frac{1}{4} \max \left\{ \sum_{i,j=1}^{n} A_{ij}(1 - x_i x_j) \colon x_i = \pm 1 \text{ for all } i \right\}. \tag{3.24}$$

Let us start with a simple 0.5-approximation algorithm for the maximum cut – one which finds a cut through at least *half* the edges of G.

Proposition 3.6.3 (The 0.5-approximation algorithm for maximum cut) *Partition the vertices of G into two sets at random, uniformly over all 2^n possible partitions. Then the expectation of the resulting cut equals*

$$0.5|E| \ge 0.5\,\text{MAX-CUT}(G),$$

where $|E|$ denotes the total number of edges of G.

Proof The random cut is generated by a symmetric Bernoulli random vector $x \sim \mathrm{Unif}\big(\{-1,1\}^n\big)$, which has independent symmetric Bernoulli coordinates. Then, in (3.23) we have $\mathbb{E}\, x_i x_j = 0$ for $i \neq j$ and $A_{ij} = 0$ for $i = j$ (since the graph has no loops). Thus, using the linearity of expectations, we get

$$\mathbb{E}\,\mathrm{CUT}(G,x) = \frac{1}{4} \sum_{i,j=1}^{n} A_{ij} = \frac{1}{2}|E|.$$

This completes the proof. ■

Exercise 3.6.4 For any $\varepsilon > 0$, give an $(0.5 - \varepsilon)$-approximation algorithm for maximum cut, which is always *guaranteed* to give a suitable cut but may have a random running time. Give a bound on the expected running time. ☞

3.6.3 Semidefinite Relaxation

Now we will do much better and give a 0.878-approximation algorithm which is due to Goemans and Williamson. It is based on a semidefinite relaxation of the NP-hard problem (3.24). It should be easy to guess what such a relaxation could be: recalling (3.21), it is natural to consider the semidefinite problem

$$\mathrm{SDP}(G) := \frac{1}{4} \max \left\{ \sum_{i,j=1}^{n} A_{ij}(1 - \langle X_i, X_j \rangle) \colon X_i \in \mathbb{R}^n,\ \|X_i\|_2 = 1 \text{ for all } i \right\}. \tag{3.25}$$

(Again – why is this a semidefinite program?)

As we will see, not only does the value $\mathrm{SDP}(G)$ approximate $\text{MAX-CUT}(G)$ to within the 0.878 factor, but we can obtain an actual partition of G (i.e., the labels x_i) which attains

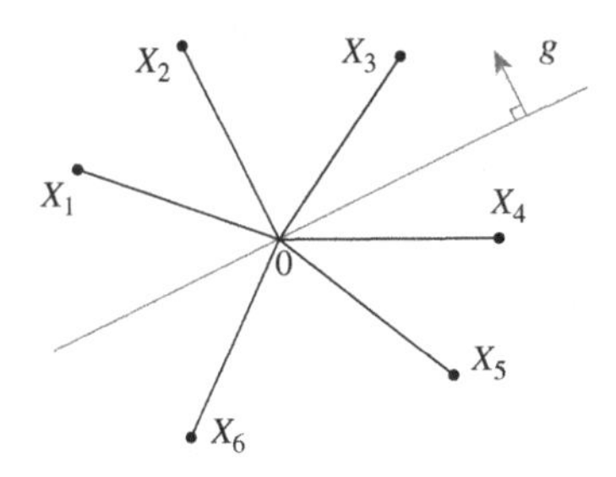

Figure 3.11 Randomized rounding of vectors $X_i \in \mathbb{R}^n$ into labels $x_i = \pm 1$. For this configuration of points X_i and a random hyperplane with normal vector g, we assign $x_1 = x_2 = x_3 = 1$ and $x_4 = x_5 = x_6 = -1$.

this value. To do this, we describe how to translate a solution (X_i) of (3.25) into labels $x_i = \pm 1$.

This can be done by the following *randomized rounding* step. Choose a random hyperplane in $\mathbb{R}^n$ passing through the origin. It cuts the set of vectors X_i into two parts; let us assign labels $x_i = 1$ to one part and $x_i = -1$ to the other part. Equivalently, we may choose a standard normal random vector

$$g \sim N(0, I_n)$$

and define

$$x_i := \operatorname{sign} \langle X_i, g \rangle, \quad i = 1, \dots, n. \tag{3.26}$$

See Figure 3.11 for an illustration.[6]

Theorem 3.6.5 (The 0.878-approximation algorithm for maximum cut) *Let G be a graph with adjacency matrix A. Let $x = (x_i)$ be the result of a randomized rounding of the solution (X_i) of the semidefinite program* (3.25). *Then*

$$\mathbb{E}\,\mathrm{CUT}(G, x) \geq 0.878\,\mathrm{SDP}(G) \geq 0.878\,\text{MAX-CUT}(G).$$

The proof of this theorem will be based on the following elementary identity. We can think of it as a more advanced version of the identity (3.6), which we used in the proof of Grothendieck's inequality, Theorem 3.5.1.

Lemma 3.6.6 (Grothendieck's identity) *Consider a random vector $g \sim N(0, I_n)$. Then, for any fixed vectors $u, v \in S^{n-1}$, we have*

$$\mathbb{E}\big(\operatorname{sign}\langle g, u\rangle \operatorname{sign}\langle g, v\rangle\big) = \frac{2}{\pi} \arcsin \langle u, v \rangle.$$

Exercise 3.6.7☕☕ Prove Grothendieck's identity. ☞

[6] In the rounding step, instead of the normal distribution we could use any other rotation invariant distribution in $\mathbb{R}^n$, for example, the uniform distribution on the sphere S^{n-1}.

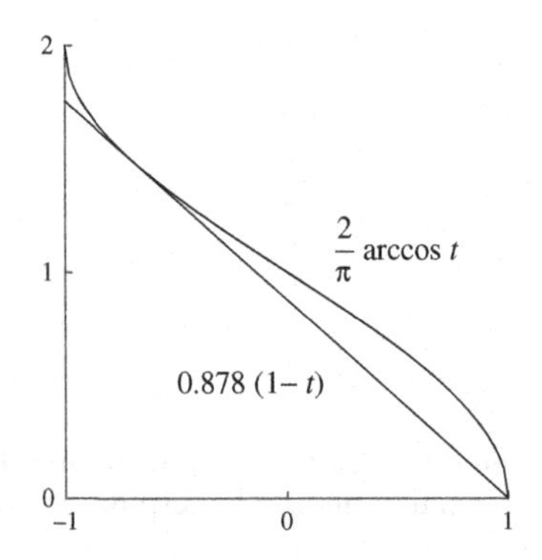

Figure 3.12 The inequality $(2/\pi)\arccos t \geq 0.878(1-t)$ holds for all $t \in [-1, 1]$.

A weak point of Grothendieck's identity is the nonlinear function arcsin, which would be hard to work with. Let us replace it with a linear function using the numeric inequality

$$1 - \frac{2}{\pi}\arcsin t = \frac{2}{\pi}\arccos t \geq 0.878(1-t), \quad t \in [-1, 1], \tag{3.27}$$

which can be easily verified using software; see Figure 3.12.

Proof of Theorem 3.6.5 By (3.23) and the linearity of expectations, we have

$$\mathbb{E}\,\mathrm{CUT}(G, x) = \frac{1}{4}\sum_{i,j=1}^{n} A_{ij}(1 - \mathbb{E}\,x_i x_j).$$

The definition of the labels x_i in the rounding step (3.26) gives

$$\begin{aligned} 1 - \mathbb{E}\,x_i x_j &= 1 - \mathbb{E}\big(\operatorname{sign}\langle X_i, g\rangle \operatorname{sign}\langle X_j, g\rangle\big) \\ &= 1 - \frac{2}{\pi}\arcsin\langle X_i, X_j\rangle \quad \text{(by Grothendieck's identity, Lemma 3.6.6)} \\ &\geq 0.878\big(1 - \langle X_i, X_j\rangle\big) \quad \text{(by (3.27))}. \end{aligned}$$

Therefore

$$\mathbb{E}\,\mathrm{CUT}(G, x) \geq 0.878 \times \frac{1}{4}\sum_{i,j=1}^{n} A_{ij}(1 - \langle X_i, X_j\rangle) = 0.878\,\mathrm{SDP}(G).$$

This proves the first inequality in the theorem. The second inequality is trivial since $\mathrm{SDP}(G) \geq \text{MAX-CUT}(G)$. (Why?) ■

3.7 Kernel Trick, and Tightening of Grothendieck's Inequality

Our proof of Grothendieck's inequality given in Section 3.5 yields a very loose bound on the absolute constant K. We now give an alternative proof that gives (almost) the best known constant $K \leq 1.783$.

Our new argument will be based on Grothendieck's identity (Lemma (3.6.6)). The main challenge in using this identity arises from the nonlinearity of the function $\arcsin(x)$. Indeed, suppose that there were no such nonlinearity, and we hypothetically had $\mathbb{E}\big(\operatorname{sign}\langle g,u\rangle \operatorname{sign}\langle g,v\rangle\big) = (2/\pi)\,\langle u,v\rangle$. Then Grothendieck's inequality would easily follow:

$$\frac{2}{\pi}\sum_{i,j} a_{ij}\,\langle u_i, v_j\rangle = \sum_{i,j} a_{ij}\,\mathbb{E}\big(\operatorname{sign}\langle g,u_i\rangle \operatorname{sign}\langle g,v_j\rangle\big) \le 1,$$

where in the last step we swapped the sum and expectation and used the assumption of Grothendieck's inequality for $x_i = \operatorname{sign}\langle g,u_i\rangle$ and $y_j = \operatorname{sign}\langle g,y_j\rangle$. This would give Grothendieck's inequality with $K \le \pi/2 \approx 1.57$.

This argument is of course wrong. To address the nonlinear form $(2/\pi)\arcsin\langle u,v\rangle$ that appears in Grothendieck's identity, we use the following remarkably powerful trick: represent $(2/\pi)\arcsin\langle u,v\rangle$ as the (linear) inner product $\langle u',v'\rangle$ of some other vectors u', v' in some Hilbert space H. In the literature on machine learning, this method is called the *kernel trick*.

We will explicitly construct nonlinear transformations $u' = \Phi(u)$, $v' = \Psi(v)$ that will do the job. Our construction is conveniently described in the language of *tensors*, which form a higher-dimensional generalization of the notion of matrices.

Definition 3.7.1 (Tensors) A tensor can be described as a multidimensional array. Thus, a kth order tensor $(a_{i_1\ldots i_k})$ is a k-dimensional array of real numbers $a_{i_1\ldots i_k}$. The canonical inner product on $\mathbb{R}^{n_1\times\cdots\times n_k}$ defines the inner product of tensors $A = (a_{i_1\ldots i_k})$ and $B = (b_{i_1\ldots i_k})$:

$$\langle A,B\rangle := \sum_{i_1,\ldots,i_k} a_{i_1\ldots i_k} b_{i_1\ldots i_k}. \tag{3.28}$$

Example 3.7.2 Scalars, vectors, and matrices are examples of tensors. As we noted in (3.19), for $m\times n$ matrices the inner product of tensors (3.28) specializes to

$$\langle A,B\rangle = \operatorname{tr}(A^{\mathsf T}B) = \sum_{i=1}^{m}\sum_{j=1}^{n} A_{ij}B_{ij}.$$

Example 3.7.3 (Rank-1 tensors) Every vector $u\in\mathbb{R}^n$ defines a kth-order *tensor product* $u\otimes\cdots\otimes u$, which is the tensor whose entries are the products of all k-tuples of the entries of u. In other words,

$$u\otimes\cdots\otimes u = u^{\otimes k} := (u_{i_1}\cdots u_{i_k}) \in \mathbb{R}^{n\times\cdots\times n}.$$

In particular, for $k=2$, the tensor product $u\otimes u$ is just the $n\times n$ matrix which is the outer product of u with itself:

$$u\otimes u = (u_i u_j)_{i,j=1}^n = uu^{\mathsf T}.$$

One can similarly define the tensor products $u\otimes v\otimes\cdots\otimes z$ for different vectors $u, v, \ldots, z$.

Exercise 3.7.4 Show that, for any vectors $u, v\in\mathbb{R}^n$ and $k\in\mathbb{N}$, we have

$$\langle u^{\otimes k}, v^{\otimes k}\rangle = \langle u,v\rangle^k.$$

This exercise shows a remarkable fact: we can represent nonlinear forms like $\langle u, v\rangle^k$ as the usual, *linear*, inner product in some other space. Formally, there exist a Hilbert space H and a transformation $\Phi\colon \mathbb{R}^n \to H$ such that

$$\langle \Phi(u), \Phi(v)\rangle = \langle u, v\rangle^k .$$

In this case, H is the space of kth-order tensors, and $\Phi(u) = u^{\otimes k}$.

In the next two exercises, we extend this observation to more general nonlinearities.

Exercise 3.7.5 ☕☕

(a) Show that there exist a Hilbert space H and a transformation $\Phi\colon \mathbb{R}^n \to H$ such that

$$\langle \Phi(u), \Phi(v)\rangle = 2\,\langle u, v\rangle^2 + 5\,\langle u, v\rangle^3 \quad \text{for all } u, v \in \mathbb{R}^n.$$ ☞

(b) More generally, consider a polynomial $f\colon \mathbb{R} \to \mathbb{R}$ with non-negative coefficients, and construct H and Φ such that

$$\langle \Phi(u), \Phi(v)\rangle = f(\langle u, v\rangle) \quad \text{for all } u, v \in \mathbb{R}^n.$$

(c) Show the same for any *real analytic function* $f\colon \mathbb{R} \to \mathbb{R}$ with non-negative coefficients, i.e., for any function that can be represented as a convergent series,

$$f(x) = \sum_{k=0}^{\infty} a_k x^k, \quad x \in \mathbb{R}, \tag{3.29}$$

and such that $a_k \geq 0$ for all k.

Exercise 3.7.6 ☕ Let $f\colon \mathbb{R} \to \mathbb{R}$ be any real analytic function (with possibly negative coefficients in (3.29)). Show that there exist a Hilbert space H and transformations $\Phi, \Psi\colon \mathbb{R}^n \to H$ such that

$$\langle \Phi(u), \Psi(v)\rangle = f(\langle u, v\rangle) \quad \text{for all } u, v \in \mathbb{R}^n.$$

Moreover, check that

$$\|\Phi(u)\|^2 = \|\Psi(u)\|^2 = \sum_{k=0}^{\infty} |a_k| \|u\|_2^{2k}.$$ ☞

Let us specialize the kernel trick to the nonlinearity $(2/\pi)\arcsin\langle u, v\rangle$ that appears in Grothendieck's identity.

Lemma 3.7.7 *There exists a Hilbert space H and transformations*[7] $\Phi, \Psi\colon S^{n-1} \to S(H)$ *such that*

$$\frac{2}{\pi}\arcsin\langle \Phi(u), \Psi(v)\rangle = \beta\,\langle u, v\rangle \quad \text{for all } u, v \in S^{n-1}, \tag{3.30}$$

where $\beta = (2/\pi)\ln(1+\sqrt{2})$.

[7] Here S^{n-1} denotes the unit Euclidean sphere in $\mathbb{R}^n$ and $S(H)$ denotes the unit sphere of the Hilbert space H.

Proof Rewrite the desired identity (3.30) as

$$\langle \Phi(u), \Psi(v) \rangle = \sin\left(\frac{\beta\pi}{2} \langle u, v \rangle\right). \tag{3.31}$$

The result of Exercise 3.7.6 gives us the Hilbert space H and the maps $\Phi, \Psi : \mathbb{R}^n \to H$ that satisfy (3.31). It remains only to determine the value of β for which Φ and Ψ map unit vectors to unit vectors. To do this, we recall the Taylor series

$$\sin t = t - \frac{t^3}{3!} + \frac{t^5}{5!} - \cdots \quad \text{and} \quad \sinh t = t + \frac{t^3}{3!} + \frac{t^5}{5!} + \cdots$$

Exercise 3.7.6 then guarantees that, for every $u \in S^{n-1}$, we have

$$\|\Phi(u)\|^2 = \|\Psi(u)\|^2 = \sinh\left(\frac{\beta\pi}{2}\right).$$

This quantity equals 1 if we set

$$\beta := \frac{2}{\pi} \operatorname{arcsinh}(1) = \frac{2}{\pi} \ln(1 + \sqrt{2}).$$

The lemma is proved. ∎

Now we are ready to prove Grothendieck's inequality (Theorem 3.5.1) with constant

$$K \le \frac{1}{\beta} = \frac{\pi}{2\ln(1+\sqrt{2})} \approx 1.783.$$

Proof of Theorem 3.5.1 We can assume without loss of generality that $u_i, v_j \in S^{N-1}$ (this is the same reduction as we made in the proof in Section 3.5). Lemma 3.7.7 gives us unit vectors $u_i' = \Phi(u_i)$ and $v_j' = \Psi(v_j)$, in some Hilbert space H, which satisfy

$$\frac{2}{\pi} \arcsin\left\langle u_i', v_j' \right\rangle = \beta \left\langle u_i, v_j \right\rangle \quad \text{for all } i, j.$$

We can again assume without loss of generality that $H = \mathbb{R}^M$ for some M. (Why?) Then

$$\begin{aligned}
\beta \sum_{i,j} a_{ij} \left\langle u_i, v_j \right\rangle &= \sum_{i,j} a_{ij} \frac{2}{\pi} \arcsin\left\langle u_i', v_j' \right\rangle \\
&= \sum_{i,j} a_{ij} \, \mathbb{E}\left(\operatorname{sign}\left\langle g, u_i' \right\rangle \operatorname{sign}\left\langle g, v_j' \right\rangle\right) \quad \text{(by Lemma 3.6.6)}, \\
&\le 1,
\end{aligned}$$

where in the last step we swapped the sum and expectation and used the assumption of Grothendieck's inequality with $x_i = \operatorname{sign}\left\langle g, u_i' \right\rangle$ and $y_j = \operatorname{sign}\left\langle g, y_j' \right\rangle$. This yields Grothendieck's inequality for $K \le 1/\beta$. ∎

3.7.1 Kernels and Feature Maps

Since the kernel trick was so successful in the proof of Grothendieck's inequality, we may ask – what other nonlinearities can be handled with the kernel trick? Let

$$K: \mathcal{X} \times \mathcal{X} \to \mathbb{R}$$

be a function of two variables on a set $\mathcal{X}$. Under what conditions on K can we find a Hilbert space H and a transformation

$$\Phi: \mathcal{X} \to H$$

such that

$$\langle \Phi(u), \Phi(v) \rangle = K(u, v) \quad \text{for all } u, v \in \mathcal{X}? \tag{3.32}$$

The answer to this question is provided by Mercer's theorem and, more precisely, the Moore–Aronszajn theorem. The necessary and sufficient condition is that K be a *positive-semidefinite kernel*, which means that, for any finite collection of points $u_1, \dots, u_N \in \mathcal{X}$, the matrix

$$\big(K(u_i, u_j)\big)_{i,j=1}^N$$

is positive-semidefinite. The map Φ is called a *feature map*, and the Hilbert space H can be constructed from the kernel K as a (unique) *reproducing kernel Hilbert space*.

Examples of positive-semidefinite kernels on $\mathbb{R}^n$ that are common in machine learning include the *Gaussian kernel* (also called the radial basis function kernel)

$$K(u, v) = \exp\Big(-\frac{\|u - v\|_2^2}{2\sigma^2}\Big), \quad u, v \in \mathbb{R}^n, \ \sigma > 0,$$

and the *polynomial kernel*

$$K(u, v) = \big(\langle u, v \rangle + r\big)^k, \quad u, v \in \mathbb{R}^n, \ r > 0, \ k \in \mathbb{N}.$$

The kernel trick (3.32), which represents a general kernel $K(u, v)$ as an inner product, is very popular in *machine learning*. It allows one to handle nonlinear models (determined by kernels K) using methods developed for linear models. In contrast with what we did in this section, in machine learning applications an explicit description of the Hilbert space H and the feature map $\Phi: \mathcal{X} \to H$ is typically not needed. Indeed, to compute the inner product $\langle \Phi(u), \Phi(v) \rangle$ in H, one does not need to know Φ: the identity (3.32) allows one to compute $K(u, v)$ instead.

3.8 Notes

Theorem 3.1.1 about the concentration of the norm of random vectors is known but difficult to locate in the existing literature. We will later prove a more general result, Theorem 6.3.2, which is valid for anisotropic random vectors. It is unknown whether the quadratic dependence on K in Theorem 3.1.1 is optimal. One may also wonder about concentration of the norm $\|X\|_2$ of random vectors X whose coordinates are not necessarily independent. In particular, for a random vector X that is uniformly distributed in a convex set K, the concentration of the norm is a central problem in geometric functional analysis; see [91, Section 2] and [35, Chapter 12].

Exercise 3.3.4 mentions the Cramér–Wold theorem. It a straightforward consequence of the uniqueness theorem for characteristic functions: see [22, Section 29].

The concept of frames introduced in Section 3.3.4 is an important extension of the notion of orthogonal bases. One can read more about frames and their applications in signal processing and data compression in, e.g., [50, 117].

Sections 3.3.5 and 3.4.4 discuss random vectors uniformly distributed in convex sets. The books [11, 35] study this topic in detail, and the surveys [180, 212] discuss the algorithmic aspects of computing the volume of convex sets in high dimensions.

Our discussion of sub-gaussian random vectors in Section 3.4 mostly follows [216]. An alternative geometric proof of Theorem 3.4.6 can be found in [13, Lemma 2.2].

Grothendieck's inequality (Theorem 3.5.1) was originally proved by A. Grothendieck in 1953 [88] with the bound on the constant $K \le \sinh(\pi/2) \approx 2.30$; a version of this original argument is presented [129, Section 2]. There are alternative proofs of Grothendieck's inequality with better and worse bounds on K; see [34] for the history. The surveys [111, 164] discuss ramifications and applications of Grothendieck's inequality in various areas of mathematics and computer science. Our first proof of Grothendieck's inequality, the proof given in Section 3.5, is similar to that in [5, Section 8.1]; it was kindly brought to the author's attention by Mark Rudelson. Our second proof, the one from Section 3.7, is due to J.-L. Krivine [118]; versions of this argument can be found e.g. in [7] and [122]. The bound on the constant

$$K \le \frac{\pi}{2\ln(1+\sqrt{2})} \approx 1.783$$

that follows from Krivine's argument is currently the best known *explicit* bound on K. It has been proved, however, that the best possible bound must be strictly smaller than Krivine's bound, but no explicit number is known [34].

Part of this chapter was about semidefinite relaxations of hard optimization problems. For an introduction to the area of convex optimization, including semidefinite programming, we refer to the books [33, 38, 122, 28]. For the use of Grothendieck's inequality in analyzing semidefinite relaxations, see [111, 7]. Our presentation of the maximum cut problem in Section 3.6 follows [38, Section 6.6] and [122, Chapter 7]. The semidefinite approach to maximum cut, which we discussed in Section 3.6.3, was pioneered in 1995 by M. Goemans and D. Williamson [81]. The approximation ratio

$$\frac{2}{\pi} \min_{0\le\theta\le\pi} \frac{\theta}{1-\cos(\theta)} \approx 0.878$$

guaranteed by the Goemans–Williamson algorithm remains the best known constant for the max-cut problem. If the unique games conjecture is true, this ratio cannot be improved, i.e., any better approximation would be NP-hard to compute [110].

In Section 3.7 we gave Krivine's proof of Grothendieck's inequality [118]. We also briefly discussed kernel methods there. To learn more about kernels, reproducing kernel Hilbert spaces, and their applications in machine learning, see e.g. the survey [100].

4

Random Matrices

We now begin the study of the non-asymptotic theory of random matrices, a study that will be continued in further chapters. Section 4.1 gives a quick reminder about singular values and matrix norms and their relationships. Section 4.2 introduces important geometric concepts – nets, covering and packing numbers, metric entropy, and discusses relations of these quantities with volume and coding. In Sections 4.4 and 4.6, we develop a basic ε*-net argument* and use it for random matrices. We first give a bound on the operator norm (Theorem 4.4.5) and then a stronger, two-sided bound on all singular values (Theorem 4.6.1) of random matrices. Three applications of random matrix theory are discussed in this chapter: a spectral clustering algorithm for recovering clusters, or communities, in complex networks (Section 4.5), covariance estimation (Section 4.7), and a spectral clustering algorithm for data presented as geometric point sets (Section 4.7.1).

4.1 Preliminaries on Matrices

You should be familiar with the notion of singular value decomposition from a basic course in linear algebra; we will recall it nevertheless. We will then introduce two matrix norms – operator and Frobenius – and discuss their relationships.

4.1.1 Singular Value Decomposition

The main object of our study will be an $m \times n$ matrix A with real entries. Recall that A can be represented using the *singular value decomposition* (SVD), which we can write as

$$A = \sum_{i=1}^{r} s_i u_i v_i^{\mathsf{T}}, \quad \text{where} \quad r = \text{rank}(A). \tag{4.1}$$

Here the non-negative numbers $s_i = s_i(A)$ are the *singular values* of A, the vectors $u_i \in \mathbb{R}^m$ are the *left singular vectors* of A, and the vectors $v_i \in \mathbb{R}^n$ are the *right singular vectors* of A.

For convenience, we often extend the sequence of singular values by setting $s_i = 0$ for $r < i \le n$, and we arrange them in a non-increasing order:

$$s_1 \ge s_2 \ge \cdots \ge s_n \ge 0.$$

The left singular vectors u_i are the orthonormal eigenvectors of AA^{T} and the right singular vectors v_i are the orthonormal eigenvectors of $A^{\mathsf{T}}A$. The singular values s_i are the square roots of the eigenvalues λ_i of both AA^{T} and $A^{\mathsf{T}}A$:

$$s_i(A) = \sqrt{\lambda_i(AA^\mathsf{T})} = \sqrt{\lambda_i(A^\mathsf{T}A)}.$$

In particular, if A is a *symmetric* matrix then the singular values of A are the absolute values of the eigenvalues λ_i of A:

$$s_i(A) = |\lambda_i(A)|,$$

and both the left and right singular vectors of A are eigenvectors of A.

The Courant–Fisher *min–max theorem* offers the following variational characterization of the eigenvalues $\lambda_i(A)$ of a symmetric matrix A, assuming they are arranged in a non-increasing order:

$$\lambda_i(A) = \max_{\dim E = i} \min_{x \in S(E)} \langle Ax, x \rangle . \tag{4.2}$$

Here the maximum is over all i-dimensional subspaces E of $\mathbb{R}^n$, the minimum is over all unit vectors $x \in E$, and $S(E)$ denotes the unit Euclidean sphere in the subspace E. For the singular values, the min–max theorem immediately implies that

$$s_i(A) = \max_{\dim E = i} \min_{x \in S(E)} \|Ax\|_2.$$

Exercise 4.1.1 Suppose that A is an invertible matrix with singular value decomposition

$$A = \sum_{i=1}^{n} s_i u_i v_i^\mathsf{T}.$$

Check that

$$A^{-1} = \sum_{i=1}^{n} \frac{1}{s_i} v_i u_i^\mathsf{T}.$$

4.1.2 Operator Norm and the Extreme Singular Values

The space of $m \times n$ matrices can be equipped with several classical norms. We mention two of them – the operator and Frobenius norms – and emphasize their connection with the spectrum of A.

When we are thinking of the space $\mathbb{R}^m$ along with the Euclidean norm $\|\cdot\|_2$ on it, we denote this Hilbert space by ℓ_2^m. The matrix A acts as a linear operator from ℓ_2^n to ℓ_2^m. Its *operator norm*, also called the *spectral norm*, is defined as

$$\|A\| := \|A : \ell_2^n \to \ell_2^m\| = \max_{x \in \mathbb{R}^n \setminus \{0\}} \frac{\|Ax\|_2}{\|x\|_2} = \max_{x \in S^{n-1}} \|Ax\|_2.$$

Equivalently, the operator norm of A can be computed by maximizing the quadratic form $\langle Ax, y \rangle$ over all unit vectors x, y:

$$\|A\| = \max_{x \in S^{n-1},\ y \in S^{m-1}} \langle Ax, y \rangle .$$

In terms of its spectrum, the operator norm of A equals the largest singular value of A:

$$s_1(A) = \|A\|.$$

(Check!)

The smallest singular value $s_n(A)$ also has a special meaning. By definition, it can only be nonzero for tall matrices, where $m \geq n$. In this case, A has full rank n if and only if $s_n(A) > 0$. Moreover, $s_n(A)$ is a quantitative measure of the *non-degeneracy* of A. Indeed,

$$s_n(A) = \frac{1}{\|A^+\|}$$

where A^+ is the Moore–Penrose pseudoinverse of A. Its norm $\|A^+\|$ is the norm of the operator A^{-1} restricted to the image of A.

4.1.3 Frobenius Norm

The *Frobenius norm*, also called *Hilbert–Schmidt* norm, of a matrix A with entries A_{ij} is defined as

$$\|A\|_F = \Big(\sum_{i=1}^{m}\sum_{j=1}^{n}|A_{ij}|^2\Big)^{1/2}.$$

Thus the Frobenius norm is the Euclidean norm on the space of matrices $\mathbb{R}^{m\times n}$. In terms of singular values, the Frobenius norm can be computed as

$$\|A\|_F = \left(\sum_{i=1}^{r} s_i(A)^2\right)^{1/2}.$$

The canonical inner product on $\mathbb{R}^{m\times n}$ can be represented in terms of matrices as

$$\langle A, B\rangle = \operatorname{tr}(A^{\mathsf{T}}B) = \sum_{i=1}^{m}\sum_{j=1}^{n} A_{ij}B_{ij}. \tag{4.3}$$

Obviously, the canonical inner product generates the canonical Euclidean norm, i.e.

$$\|A\|_F^2 = \langle A, A\rangle .$$

Let us now compare the operator and the Frobenius norm. If we look at the vector $s = (s_1, \ldots, s_r)$ of singular values of A, these norms become the ℓ_∞ and ℓ_2 norms, respectively:

$$\|A\| = \|s\|_\infty, \quad \|A\|_F = \|s\|_2.$$

Using the inequality $\|s\|_\infty \leq \|s\|_2 \leq \sqrt{r}\,\|s\|_\infty$ for $s \in \mathbb{R}^n$ (check it!), we obtain the best possible relation between the operator and Frobenius norms:

$$\|A\| \leq \|A\|_F \leq \sqrt{r}\,\|A\|. \tag{4.4}$$

Exercise 4.1.2 ☕☕ Prove the following bound on the singular values s_i of any matrix A:

$$s_i \leq \frac{1}{\sqrt{i}}\|A\|_F.$$

4.1.4 Low-Rank Approximation

Suppose we want to approximate a given matrix A of rank r by a matrix A_k that has a given lower rank $k < r$. What is the best choice for A_k? In other words, what matrix A_k of rank k minimizes the distance to A? This distance can be measured by the operator norm or by the Frobenius norm.

In either case, the *Eckart–Young–Mirsky theorem* gives the answer to this low-rank approximation problem. It states that the minimizer A_k is obtained by truncating the singular value decomposition of A at the kth term:

$$A_k = \sum_{i=1}^{k} s_i u_i v_i^{\mathsf{T}}.$$

In other words, the Eckart–Young–Mirsky theorem states that

$$\|A - A_k\| = \min_{\operatorname{rank}(A') \le k} \|A - A'\|.$$

A similar statement holds for the Frobenius norm (and, in fact, for any unitarily invariant norm). The matrix A_k is often called the *best rank-k approximation* of A.

Exercise 4.1.3 (Best rank-k approximation) Let A_k be the best rank-k approximation of a matrix A. Express $\|A - A_k\|^2$ and $\|A - A_k\|_F^2$ in terms of the singular values s_i of A.

4.1.5 Approximate Isometries

The extreme singular values $s_1(A)$ and $s_n(A)$ have an important geometric meaning. They are respectively the smallest number M and the largest number m that make the following inequality true:

$$m\|x\|_2 \le \|Ax\|_2 \le M\|x\|_2 \quad \text{for all } x \in \mathbb{R}^n. \tag{4.5}$$

(Check!) Applying this inequality for $x - y$ instead of x and with the best bounds, we can rewrite it as

$$s_n(A)\|x - y\|_2 \le \|Ax - Ay\|_2 \le s_1(A)\|x - y\|_2 \quad \text{for all } x \in \mathbb{R}^n.$$

This means that the matrix A, acting as an operator from $\mathbb{R}^n$ to $\mathbb{R}^m$, can only change the distance between any points by a factor that lies between $s_n(A)$ and $s_1(A)$. Thus the extreme singular values control the *distortion* of the geometry of $\mathbb{R}^n$ under the action of A.

The best possible matrices in this sense, which preserve distances exactly, are called *isometries*. Let us recall their characterization, which can be proved using elementary linear algebra.

Exercise 4.1.4 (Isometries) Let A be an $m \times n$ matrix with $m \ge n$. Prove that the following statements are equivalent.

(a) $A^{\mathsf{T}} A = I_n$.
(b) $P := AA^{\mathsf{T}}$ is an *orthogonal projection*[1] in $\mathbb{R}^m$ onto a subspace of dimension n.

[1] Recall that P is a projection if $P^2 = P$, and P is called orthogonal if the image and kernel of P are orthogonal subspaces.

(c) A is an *isometry*, or isometric embedding of $\mathbb{R}^n$ into $\mathbb{R}^m$, which means that

$$\|Ax\|_2 = \|x\|_2 \quad \text{for all } x \in \mathbb{R}^n.$$

(d) All singular values of A equal 1; equivalently

$$s_n(A) = s_1(A) = 1.$$

Quite often the conditions of Exercise 4.1.4 hold only approximately, in which case we regard A as an *approximate isometry*.

Lemma 4.1.5 (Approximate isometries) *Let A be an $m \times n$ matrix and $\delta > 0$. Suppose that*

$$\|A^{\mathsf T} A - I_n\| \le \max(\delta, \delta^2).$$

Then

$$(1-\delta)\|x\|_2 \le \|Ax\|_2 \le (1+\delta)\|x\|_2 \quad \textit{for all } x \in \mathbb{R}^n. \tag{4.6}$$

Consequently, all singular values of A lie between $1-\delta$ and $1+\delta$:

$$1-\delta \le s_n(A) \le s_1(A) \le 1+\delta. \tag{4.7}$$

Proof To prove (4.6), we may assume without loss of generality that $\|x\|_2 = 1$. (Why?) Then, using the assumption, we get

$$\max(\delta, \delta^2) \ge \left|\langle (A^{\mathsf T} A - I_n)x, x\rangle\right| = \left|\|Ax\|_2^2 - 1\right|.$$

Applying the elementary inequality

$$\max(|z-1|, |z-1|^2) \le |z^2 - 1|, \quad z \ge 0, \tag{4.8}$$

for $z = \|Ax\|_2$, we conclude that

$$|\|Ax\|_2 - 1| \le \delta.$$

This proves (4.6), which in turn implies (4.7), as we saw at the beginning of this section. ∎

Exercise 4.1.6 (Approximate isometries) Prove the following converse to Lemma 4.1.5: if (4.7) holds, then

$$\|A^{\mathsf T} A - I_n\| \le 3\max(\delta, \delta^2).$$

Remark 4.1.7 (Projections vs. isometries) Consider an $n \times m$ matrix Q. Then

$$QQ^{\mathsf T} = I_n$$

if and only if

$$P := Q^{\mathsf T} Q$$

is an orthogonal projection in $\mathbb{R}^m$ onto a subspace of dimension n. (This can be checked directly or deduced from Exercise 4.1.4 by taking $A = Q^{\mathsf T}$.) When this is the case, the matrix Q itself is often called a *projection* from $\mathbb{R}^m$ onto $\mathbb{R}^n$.

Note that A is an isometric embedding of $\mathbb{R}^n$ into $\mathbb{R}^m$ if and only if A^T is a projection from $\mathbb{R}^m$ onto $\mathbb{R}^n$. These remarks can be also made for an approximate isometry A; the transpose A^T in this case is an *approximate projection.*

Exercise 4.1.8 (Isometries and projections from unitary matrices) Canonical examples of isometries and projections can be constructed from a fixed unitary matrix U. Check that any sub-matrix of U obtained by selecting a subset of columns is an isometry and any sub-matrix obtained by selecting a subset of rows is a projection.

4.2 Nets, Covering Numbers, and Packing Numbers

We are going to develop a simple but powerful method – an ε-net argument – and illustrate its usefulness for the analysis of random matrices. In this section, we recall the concept of an *ε-net*, which you may have met in a course on real analysis, and we relate it to some other basic notions – covering, packing, entropy, volume, and coding.

Definition 4.2.1 (ε-Net) Let (T, d) be a metric space. Consider a subset $K \subset T$ and let $\varepsilon > 0$. A subset $\mathcal{N} \subseteq K$ is called an *ε-net of K* if every point in K is within a distance ε of some point of $\mathcal{N}$, i.e.

$$\forall x \in K \ \exists x_0 \in \mathcal{N} : d(x, x_0) \le \varepsilon.$$

Equivalently, $\mathcal{N}$ is an ε-net of K if and only if K can be covered by balls with centers in $\mathcal{N}$ and radii ε; see Figure 4.1(a).

If you ever feel confused by too much generality, it might be helpful to keep in mind an important example. Let $T = \mathbb{R}^n$ with d the Euclidean distance, i.e.,

$$d(x, y) = \|x - y\|_2, \quad x, y \in \mathbb{R}^n. \tag{4.9}$$

In this case, we can cover a subset $K \subset \mathbb{R}^n$ by *round balls*, as shown in Figure 4.1a. We have already seen an example of such a covering in Corollary 0.0.4, where K was a polytope.

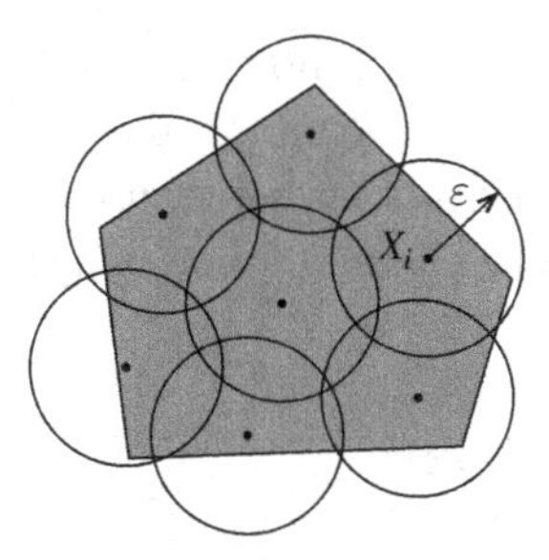

(a) This covering of a pentagon K by seven ε-balls shows that $\mathcal{N}(K,\varepsilon) \le 7$.

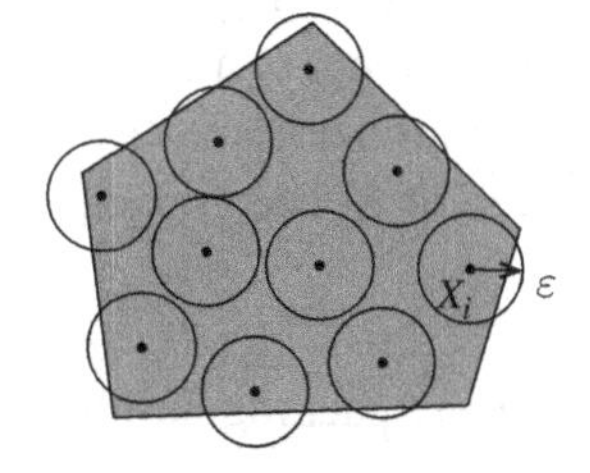

(b) This packing of a pentagon K by ten ε-balls shows that $\mathcal{P}(K,\varepsilon) \ge 10$.

Figure 4.1 Packing and covering.

Definition 4.2.2 (Covering numbers) The smallest possible cardinality of an ε-net of K is called the *covering number* of K and is denoted $\mathcal{N}(K, d, \varepsilon)$. Equivalently, $\mathcal{N}(K, d, \varepsilon)$ is the smallest number of closed balls with centers in K and radii ε whose union covers K.

Remark 4.2.3 (Compactness) An important result in real analysis states that a subset K of a complete metric space (T, d) is *precompact* (i.e., the closure of K is compact) if and only if

$$\mathcal{N}(K, d, \varepsilon) < \infty \quad \text{for every } \varepsilon > 0.$$

Thus we can think of the magnitude $\mathcal{N}(K, d, \varepsilon)$ as a quantitative measure of the compactness of K.

Closely related to covering is the notion of *packing*.

Definition 4.2.4 (Packing numbers) A subset $\mathcal{N}$ of a metric space (T, d) is *ε-separated* if $d(x, y) > \varepsilon$ for all distinct points $x, y \in \mathcal{N}$. The largest possible cardinality of an ε-separated subset of a given set $K \subset T$ is called the *packing number* of K and is denoted $\mathcal{P}(K, d, \varepsilon)$.

Exercise 4.2.5 (Packing the balls into K)☕☕

(a) Suppose that T is a normed space. Prove that $\mathcal{P}(K, d, \varepsilon)$ is the largest number of closed disjoint balls with centers in K and radii $\varepsilon/2$. See Figure 4.1b for an illustration.
(b) Show by example that the previous statement may be false for a general metric space T.

Lemma 4.2.6 (Nets from separated sets) *Let $\mathcal{N}$ be a maximal*[2] *ε-separated subset of K. Then $\mathcal{N}$ is an ε-net of K.*

Proof Let $x \in K$; we want to show that there exists $x_0 \in \mathcal{N}$ such that $d(x, x_0) \leq \varepsilon$. If $x \in \mathcal{N}$, the conclusion is trivial on choosing $x_0 = x$. Suppose now that $x \notin \mathcal{N}$. The maximality assumption implies that $\mathcal{N} \cup \{x_0\}$ is not ε-separated. But this means precisely that $d(x, x_0) \leq \varepsilon$ for some $x_0 \in \mathcal{N}$. ∎

Remark 4.2.7 (Constructing a net) Lemma 4.2.6 leads to the following simple algorithm for constructing an ε-net of a given set K. Choose a point $x_1 \in K$ arbitrarily, choose a point $x_2 \in K$ which is farther than ε from x_1, choose x_3 so that it is farther than ε from both x_1 and x_2, and so on. If K is compact, the algorithm terminates in finite time (why?) and gives an ε-net of K.

The covering and packing numbers are essentially equivalent:

Lemma 4.2.8 (Equivalence of covering and packing numbers) *For any set $K \subset T$ and any $\varepsilon > 0$, we have*

$$\mathcal{P}(K, d, 2\varepsilon) \leq \mathcal{N}(K, d, \varepsilon) \leq \mathcal{P}(K, d, \varepsilon).$$

[2] Here by "maximal" we mean that adding any new point to $\mathcal{N}$ destroys the separation property.

Proof The upper bound follows from Lemma 4.2.6. (How?)

To prove the lower bound, choose a 2ε-separated subset $\mathcal{P} = \{x_i\}$ in K and an ε-net $\mathcal{N} = \{y_j\}$ of K. By the definition of a net, each point x_i belongs a closed ε-ball centered at some point y_j. Moreover, since any closed ε-ball cannot contain a pair of 2ε-separated points, each ε-ball centered at y_j may contain at most one point x_i. The pigeonhole principle then yields $|\mathcal{P}| \leq |\mathcal{N}|$. Since this happens for arbitrary packing number $\mathcal{P}$ and covering number $\mathcal{N}$, the lower bound in the lemma is proved. ■

Exercise 4.2.9 (Allowing the centers to be outside K) In our definition of the covering numbers of K, we required that the centers x_i of the balls $B(x_i, \varepsilon)$ that form a covering lie in K. Relaxing this condition, define the *exterior covering number* $\mathcal{N}^{\text{ext}}(K, d, \varepsilon)$ similarly but without requiring that $x_i \in K$. Prove that

$$\mathcal{N}^{\text{ext}}(K, d, \varepsilon) \leq \mathcal{N}(K, d, \varepsilon) \leq \mathcal{N}^{\text{ext}}(K, d, \varepsilon/2).$$

Exercise 4.2.10 (Monotonicity) Give a counterexample to the following monotonicity property:

$$L \subset K \quad \text{implies} \quad \mathcal{N}(L, d, \varepsilon) \leq \mathcal{N}(K, d, \varepsilon).$$

Prove an approximate version of monotonicity:

$$L \subset K \quad \text{implies} \quad \mathcal{N}(L, d, \varepsilon) \leq \mathcal{N}(K, d, \varepsilon/2).$$

4.2.1 Covering Numbers and Volume

Let us now specialize our study of covering numbers to the most important example, where $T = \mathbb{R}^n$ with Euclidean metric

$$d(x, y) = \|x - y\|_2,$$

as in (4.9). To ease the notation, we often omit the metric when it is understood from the context, thus writing

$$\mathcal{N}(K, \varepsilon) = \mathcal{N}(K, d, \varepsilon).$$

If the covering numbers measure the size of K, how are they related to the most classical measure of size, the volume of K in $\mathbb{R}^n$? There could not be a full equivalence between these two quantities, since "flat" sets have zero volume but nonzero covering numbers.

Still, there is a useful partial equivalence, which is often quite sharp. It is based on the notion of the *Minkowski sum* of sets in $\mathbb{R}^n$.

Definition 4.2.11 (Minkowski sum) Let A and B be subsets of $\mathbb{R}^n$. The *Minkowski sum* $A + B$ is defined as

$$A + B := \{a + b \colon a \in A,\ b \in B\}.$$

Figure 4.2 shows an example of the Minkowski sum of two sets on the plane.

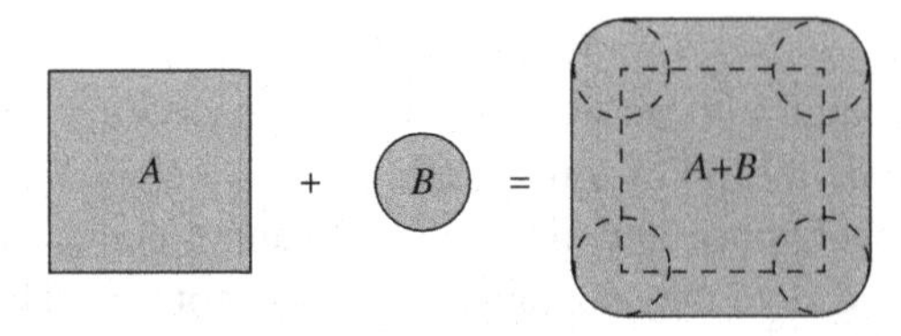

Figure 4.2 The Minkowski sum of a square and a circle is a square with rounded corners.

Proposition 4.2.12 (Covering numbers and volume) *Let K be a subset of $\mathbb{R}^n$ and $\varepsilon > 0$. Then*

$$\frac{|K|}{|\varepsilon B_2^n|} \le \mathcal{N}(K, \varepsilon) \le \mathcal{P}(K, \varepsilon) \le \frac{|(K + (\varepsilon/2)B_2^n)|}{|(\varepsilon/2)B_2^n|}.$$

Here $|\cdot|$ denotes the volume in $\mathbb{R}^n$, B_2^n denotes the unit Euclidean ball in $\mathbb{R}^n$;[3] *so εB_2^n is a Euclidean ball with radius ε.*

Proof The middle inequality follows from Lemma 4.2.8, so all we need to prove are the left and right bounds.

Lower bound. Let $N := \mathcal{N}(K, \varepsilon)$. Then K can be covered by N balls with radii ε. Comparing the volumes, we obtain

$$|K| \le N\,|\varepsilon B_2^n|.$$

Dividing both sides by $|\varepsilon B_2^n|$ yields the lower bound.

Upper bound. Let $N := \mathcal{P}(K, \varepsilon)$. Then one can construct N closed disjoint balls $B(x_i, \varepsilon/2)$ with centers $x_i \in K$ and radii $\varepsilon/2$ (see Exercise 4.2.5). While these balls may not need to fit entirely into K (see Figure 4.1b), they do fit into a slightly inflated set, namely $K + (\varepsilon/2)B_2^n$. (Why?) Comparing the volumes, we obtain

$$N\,|(\varepsilon/2)B_2^n| \le |K + (\varepsilon/2)B_2^n|,$$

which leads to the upper bound in the proposition. ∎

An important consequence of the volumetric bound (4.10) is that the covering (and thus packing) numbers of the Euclidean ball, as well as many other sets, are *exponential* in the dimension n. Let us check this.

Corollary 4.2.13 (Covering numbers of the Euclidean ball) *The covering numbers of the unit Euclidean ball B_2^n satisfy the following for any $\varepsilon > 0$:*

$$\left(\frac{1}{\varepsilon}\right)^n \le \mathcal{N}(B_2^n, \varepsilon) \le \left(\frac{2}{\varepsilon} + 1\right)^n.$$

The same upper bound is true for the unit Euclidean sphere S^{n-1}.

[3] Thus $B_2^n = \{x \in \mathbb{R}^n : \|x\|_2 \le 1\}$.

Proof The lower bound follows immediately from Proposition 4.2.12, since the volume in $\mathbb{R}^n$ scales as follows:

$$|\varepsilon B_2^n| = \varepsilon^n |B_2^n|.$$

The upper bound follows from Proposition 4.2.12, too:

$$\mathcal{N}(B_2^n, \varepsilon) \le \frac{|(1+\varepsilon/2)B_2^n|}{|(\varepsilon/2)B_2^n|} = \frac{(1+\varepsilon/2)^n}{(\varepsilon/2)^n} = \Big(\frac{2}{\varepsilon}+1\Big)^n.$$

The upper bound for the sphere can be proved in the same way. ■

To simplify the bound, note that in the nontrivial range $\varepsilon \in (0, 1]$ we have

$$\Big(\frac{1}{\varepsilon}\Big)^n \le \mathcal{N}(B_2^n, \varepsilon) \le \Big(\frac{3}{\varepsilon}\Big)^n. \tag{4.10}$$

In the trivial range, where $\varepsilon > 1$, the unit ball can be covered by just one ε-ball, so $N(B_2^n, \varepsilon) = 1$.

The volumetric argument we just gave works well in many other situations. Let us give an important example.

Definition 4.2.14 (Hamming cube) The Hamming cube $\{0, 1\}^n$ consists of all binary strings of length n. The *Hamming distance* $d_H(x, y)$ between two binary strings is defined as the number of bits where x and y disagree, i.e.

$$d_H(x, y) := \#\big\{i : x(i) \ne y(i)\big\}, \quad x, y \in \{0, 1\}^n.$$

Endowed with this metric, the Hamming cube is a metric space $(\{0, 1\}^n, d_H)$, which is sometimes called the *Hamming space*.

Exercise 4.2.15☕ Check that d_H is indeed a metric.

Exercise 4.2.16 (Covering and packing numbers of the Hamming cube)☕☕☕ Let $K = \{0, 1\}^n$. Prove that, for every integer $m \in [0, n]$, we have

$$\frac{2^n}{\sum_{k=0}^{m} \binom{n}{k}} \le \mathcal{N}(K, d_H, m) \le \mathcal{P}(K, d_H, m) \le \frac{2^n}{\sum_{k=0}^{\lfloor m/2 \rfloor} \binom{n}{k}}$$ ☞

To make these bounds easier to compute, one can use the bounds for binomial sums from Exercise 0.0.5.

4.3 Application: Error Correcting Codes

Covering and packing arguments frequently appear in applications to *coding theory*. Here we give two examples that relate the covering and packing numbers to complexity and error correction.

4.3.1 Metric Entropy and Complexity

Intuitively, the covering and packing numbers measure the *complexity* of a set K. The logarithm of the covering number $\log_2 \mathcal{N}(K, \varepsilon)$ is often called the *metric entropy* of K. As we will see now, the metric entropy is equivalent to the number of bits needed to encode the points in K.

Proposition 4.3.1 (Metric entropy and coding) *Let (T, d) be a metric space, and consider a subset $K \subset T$. Let $\mathcal{C}(K, d, \varepsilon)$ denote the smallest number of bits sufficient to specify every point $x \in K$ with accuracy ε in the metric d. Then*

$$\log_2 \mathcal{N}(K, d, \varepsilon) \le \mathcal{C}(K, d, \varepsilon) \le \log_2 \mathcal{N}(K, d, \varepsilon/2).$$

Proof Lower bound. Assume $\mathcal{C}(K, d, \varepsilon) \le N$. This means that there exists a transformation (an "encoding") of the points $x \in K$ into bit strings of length N which specifies every point with accuracy ε. Such a transformation induces a partition of K into at most 2^N subsets, which are obtained by grouping the points represented by the same bit string; see Figure 4.3 for an illustration. Each subset must have diameter[4] at most ε, and thus it can be covered by a ball centered in K and with radius ε. (Why?) Thus K can be covered by at most 2^N balls with radii ε. This implies that $\mathcal{N}(K, d, \varepsilon) \le 2^N$. Taking logarithms on both sides, we obtain the lower bound in the proposition.

Upper bound. Assume that $\log_2 \mathcal{N}(K, d, \varepsilon/2) \le N$; this means that there exists an $(\varepsilon/2)$-net $\mathcal{N}$ of K with cardinality $|\mathcal{N}| \le 2^N$. To every point $x \in K$, let us assign a point $x_0 \in \mathcal{N}$ that is closest to x. Since there are at most 2^N such points, N bits are sufficient to specify the point x_0. It remains to note that the encoding $x \mapsto x_0$ represents points in K with accuracy ε. Indeed, if both x and y are encoded by the same x_0 then, by the triangle inequality,

$$d(x, y) \le d(x, x_0) + d(y, x_0) \le \frac{\varepsilon}{2} + \frac{\varepsilon}{2} = \varepsilon.$$

This shows that $\mathcal{C}(K, d, \varepsilon) \le N$. This completes the proof. ∎

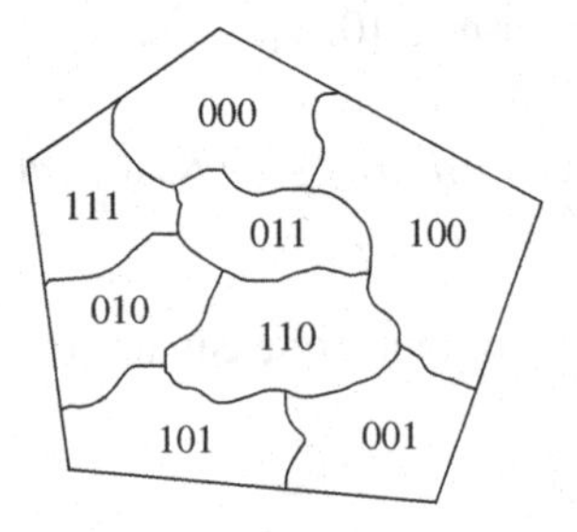

Figure 4.3 Encoding points in K as N-bit strings induces a partition of K into at most 2^N subsets.

[4] If (T, d) is a metric space and $K \subset T$, the diameter of the set K is defined as $\text{diam}(K) := \sup\{d(x, y) : x, y \in K\}$.

4.3.2 Error Correcting Codes

Suppose Alice wants to send Bob a message that consists of k letters, such as

$$x := \text{“}\textit{fill the glass}\text{”}.$$

Suppose further that an adversary may corrupt Alice's message by changing at most r letters in it. For example, Bob may receive

$$y := \text{“}\textit{bill the class}\text{”}$$

if $r = 2$. Is there a way to protect the communication channel between Alice and Bob, a method that can correct adversarial errors?

A common approach relies on using *redundancy*. Alice would encode her k-letter message into a longer, n-letter, message for some $n > k$, hoping that the extra information would help Bob get her message right despite any r errors.

Example 4.3.2 (Repetition code) Alice may just repeat her message several times, thus sending to Bob

$$E(x) := \text{“}\textit{fill the glass fill the glass fill the glass fill the glass fill the glass}\text{”}.$$

Bill could then use the *majority decoding*: to determine the value of any particular letter, he would look at the received copies of it in $E(x)$ and choose the value that occurs most frequently. If the original message x is repeated $2r + 1$ times, then the majority decoding recovers x exactly even when r letters of $E(x)$ are corrupted. (Why?)

The problem with majority decoding is that it is very inefficient: it uses

$$n = (2r + 1)k \tag{4.11}$$

letters to encode a k-letter message. As we will see shortly, there exist error correcting codes with much smaller n values.

But first let us formalize the notion of an error correcting code – an encoding of k-letter strings into n-letter strings that can correct r errors. For convenience, instead of using the English alphabet we shall work with the binary alphabet consisting of two letters 0 and 1.

Definition 4.3.3 (Error correcting code) Fix integers k, n, and r. The two maps

$$E\colon \{0,1\}^k \to \{0,1\}^n \quad \text{and} \quad D\colon \{0,1\}^n \to \{0,1\}^k$$

are called *encoding* and *decoding* maps and can correct r errors if we have

$$D(y) = x$$

for every word $x \in \{0,1\}^n$ and every string $y \in \{0,1\}^k$ that differs from $E(x)$ in at most r bits. The encoding map E is called an *error correcting code*; its image $E(\{0,1\}^k)$ is called a *codebook* (and very often the image itself is called the *error correcting code*); the elements $E(x)$ of the image are called *codewords*.

We now relate error correction to the packing numbers of the Hamming cube $(\{0,1\}^n, d_H)$, where d_H is the Hamming metric introduced in Definition 4.2.14.

Lemma 4.3.4 (Error correction and packing) *Assume that positive integers k, n, and r are such that*

$$\log_2 \mathcal{P}(\{0,1\}^n, d_H, 2r) \geq k.$$

Then there exists an error correcting code that encodes k-bit strings into n-bit strings and can correct r errors.

Proof By assumption, there exists a subset $\mathcal{N} \subset \{0,1\}^n$ with cardinality $|\mathcal{N}| = 2^k$ and such that the closed balls centered at the points in $\mathcal{N}$ and with radii r are disjoint. (Why?) We then define the encoding and decoding maps as follows: choose $E\colon \{0,1\}^k \to \mathcal{N}$ to be an arbitrary one-to-one map and $D\colon \{0,1\}^n \to \{0,1\}^k$ to be a nearest-neighbor decoder.[5]

Now, if $y \in \{0,1\}^n$ differs from $E(x)$ in at most r bits, y lies in the closed ball centered at $E(x)$ and with radius r. Since such balls are disjoint by construction, y must be strictly closer to $E(x)$ than to any other codeword $E(x')$ in $\mathcal{N}$. Thus the nearest-neighbor decoding decodes y correctly, i.e. $D(y) = x$. This completes the proof. ■

Let us substitute into Lemma 4.3.4 the bounds on the packing numbers of the Hamming cube from Exercise 4.2.16.

Theorem 4.3.5 (Guarantees for an error correcting code) *Assume that positive integers k, n, and r are such that*

$$n \geq k + 2r \log_2 \left(\frac{en}{2r}\right).$$

Then there exists an error correcting code that encodes k-bit strings into n-bit strings and can correct r errors.

Proof Passing from packing to covering numbers, using Lemma 4.2.8, and then using the bounds on the covering numbers from Exercise 4.2.16 (and simplifying using Exercise 0.0.5), we get

$$\mathcal{P}(\{0,1\}^n, d_H, 2r) \geq \mathcal{N}(\{0,1\}^n, d_H, 2r) \geq 2^n \left(\frac{2r}{en}\right)^{2r}.$$

By assumption, this quantity is further bounded below by 2^k. An application of Lemma 4.3.4 completes the proof. ■

Informally, Theorem 4.3.5 shows that we can correct r errors if we make the *information overhead* $n-k$ almost linear in r:

$$n - k \asymp r \log\left(\frac{n}{r}\right).$$

This overhead is much smaller than for the repetition code (4.11). For example, to correct two errors in Alice's 12-letter message *"fill the glass"*, encoding it into a 30-letter codeword would suffice.

[5] Formally, we set $D(y) = x_0$, where $E(x_0)$ is the closest codeword in $\mathcal{N}$ to y.

Remark 4.3.6 (Rate) The guarantees of a given error correcting code are traditionally expressed in terms of the tradeoff between the *rate* and the *fraction of errors*, defined respectively as

$$R := \frac{k}{n} \quad \text{and} \quad \delta := \frac{r}{n}.$$

Theorem 4.3.5 states that there exist error correcting codes with rate as high as

$$R \geq 1 - f(2\delta)$$

where $f(t) = t \log_2(e/t)$.

Exercise 4.3.7 (Optimality)☕☕☕

(a) Prove the converse to the statement of Lemma 4.3.4.
(b) Deduce a converse to Theorem 4.3.5. Conclude that for any error correcting code that encodes k-bit strings into n-bit strings and can correct r errors, the rate must be

$$R \leq 1 - f(\delta)$$

where $f(t) = t \log_2(e/t)$ as before.

4.4 Upper Bounds on Random Sub-Gaussian Matrices

We are now ready to begin to study the non-asymptotic theory of random matrices. Random matrix theory is concerned with $m \times n$ matrices A with random entries. The central questions of this theory are about the distributions of the singular values, eigenvalues (if A is symmetric), and eigenvectors of A.

Theorem 4.4.5 will give a first bound on the operator norm (equivalently, on the largest singular value) of a random matrix with independent sub-gaussian entries. It is neither the sharpest nor the most general result; it will be sharpened and extended in Sections 4.6 and 6.5.

But before we do this, let us pause to learn how ε-nets can help us compute the operator norm of a matrix.

4.4.1 Computing the Norm on a Net

The notion of ε-nets can help us to simplify various problems involving high-dimensional sets. One such problem is the computation of the operator norm of an $m \times n$ matrix A. The operator norm was defined in Section 4.1.2 as

$$\|A\| = \max_{x \in S^{n-1}} \|Ax\|_2.$$

Thus, to evaluate $\|A\|$ one needs to bound $\|Ax\|_2$, uniformly over the sphere S^{n-1}. We will show that, instead of the entire sphere, it is enough to gain control just over an ε-net of the sphere (in the Euclidean metric).

Lemma 4.4.1 (Computing the operator norm on a net) *Let A be an $m \times n$ matrix and $\varepsilon \in [0, 1)$. Then, for any ε-net $\mathcal{N}$ of the sphere S^{n-1}, we have*

$$\sup_{x \in \mathcal{N}} \|Ax\|_2 \leq \|A\| \leq \frac{1}{1 - \varepsilon} \sup_{x \in \mathcal{N}} \|Ax\|_2$$

Proof The lower bound in the conclusion is trivial since $\mathcal{N} \subset S^{n-1}$. To prove the upper bound, fix a vector $x \in S^{n-1}$ for which

$$\|A\| = \|Ax\|_2$$

and choose an $x_0 \in \mathcal{N}$ that approximates x, so that

$$\|x - x_0\|_2 \le \varepsilon.$$

By the definition of the operator norm, this implies that

$$\|Ax - Ax_0\|_2 = \|A(x - x_0)\|_2 \le \|A\| \|x - x_0\|_2 \le \varepsilon \|A\|.$$

Using the triangle inequality, we find that

$$\|Ax_0\|_2 \ge \|Ax\|_2 - \|Ax - Ax_0\|_2 \ge \|A\| - \varepsilon \|A\| = (1 - \varepsilon)\|A\|.$$

Dividing both sides of this inequality by $1 - \varepsilon$, we complete the proof. ∎

Exercise 4.4.2 Let $x \in \mathbb{R}^n$ and let $\mathcal{N}$ be an ε-net of the sphere S^{n-1}. Show that

$$\sup_{y \in \mathcal{N}} \langle x, y \rangle \le \|x\|_2 \le \frac{1}{1 - \varepsilon} \sup_{y \in \mathcal{N}} \langle x, y \rangle .$$

Recall from Section 4.1.2 that the operator norm of A can be computed by maximizing a quadratic form:

$$\|A\| = \max_{x \in S^{n-1},\ y \in S^{m-1}} \langle Ax, y \rangle .$$

Moreover, for symmetric matrices one can take $x = y$ in this formula. The following exercise shows that instead of controlling the quadratic form on the spheres, it suffices to have control just over the ε-nets.

Exercise 4.4.3 (Quadratic form on a net) Let A be an $m \times n$ matrix and $\varepsilon \in [0, 1/2)$.

(a) Show that, for any ε-net $\mathcal{N}$ of the sphere S^{n-1} and any ε-net $\mathcal{M}$ of the sphere S^{m-1}, we have

$$\sup_{x \in \mathcal{N},\ y \in \mathcal{M}} \langle Ax, y \rangle \le \|A\| \le \frac{1}{1 - 2\varepsilon} \sup_{x \in \mathcal{N},\ y \in \mathcal{M}} \langle Ax, y \rangle .$$

(b) Moreover, if $m = n$ and A is symmetric, show that

$$\sup_{x \in \mathcal{N}} |\langle Ax, x \rangle| \le \|A\| \le \frac{1}{1 - 2\varepsilon} \sup_{x \in \mathcal{N}} |\langle Ax, x \rangle| .$$

☞

Exercise 4.4.4 (Deviation of the norm on a net) Let A be an $m \times n$ matrix, $\mu \in \mathbb{R}$ and $\varepsilon \in [0, 1/2)$. Show that, for any ε-net $\mathcal{N}$ of the sphere S^{n-1}, we have

$$\sup_{x \in S^{n-1}} |\|Ax\|_2 - \mu| \le \frac{C}{1 - 2\varepsilon} \sup_{x \in \mathcal{N}} |\|Ax\|_2 - \mu| .$$

☞

4.4.2 The Norms of Sub-Gaussian Random Matrices

We are ready for the first result on random matrices. The following theorem states that the norm of an $m \times n$ random matrix A with independent sub-gaussian entries satisfies

$$\|A\| \lesssim \sqrt{m} + \sqrt{n}$$

with high probability.

Theorem 4.4.5 (Norm of matrices with sub-gaussian entries) *Let A be an $m \times n$ random matrix whose entries A_{ij} are independent mean-zero sub-gaussian random variables. Then, for any $t > 0$ we have*[6]

$$\|A\| \le CK\big(\sqrt{m} + \sqrt{n} + t\big)$$

with probability at least $1 - 2\exp(-t^2)$. Here $K = \max_{i,j} \|A_{ij}\|_{\psi_2}$.

Proof This proof is an example of an *ε-net argument*. We need to bound $\langle Ax, y\rangle$ for all vectors x and y on the unit sphere. To this end, we will discretize the sphere using a net (this is the approximation step), establish a tight control of $\langle Ax, y\rangle$ for fixed vectors x and y from the net (the concentration step), and finish by taking a union bound over all x and y in the net.

Step 1: Approximation. Choose $\varepsilon = 1/4$. Using Corollary 4.2.13, we can find an ε-net $\mathcal{N}$ of the sphere S^{n-1} and ε-net $\mathcal{M}$ of the sphere S^{m-1} with cardinalities

$$|\mathcal{N}| \le 9^n \quad \text{and} \quad |\mathcal{M}| \le 9^m. \tag{4.12}$$

By Exercise 4.4.3, the operator norm of A can be bounded using these nets as follows:

$$\|A\| \le 2 \max_{x\in\mathcal{N},\, y\in\mathcal{M}} \langle Ax, y\rangle\,. \tag{4.13}$$

Step 2: Concentration. Fix $x \in \mathcal{N}$ and $y \in \mathcal{M}$. Then the quadratic form

$$\langle Ax, y\rangle = \sum_{i=1}^{n}\sum_{j=1}^{m} A_{ij}x_i y_j$$

is a sum of independent sub-gaussian random variables. Proposition 2.6.1 states that the sum is sub-gaussian and

$$\begin{aligned}\|\langle Ax, y\rangle\|_{\psi_2}^2 &\le C\sum_{i=1}^{n}\sum_{j=1}^{m}\|A_{ij}x_i y_j\|_{\psi_2}^2 \le CK^2\sum_{i=1}^{n}\sum_{j=1}^{m} x_i^2 y_j^2\\ &= CK^2\Big(\sum_{i=1}^{n} x_i^2\Big)\Big(\sum_{j=1}^{m} y_i^2\Big) = CK^2.\end{aligned}$$

Recalling (2.14), we can restate this as the tail bound

$$\mathbb{P}\left\{\langle Ax, y\rangle \ge u\right\} \le 2\exp(-cu^2/K^2), \quad u \ge 0. \tag{4.14}$$

[6] In results like this, C and c will always denote positive absolute constants.

Step 3: Union bound. Next, we unfix x and y using a union bound. Suppose that the event $\max_{x\in\mathcal{N},\, y\in\mathcal{M}} \langle Ax, y\rangle \geq u$ occurs. Then there exist $x \in \mathcal{N}$ and $y \in \mathcal{M}$ such that $\langle Ax, y\rangle \geq u$. Thus the union bound yields

$$\mathbb{P}\left\{\max_{x\in\mathcal{N},\, y\in\mathcal{M}} \langle Ax, y\rangle \geq u\right\} \leq \sum_{x\in\mathcal{N},\, y\in\mathcal{M}} \mathbb{P}\left\{\langle Ax, y\rangle \geq u\right\}.$$

Using the tail bound (4.14) and the estimate (4.12) on the sizes of $\mathcal{N}$ and $\mathcal{M}$, we bound the probability above by

$$9^{n+m}\, 2\exp(-cu^2/K^2). \tag{4.15}$$

Choose

$$u = CK(\sqrt{n} + \sqrt{m} + t). \tag{4.16}$$

Then $u^2 \geq C^2K^2(n+m+t^2)$ and, if the constant C is chosen sufficiently large, the exponent in (4.15) will also be sufficiently large, say $cu^2/K^2 \geq 3(n+m) + t^2$. Thus

$$\mathbb{P}\left\{\max_{x\in\mathcal{N},\, y\in\mathcal{M}} \langle Ax, y\rangle \geq u\right\} \leq 9^{n+m}\, 2\exp\left(-3(n+m) - t^2\right) \leq 2\exp(-t^2).$$

Finally, combining this with (4.13), we conclude that

$$\mathbb{P}\left\{\|A\| \geq 2u\right\} \leq 2\exp(-t^2).$$

Recalling our choice of u in (4.16), we complete the proof. ■

Exercise 4.4.6 (Expected norm) Deduce from Theorem 4.4.5 that

$$\mathbb{E}\,\|A\| \leq CK\big(\sqrt{m} + \sqrt{n}\big).$$

Exercise 4.4.7 (Optimality) Suppose that in Theorem 4.4.5 the entries A_{ij} have unit variances. Prove that

$$\mathbb{E}\,\|A\| \geq C\big(\sqrt{m} + \sqrt{n}\big).$$

Theorem 4.4.5 can be easily extended for symmetric matrices, and the bound for them is

$$\|A\| \lesssim \sqrt{n}$$

with high probability.

Corollary 4.4.8 (Norm of symmetric matrices with sub-gaussian entries) *Let A be an $n\times n$ symmetric random matrix whose entries A_{ij} on and above the diagonal are independent mean-zero sub-gaussian random variables. Then, for any $t > 0$, we have*

$$\|A\| \leq CK\big(\sqrt{n} + t\big)$$

with probability at least $1 - 4\exp(-t^2)$. Here $K = \max_{i,j} \|A_{ij}\|_{\psi_2}$.

Proof Decompose A into an upper-triangular part A^+ and a lower-triangular part A^-. It does not matter where the diagonal goes; let us include it into A^+ to be specific. Then

$$A = A^+ + A^-.$$

Theorem 4.4.5 applies for each part, A^+ or A^-, separately. By a union bound, we have simultaneously

$$\|A^+\| \le CK(\sqrt{n} + t) \quad \text{and} \quad \|A^-\| \le CK(\sqrt{n} + t)$$

with probability at least $1 - 4\exp(-t^2)$. Since by the triangle inequality $\|A\| \le \|A^+\| + \|A^-\|$, the proof is complete. ∎

4.5 Application: Community Detection in Networks

The results of random matrix theory are useful in many applications. Here we give an illustration in the analysis of networks.

Real-world networks tend to have *communities* – clusters of tightly connected vertices. Finding the communities accurately and efficiently is one of the main problems in network analysis, known as the *community detection problem.*

4.5.1 Stochastic Block Model

We will try to solve the community detection problem for a basic probabilistic model, that of a network with two communities. It is a simple extension of the Erdös–Rényi model of random graphs, which we described in Section 2.4.

Definition 4.5.1 (Stochastic block model) Divide n vertices into two sets ("communities") of sizes $n/2$ each. Construct a random graph G by connecting every pair of vertices independently with probability p if they belong to the same community and q if they belong to different communities. This distribution on graphs is called the *stochastic block model* and is denoted $G(n, p, q)$.[7]

In the partial case where $p = q$ we obtain the Erdös–Rényi model $G(n, p)$. But we assume that $p > q$ here. In this case, edges are more likely to occur within than across communities. This gives the network a community structure; see Figure 4.4.

4.5.2 Expected Adjacency Matrix

It is convenient to identify a graph G with its adjacency matrix A, which we introduced in Definition 3.6.2. For a random graph $G \sim G(n, p, q)$, the adjacency matrix A is a *random matrix*, and we will examine A using the tools developed earlier in this chapter.

It is enlightening to split A into deterministic and random parts,

$$A = D + R,$$

[7] The term *stochastic block model* can also refer to a more general model of random graphs with multiple communities of variable sizes.

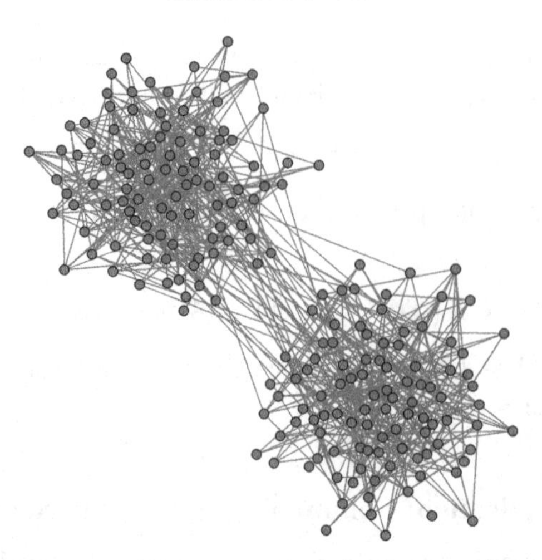

Figure 4.4 A random graph generated according to the stochastic block model $G(n, p, q)$ with $n = 200$, $p = 1/20$, and $q = 1/200$.

where D is the expectation of A. We may think of D as an informative part (the "signal") and R as "noise".

To see why D is informative, let us compute its eigenstructure. The entries A_{ij} have a Bernoulli distribution; they are either $\mathrm{Ber}(p)$ or $\mathrm{Ber}(q)$ depending on the community membership of vertices i and j. Thus the entries of D are either p or q, depending on the membership. For illustration, if we group the vertices that belong to the same community together, then for $n = 4$ the matrix D will look like this:

$$D = \mathbb{E}\, A = \left[\begin{array}{cc|cc} p & p & q & q \\ p & p & q & q \\ \hline q & q & p & p \\ q & q & p & p \end{array}\right].$$

Exercise 4.5.2 Check that the matrix D has rank 2 and that the nonzero eigenvalues λ_i and the corresponding eigenvectors u_i are

$$\lambda_1 = \left(\frac{p+q}{2}\right)n, \quad u_1 = \left[\begin{array}{c} 1 \\ 1 \\ \hline 1 \\ 1 \end{array}\right]; \quad \lambda_2 = \left(\frac{p-q}{2}\right)n, \quad u_2 = \left[\begin{array}{c} 1 \\ 1 \\ \hline -1 \\ -1 \end{array}\right]. \tag{4.17}$$

The important object here is the second eigenvector, u_2. It contains all the information about the community structure. If we knew u_2, we could identify the communities precisely from the sizes of the coefficients of u_2.

But we do not know $D = \mathbb{E}\, A$ and so we do not have access to u_2. Instead, we know $A = D + R$, a noisy version of D. The level (magnitude) of the signal D is

$$\|D\| = \lambda_1 \asymp n,$$

while the level of the noise R can be estimated using Corollary 4.4.8:

$$\|R\| \leq C\sqrt{n} \quad \text{with probability at least } 1 - 4e^{-n}. \tag{4.18}$$

Thus, for large n, the noise R is much smaller than the signal D. In other words, A is close to D, and thus we should be able to use A instead of D to extract community information. This can be justified using the classical perturbation theory for matrices.

4.5.3 Perturbation Theory

Perturbation theory describes how the eigenvalues and eigenvectors of a matrix change under perturbations of the matrix. For the eigenvalues, we have

Theorem 4.5.3 (Weyl's inequality) *For any symmetric matrices S and T with the same dimensions, we have*

$$\max_i |\lambda_i(S) - \lambda_i(T)| \leq \|S - T\|.$$

Thus, the operator norm determines the stability of the spectrum.

Exercise 4.5.4☕☕ Deduce Weyl's inequality from the Courant–Fisher min–max characterization of eigenvalues (4.2).

A similar result holds for eigenvectors, but we need to be careful to track the same eigenvector before and after the perturbation. If the eigenvalues $\lambda_i(S)$ and $\lambda_{i+1}(S)$ are too close to each other, the perturbation can swap their order and force us to compare the wrong eigenvectors. To prevent this from happening, we will assume that the eigenvalues of S are well separated.

Theorem 4.5.5 (Davis–Kahan) *Let S and T be symmetric matrices with the same dimensions. Fix i and assume that the ith largest eigenvalue of S is well separated from the rest of the spectrum:*

$$\min_{j:\, j \neq i} |\lambda_i(S) - \lambda_j(S)| = \delta > 0.$$

Then the angle between the eigenvectors of S and T corresponding to the ith largest eigenvalues (as a number between 0 and $\pi/2$) satisfies

$$\sin \angle\,(v_i(S),\ v_i(T)) \leq \frac{2\|S - T\|}{\delta}.$$

We do not prove the Davis–Kahan theorem here.

The conclusion of the Davis–Kahan theorem implies that the *unit* eigenvectors $v_i(S)$ and $v_i(T)$ are close to each other up to a sign, namely

$$\exists \theta \in \{-1, 1\}: \quad \|v_i(S) - \theta v_i(T)\|_2 \leq \frac{2^{3/2}\|S - T\|}{\delta}. \tag{4.19}$$

(Check!)

4.5.4 Spectral Clustering

Returning to the community detection problem, let us apply the Davis–Kahan theorem for $S = D$ and $T = A = D + R$ and for the second largest eigenvalue. We need to check that λ_2 is well separated from the rest of the spectrum of D, that is, from 0 and λ_1. The distance is

$$\delta = \min(\lambda_2, \lambda_1 - \lambda_2) = \min\left(\frac{p-q}{2}, q\right) n =: \mu n.$$

Recalling the bound (4.18) on $R = T - S$ and applying (4.19), we can bound the distance between the unit eigenvectors of D and A. It follows that there exists a sign $\theta \in \{-1, 1\}$ such that

$$\|v_2(D) - \theta v_2(A)\|_2 \leq \frac{C\sqrt{n}}{\mu n} = \frac{C}{\mu\sqrt{n}}$$

with probability at least $1 - 4e^{-n}$. We have already computed the eigenvectors $u_i(D)$ of D in (4.17), but there they had norm $\sqrt{n}$. So, multiplying both sides by $\sqrt{n}$, we obtain in this normalization that

$$\|u_2(D) - \theta u_2(A)\|_2 \leq \frac{C}{\mu}.$$

It follows that the *signs* of most coefficients of $\theta v_2(A)$ and $v_2(D)$ must agree. Indeed, we know that

$$\sum_{j=1}^{n} |u_2(D)_j - \theta u_2(A)_j|^2 \leq \frac{C}{\mu^2}. \tag{4.20}$$

and we also know from (4.17) that the coefficients $u_2(D)_j$ are all ± 1. So, every coefficient j for which the signs of $\theta v_2(A)_j$ and $v_2(D)_j$ disagree contributes at least 1 to the sum in (4.20). Thus the number of disagreeing signs must be bounded by

$$\frac{C}{\mu^2}.$$

Summarizing, we can use the vector $v_2(A)$ to accurately estimate the vector $v_2 = v_2(D)$ in (4.17) whose signs identify the two communities. This method for community detection is usually called *spectral clustering*. Let us explicitly state this method and the guarantees that we have just obtained.

Spectral Clustering Algorithm

Input: graph G

Output: a partition of the vertices of G into two communities

1: Compute the adjacency matrix A of the graph.

2: Compute the eigenvector $v_2(A)$ corresponding to the second largest eigenvalue of A.

3: Partition the vertices into two communities on the basis of the signs of the coefficients of $v_2(A)$. (To be specific, if $v_2(A)_j > 0$ then put vertex j into the first community, otherwise into the second.)

Theorem 4.5.6 (Spectral clustering for the stochastic block model) *Let $G \sim G(n, p, q)$ with $p > q$, and $\min(q, p-q) = \mu > 0$. Then, with probability at least $1 - 4e^{-n}$, the spectral clustering algorithm identifies the communities of G correctly up to C/μ^2 misclassified vertices.*

Summarizing, the spectral clustering algorithm correctly classifies all except a *constant* number of vertices, provided that the random graph is dense enough ($q \geq$ const) and that the probabilities of within- and across-community edges are well separated ($p - q \geq$ const).

4.6 Two-Sided Bounds on Sub-Gaussian Matrices

Let us return to Theorem 4.4.5, which gives an upper bound on the spectrum of an $m \times n$ matrix A with independent sub-gaussian entries. It essentially states that

$$s_1(A) \leq C(\sqrt{m} + \sqrt{n})$$

with high probability. We will now improve this result in two important ways.

First, we are going to prove sharper and *two-sided* bounds on the entire spectrum of A:

$$\sqrt{m} - C\sqrt{n} \leq s_i(A) \leq \sqrt{m} + C\sqrt{n}.$$

In other words, we will show that a tall random matrix (with $m \gg n$) is an *approximate isometry* in the sense of Section 4.1.5.

Second, the independence of entries is going to be relaxed to just the *independence of rows*. Thus we assume that the rows of A are sub-gaussian random vectors. (We studied such vectors in Section 3.4.) This relaxation of independence is important in some applications to data science, where the rows of A could be samples from a high-dimensional distribution. The samples are usually independent, and so are the rows of A. But there is no reason to assume independence of the columns of A, since the coordinates of the distribution (the "parameters") are usually not independent.

Theorem 4.6.1 (Two-sided bound on sub-gaussian matrices) *Let A be an $m \times n$ matrix whose rows A_i are independent mean-zero sub-gaussian isotropic random vectors in $\mathbb{R}^n$. Then, for any $t \geq 0$ we have*

$$\sqrt{m} - CK^2(\sqrt{n} + t) \leq s_n(A) \leq s_1(A) \leq \sqrt{m} + CK^2(\sqrt{n} + t), \tag{4.21}$$

with probability at least $1 - 2\exp(-t^2)$. Here $K = \max_i \|A_i\|_{\psi_2}$.

We will prove a slightly stronger conclusion than (4.21), namely that

$$\left\| \frac{1}{m} A^\mathsf{T} A - I_n \right\| \leq K^2 \max(\delta, \delta^2) \quad \text{where} \quad \delta = C\Big(\sqrt{\frac{n}{m}} + \frac{t}{\sqrt{m}}\Big). \tag{4.22}$$

Using Lemma 4.1.5, one can quickly check that (4.22) indeed implies (4.21). (Do this!)

Proof We will prove (4.22) using an *ε-net argument*. This will be similar to the proof of Theorem 4.4.5, but we now use Bernstein's concentration inequality instead of Hoeffding's.

Step 1: Approximation. Using Corollary 4.2.13, we can find a 1/4-net $\mathcal{N}$ of the unit sphere S^{n-1} with cardinality

$$|\mathcal{N}| \le 9^n.$$

Using Lemma 4.4.1, we can evaluate the operator norm in (4.22) on $\mathcal{N}$:

$$\left\| \frac{1}{m} A^{\mathsf{T}} A - I_n \right\| \le 2 \max_{x \in \mathcal{N}} \left| \left\langle \left(\frac{1}{m} A^{\mathsf{T}} A - I_n \right) x \right\rangle x \right| = 2 \max_{x \in \mathcal{N}} \left| \frac{1}{m} \|Ax\|_2^2 - 1 \right|.$$

To complete the proof of (4.22) it suffices to show that, with the required probability,

$$\max_{x \in \mathcal{N}} \left| \frac{1}{m} \|Ax\|_2^2 - 1 \right| \le \frac{\varepsilon}{2} \quad \text{where} \quad \varepsilon := K^2 \max(\delta, \delta^2).$$

Step 2: Concentration. Fix $x \in S^{n-1}$ and express $\|Ax\|_2^2$ as a sum of independent random variables:

$$\|Ax\|_2^2 = \sum_{i=1}^{m} \langle A_i, x \rangle^2 =: \sum_{i=1}^{m} X_i^2 \tag{4.23}$$

where the A_i denote the rows of A. By assumption, the A_i are independent, isotropic, and sub-gaussian random vectors with $\|A_i\|_{\psi_2} \le K$. Thus the $X_i = \langle A_i, x \rangle$ are independent sub-gaussian random variables with $\mathbb{E}\, X_i^2 = 1$ and $\|X_i\|_{\psi_2} \le K$. Therefore the $X_i^2 - 1$ are independent, mean-zero, and sub-exponential random variables, with

$$\|X_i^2 - 1\|_{\psi_1} \le C K^2.$$

(Check this; we did a similar computation in the proof of Theorem 3.1.1.) Thus we can use Bernstein's inequality (Corollary 2.8.3) and obtain

$$\begin{aligned}
\mathbb{P} \left\{ \left| \frac{1}{m} \|Ax\|_2^2 - 1 \right| \ge \frac{\varepsilon}{2} \right\} &= \mathbb{P} \left\{ \left| \frac{1}{m} \sum_{i=1}^{m} X_i^2 - 1 \right| \ge \frac{\varepsilon}{2} \right\} \\
&\le 2 \exp\left(- c_1 \min\left(\frac{\varepsilon^2}{K^4}, \frac{\varepsilon}{K^2} \right) m \right) \\
&= 2 \exp\left(- c_1 \delta^2 m \right) \quad \left(\text{since } \frac{\varepsilon}{K^2} = \max(\delta, \delta^2)\right) \\
&\le 2 \exp\left(- c_1 C^2 (n + t^2) \right).
\end{aligned}$$

The last bound follows from the definition of δ in (4.22) and using the inequality $(a+b)^2 \ge a^2 + b^2$ for $a, b \ge 0$.

Step 3: Union bound. Now we can unfix $x \in \mathcal{N}$ using a union bound. Recalling that $\mathcal{N}$ has cardinality bounded by 9^n, we obtain

$$\mathbb{P} \left\{ \max_{x \in \mathcal{N}} \left| \frac{1}{m} \|Ax\|_2^2 - 1 \right| \ge \frac{\varepsilon}{2} \right\} \le 9^n \, 2 \exp\left(- c_1 C^2 (n + t^2) \right) \le 2 \exp(-t^2),$$

if we choose the absolute constant C in (4.22) large enough. As we noted in step 1, this completes the proof of the theorem. ■

Exercise 4.6.2 Deduce from (4.22) that

$$\mathbb{E}\left\|\frac{1}{m}A^{\mathsf{T}}A - I_n\right\| \leq CK^2\Big(\sqrt{\frac{n}{m}} + \frac{n}{m}\Big).$$

Exercise 4.6.3 Deduce from Theorem 4.6.1 the following bounds on the expectation:

$$\sqrt{m} - CK^2\sqrt{n} \leq \mathbb{E}\, s_n(A) \leq \mathbb{E}\, s_1(A) \leq \sqrt{m} + CK^2\sqrt{n}.$$

Exercise 4.6.4 Give a simpler proof of Theorem 4.6.1, using Theorem 3.1.1 to obtain a concentration bound for $\|Ax\|_2$ and Exercise 4.4.4 to reduce to a union bound over a net.

4.7 Application: Covariance Estimation and Clustering

Suppose that we are analyzing some high-dimensional data which are represented as points $X_1, \ldots, X_m$ sampled from an unknown distribution in $\mathbb{R}^n$. One of the most basic data exploration tools is principal component analysis (PCA), which we discussed briefly in Section 3.2.1.

Since we do not have access to the full distribution but only to the finite sample $\{X_1, \ldots, X_m\}$, we can only expect to compute the covariance matrix of the underlying distribution approximately. If we can do so, the Davis–Kahan theorem 4.5.5 would allow us to estimate the principal components of the underlying distribution, which are the eigenvectors of the covariance matrix.

So, how can we estimate the covariance matrix from the data? Let X denote the random vector drawn from the (unknown) distribution. Assume for simplicity that X has zero mean, and let us denote its covariance matrix by

$$\Sigma = \mathbb{E}\, XX^{\mathsf{T}}.$$

(Actually, our analysis will not require a zero mean, in which case Σ is simply the second moment matrix of X, as we explained in Section 3.2.)

To estimate Σ we can use the *sample covariance* matrix Σ_m, which is computed from the sample $X_1, \ldots, X_m$ as follows:

$$\Sigma_m = \frac{1}{m}\sum_{i=1}^{m} X_i X_i^{\mathsf{T}}.$$

In other words, to compute Σ we replace the expectation over the entire distribution (the "population expectation") by the average over the sample (the "sample expectation").

Since X_i and X are identically distributed, our estimate is unbiased, that is,

$$\mathbb{E}\, \Sigma_m = \Sigma.$$

Then the law of large numbers (Theorem 1.3.1) applied to each entry of Σ yields

$$\Sigma_m \to \Sigma \quad \text{almost surely}$$

as the sample size m increases to infinity. This leads to the quantitative question: how large must the sample size m be to guarantee that

$$\Sigma_m \approx \Sigma$$

with high probability? For dimensional reasons, we need at least $m \gtrsim n$ sample points. (Why?) And we now show that $m \asymp n$ sample points suffice.

Theorem 4.7.1 (Covariance estimation) *Let X be a sub-gaussian random vector in $\mathbb{R}^n$. More precisely, assume that there exists $K \geq 1$ such that* [8]

$$\| \langle X, x \rangle \|_{\psi_2} \leq K \| \langle X, x \rangle \|_{L^2} \quad \textit{for any } x \in \mathbb{R}^n. \tag{4.24}$$

Then, for every positive integer m, we have

$$\mathbb{E} \|\Sigma_m - \Sigma\| \leq CK^2 \Big(\sqrt{\frac{n}{m}} + \frac{n}{m} \Big) \|\Sigma\|.$$

Proof Let us first bring the random vectors $X, X_1, \dots, X_m$ to the isotropic position. There exist independent isotropic random vectors $Z, Z_1, \dots, Z_m$ such that

$$X = \Sigma^{1/2} Z \quad \text{and} \quad X_i = \Sigma^{1/2} Z_i.$$

(We checked this in Exercise 3.2.2.) The sub-gaussian assumption (4.24) then implies that

$$\|Z\|_{\psi_2} \leq K \quad \text{and} \quad \|Z_i\|_{\psi_2} \leq K.$$

(Check!) Then

$$\|\Sigma_m - \Sigma\| = \|\Sigma^{1/2} R_m \Sigma^{1/2}\| \leq \|R_m\| \, \|\Sigma\| \quad \text{where} \quad R_m := \frac{1}{m} \sum_{i=1}^m Z_i Z_i^\mathsf{T} - I_n. \tag{4.25}$$

Consider the $m \times n$ random matrix A whose rows are Z_i^T. Then

$$\frac{1}{m} A^\mathsf{T} A - I_n = \frac{1}{m} \sum_{i=1}^m Z_i Z_i^\mathsf{T} - I_n = R_m.$$

We can apply Theorem 4.6.1 for A and get

$$\mathbb{E} \|R_m\| \leq CK^2 \Big(\sqrt{\frac{n}{m}} + \frac{n}{m} \Big).$$

(See Exercise 4.6.2.) Substituting this into (4.25), we complete the proof. ■

Remark 4.7.2 (Sample complexity) Theorem 4.7.1 implies that for any $\varepsilon \in (0, 1)$, we are guaranteed to have covariance estimation with a good relative error,

$$\mathbb{E} \|\Sigma_m - \Sigma\| \leq \varepsilon \|\Sigma\|,$$

[8] Here we use the notation for the L^2 norm of random variables from Section 1.1, namely $\| \langle X, x \rangle \|_{L^2}^2 = \mathbb{E} \langle X, x \rangle^2 = \langle \Sigma x, x \rangle$.

if we take a sample of size

$$m \asymp \varepsilon^{-2} n.$$

In other words, the covariance matrix can be estimated accurately by the sample covariance matrix *if the sample size m is proportional to the dimension n.*

Exercise 4.7.3 (Tail bound) Our argument also implies the following high-probability guarantee. Check that for any $u \geq 0$, we have

$$\|\Sigma_m - \Sigma\| \leq CK^2\Big(\sqrt{\frac{n+u}{m}} + \frac{n+u}{m}\Big)\,\|\Sigma\|$$

with probability at least $1 - 2e^{-u}$.

4.7.1 Application: Clustering of Point Sets

We are going to illustrate Theorem 4.7.1 with an application to clustering. As in Section 4.5, we try to identify clusters in the data. But the nature of the data will be different – instead of networks, we will now be working with point sets in $\mathbb{R}^n$. The general goal is to partition a given set of points into few clusters. What exactly constitutes a cluster is not well defined in data science. But common sense suggests that the points in the same cluster should tend to be closer to each other than the points taken from different clusters.

Just as we did for networks, we will design a basic probabilistic model of point sets in $\mathbb{R}^n$ with two communities and study the clustering problem for that model.

Definition 4.7.4 (Gaussian mixture model) Generate m random points in $\mathbb{R}^n$ as follows. Flip a fair coin; if we get heads, draw a point from $N(\mu, I_n)$ and if we get tails, from $N(-\mu, I_n)$. This distribution of points is called the Gaussian mixture model with means μ and $-\mu$.

Equivalently, we may consider a random vector

$$X = \theta\mu + g$$

where θ is a symmetric Bernoulli random variable, $g \in N(0, I_n)$, and θ and g are independent. Draw a sample $X_1, \ldots, X_m$ of independent random vectors that are distributed identically to X. Then the sample will be distributed according to the Gaussian mixture model; see Figure 4.5 for an illustration.

Suppose we are given a sample of m points drawn according to the Gaussian mixture model. Our goal is to identify which points belong to which cluster. To this end, we can use a variant of the *spectral clustering* algorithm that we introduced for networks in Section 3.2.1.

To see why a spectral method has a chance of working here, note that the distribution of X is not isotropic but rather stretched in the direction of μ. (This is the horizontal direction in Figure 4.5.) Thus, we can approximately compute μ by computing the first principal component of the data. Next, we can project the data points onto the line spanned by μ, and

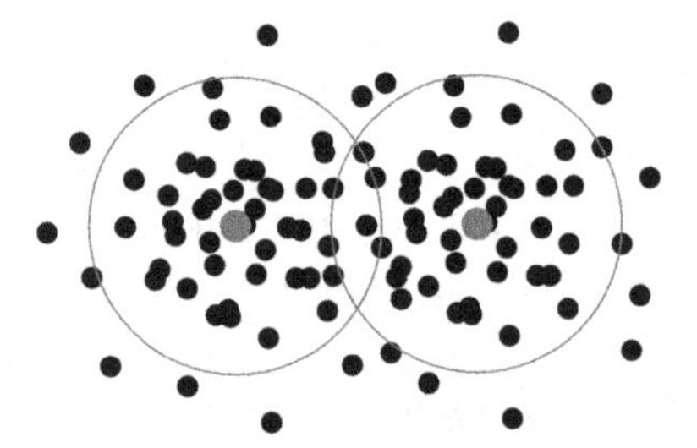

Figure 4.5 A simulation of points generated according to the Gaussian mixture model, which has two clusters with different means.

thus classify them – by just looking at on which side of the origin the projections lie. This leads to the following algorithm.

Spectral Clustering Algorithm

Input: points $X_1, \dots, X_m$ in $\mathbb{R}^n$

Output: a partition of the points into two clusters

1: Compute the sample covariance matrix $\Sigma_m = m^{-1} \sum_{i=1}^m X_i X_i^\mathsf{T}$.

2: Compute the eigenvector $v = v_1(\Sigma_m)$ corresponding to the largest eigenvalue of Σ_m.

3: Partition the vertices into two communities on the basis of the signs of the inner product of v with the data points. (To be specific, if $\langle v, X_i \rangle > 0$ put point X_i into the first community, otherwise in the second.)

Theorem 4.7.5 (Guarantees of spectral clustering of the Gaussian mixture model) *Let $X_1, \dots, X_m$ be points in $\mathbb{R}^n$ drawn from the Gaussian mixture model as above, i.e. there are two communities with means μ and $-\mu$.*

Let $\varepsilon > 0$ be such that $\|\mu\|_2 \ge C\sqrt{\log(1/\varepsilon)}$. Suppose the sample size satisfies

$$m \ge \left(\frac{n}{\|\mu\|_2} \right)^c$$

where $c > 0$ is an appropriate absolute constant.

Then, with probability at least $1 - 4e^{-n}$, the above spectral clustering algorithm identifies the communities correctly up to εm misclassified points.

Exercise 4.7.6 (Spectral clustering of the Gaussian mixture model)☕☕☕ Prove Theorem 4.7.5 for the spectral clustering algorithm applied to the Gaussian mixture model. Proceed as follows.

(a) Compute the covariance matrix Σ of X; note that the eigenvector corresponding to the largest eigenvalue is parallel to μ.

(b) Use results about covariance estimation to show that the sample covariance matrix Σ_m is close to Σ if the sample size m is relatively large.

(c) Use the Davis–Kahan theorem 4.5.5 to deduce that the first eigenvector $v = v_1(\Sigma_m)$ is close to the direction of μ.

(d) Conclude that the signs of the $\langle \mu, X_i \rangle$ predict well to which community X_i belongs.
(e) Since $v \approx \mu$, conclude the same for v.

4.8 Notes

The notions of covering and packing numbers and metric entropy introduced in Section 4.2 are studied thoroughly in asymptotic geometric analysis. Most of the material covered in that section can be found in standard sources such as [11, Chapter 4] and [164].

In Section 4.3.2 we gave some basic results about error correcting codes. The book [210] offers a more systematic introduction to coding theory. Theorem 4.3.5 is a simplified version of the landmark *Gilbert–Varshamov bound* on the rates of error correcting codes. Our proof of this result relies on a bound on the binomial sum from Exercise 0.0.5. A slight tightening of the binomial sum bound leads to the following improved bound on the rate in Remark 4.3.6: there exist codes with rate

$$R \geq 1 - h(2\delta) - o(1),$$

where

$$h(x) = -x \log_2(x) + (1-x) \log_2(1-x)$$

is the *binary entropy function*. This result is known as the *Gilbert–Varshamov bound*. One can tighten up the result of Exercise 4.3.7 similarly and prove that, for any error correcting code, the rate is bounded as

$$R \leq 1 - h(\delta).$$

This result is known as the *Hamming bound*.

Our introduction to non-asymptotic random matrix theory in Sections 4.4 and 4.6 mostly follows [216].

In Section 4.5 we gave an application of random matrix theory to networks. For a comprehensive introduction to the interdisciplinary area of network analysis, see e.g. the book [154]. Stochastic block models (Definition 4.5.1) were introduced in [101]. The community detection problem in stochastic block models has attracted a lot of attention: see the book [154], the survey [75], papers including [137, 221, 153, 94, 1, 26, 54, 124, 92, 106] and the references therein.

In Section 4.7 we discussed covariance estimation following [216]; more general results will appear in Section 9.2.3. The covariance estimation problem has been studied extensively in high-dimensional statistics; see e.g. [216, 170, 115, 42, 127, 52] and the references therein.

In Section 4.7.1 we gave an application to the clustering of Gaussian mixture models. This problem has been well studied in statistics and computer science communities; see e.g. [149, Chapter 6] and [109, 150, 18, 102, 10, 87].

5

Concentration Without Independence

The approach to concentration inequalities that we have developed so far relies crucially on the independence of the random variables. We now pursue some alternative approaches to concentration which are not based on independence. In Section 5.1, we demonstrate how to derive the concentration from isoperimetric inequalities. We first do this for the example of the Euclidean sphere and then discuss other natural settings in Section 5.2.

In Section 5.3 we use concentration on the sphere to derive the classical Johnson–Lindenstrauss lemma, a basic result about dimension reduction for high-dimensional data.

Section 5.4 introduces matrix concentration inequalities. We prove the matrix Bernstein inequality, a remarkably general extension of the classical Bernstein inequality from Section 2.8 for random matrices. We then give two applications in Sections 5.5 and 5.6, extending our analysis for community detection and covariance estimation problems to sparse networks and fairly general distributions in $\mathbb{R}^n$.

5.1 Concentration of Lipschitz Functions for the Sphere

Consider a Gaussian random vector $X \sim N(0, I_n)$ and a function $f\colon \mathbb{R}^n \to \mathbb{R}$. When does the random vector $f(X)$ concentrate about its mean, i.e.,

$$f(X) \approx \mathbb{E}\, f(X) \quad \text{with high probability?}$$

This question is easy for *linear functions* f. Indeed, in this case $f(X)$ has a normal distribution, and it concentrates around its mean well (recall Exercise 3.3.3 and Proposition 2.1.2).

We now study the concentration of *nonlinear* functions $f(X)$ of random vectors X. We cannot expect to have good concentration for completely arbitrary f (why?). But if f does not oscillate too wildly, we might expect concentration. The concept of Lipschitz functions, which we introduce now, will help us to rule out rigorously functions that have wild oscillations.

5.1.1 Lipschitz Functions

Definition 5.1.1 (Lipschitz functions) Let (X, d_X) and (Y, d_Y) be metric spaces. A function $f\colon X \to Y$ is called *Lipschitz* if there exists $L \in \mathbb{R}$ such that

$$d_Y(f(u), f(v)) \le L d_X(u, v) \quad \text{for every } u, v \in X.$$

The infimum of all L in this definition is called the *Lipschitz norm* of f and is denoted $\|f\|_{\mathrm{Lip}}$.

In other words, Lipschitz functions may not blow up distances between points too much. Lipschitz functions with $\|f\|_{\text{Lip}} \le 1$ are usually called *contractions*, since they may only shrink distances.

Lipschitz functions form an intermediate class between uniformly continuous and differentiable functions:

Exercise 5.1.2 (Continuity, differentiability, and Lipschitz functions) Prove the following statements.

(a) Every Lipschitz function is uniformly continuous.
(b) Every differentiable function $f: \mathbb{R}^n \to \mathbb{R}$ is Lipschitz, and

$$\|f\|_{\text{Lip}} \le \sup_{x \in \mathbb{R}^n} \|\nabla f(x)\|_2.$$

(c) Give an example of a non-Lipschitz but uniformly continuous function $f: [-1, 1] \to \mathbb{R}$.
(d) Give an example of a non-differentiable but Lipschitz function $f: [-1, 1] \to \mathbb{R}$.

Here are a few useful examples of Lipschitz functions on $\mathbb{R}^n$.

Exercise 5.1.3 (Linear functionals and norms as Lipschitz functions) Prove the following statements.

(a) For a fixed $\theta \in \mathbb{R}^n$, the linear functional

$$f(x) = \langle x, \theta \rangle$$

is a Lipschitz function on $\mathbb{R}^n$, and $\|f\|_{\text{Lip}} = \|\theta\|_2$.
(b) More generally, an $m \times n$ matrix A acting as a linear operator

$$A: (\mathbb{R}^n, \|\cdot\|_2) \to (\mathbb{R}^m, \|\cdot\|_2)$$

is Lipschitz, and $\|A\|_{\text{Lip}} = \|A\|$.
(c) Any norm $f(x) = \|x\|$ on $(\mathbb{R}^n, \|\cdot\|_2)$ is a Lipschitz function. The Lipschitz norm of f is the smallest L that satisfies

$$\|x\| \le L\|x\|_2 \quad \text{for all } x \in \mathbb{R}^n.$$

5.1.2 Concentration via Isoperimetric Inequalities

The main result of this section is that any Lipschitz function on the Euclidean sphere $S^{n-1} = \{x \in \mathbb{R}^n: \|x\|_2 = 1\}$ concentrates well.

Theorem 5.1.4 (Concentration of Lipschitz functions on the sphere) *Consider a random vector $X \sim \text{Unif}(\sqrt{n}S^{n-1})$, i.e., X is uniformly distributed on the Euclidean sphere of radius $\sqrt{n}$. Consider a Lipschitz function*[1] *$f: \sqrt{n}S^{n-1} \to \mathbb{R}$. Then*

[1] This theorem is valid for both the geodesic metric on the sphere (where $d(x, y)$ is the length of the shortest arc connecting x and y) and the Euclidean metric $d(x, y) = \|x - y\|_2$. We will prove the theorem for the Euclidean metric; Exercise 5.1.11 extends it to the geodesic metric.

$$\|f(X) - \mathbb{E} f(X)\|_{\psi_2} \le C\|f\|_{\mathrm{Lip}}.$$

Using the definition of the sub-gaussian norm, the conclusion of Theorem 5.1.4 can be stated as follows: for every $t \ge 0$, we have

$$\mathbb{P}\left\{|f(X) - \mathbb{E} f(X)| \ge t\right\} \le 2\exp\Big(-\frac{ct^2}{\|f\|_{\mathrm{Lip}}^2}\Big).$$

Let us set out a strategy to prove Theorem 5.1.4. We already proved it for linear functions. Indeed, Theorem 3.4.6 states that $X \sim \mathrm{Unif}(\sqrt{n}S^{n-1})$ is a sub-gaussian random vector, and this by definition means that any linear function of X is a sub-gaussian random variable.

To prove Theorem 5.1.4 in full generality, we will argue that any nonlinear Lipschitz function must concentrate at least as strongly as a linear function. To show this, instead of comparing nonlinear with linear functions directly, we will compare the areas of their *sub-level sets* – the subsets of the sphere of the form $\{x\colon f(x) \le a\}$. The sub-level sets of linear functions are obviously spherical caps. We can compare the areas of general sets and spherical caps using a remarkable geometric principle – an *isoperimetric inequality*.

The most familiar form of an isoperimetric inequality applies to subsets of $\mathbb{R}^3$ (and also in $\mathbb{R}^n$):

Theorem 5.1.5 (Isoperimetric inequality on $\mathbb{R}^n$) *Among all subsets $A \subset \mathbb{R}^n$ with given volume, Euclidean balls have minimal area. Moreover, for any $\varepsilon > 0$, Euclidean balls minimize the volume of the ε-neighborhood of A, defined as*[2]

$$A_\varepsilon := \left\{x \in \mathbb{R}^n\colon \exists y \in A \text{ such that } \|x - y\|_2 \le \varepsilon\right\} = A + \varepsilon B_2^n.$$

Figure 5.1 illustrates the isoperimetric inequality. Note that the "moreover" part of Theorem 5.1.5 implies the first part: to see this, let $\varepsilon \to 0$.

A similar isoperimetric inequality holds for subsets of the sphere S^{n-1}, and in this case the minimizers are *spherical caps* – the neighborhoods of a single point.[3] To state this principle, we denote by σ_{n-1} the normalized area on the sphere S^{n-1} (i.e. the $(n-1)$-dimensional Lebesgue measure).

Theorem 5.1.6 (Isoperimetric inequality on the sphere) *Let $\varepsilon > 0$. Then, among all sets $A \subset S^{n-1}$ with given area $\sigma_{n-1}(A)$, the spherical caps minimize the area of the neighborhood $\sigma_{n-1}(A_\varepsilon)$, where*

Figure 5.1 The isoperimetric inequality in $\mathbb{R}^n$ states that among all sets A of given volume, the Euclidean balls minimize the volume of the ε-neighborhood A_ε.

[2] Here we use the notation for the Minkowski sum introduced in Definition 4.2.11.

[3] More formally, a closed spherical cap centered at a point $a \in S^{n-1}$ and with radius ε can be defined as $C(a, \varepsilon) = \{x \in S^{n-1}\colon \|x - a\|_2 \le \varepsilon\}$.

$$A_\varepsilon := \left\{x \in S^{n-1} : \exists y \in A \text{ such that } \|x - y\|_2 \le \varepsilon\right\}.$$

We do not prove isoperimetric inequalities (Theorems 5.1.5 and 5.1.6) in this book; the bibliography notes for this chapter refer to several proofs of these results.

5.1.3 Blow-Up of Sets on the Sphere

The isoperimetric inequality implies a remarkable phenomenon that may sound counterintuitive: if a set A makes up at least *half* the sphere (in terms of area) then the neighborhood A_ε will make up *most* of the sphere. We now state and prove this "blow-up" phenomenon, and then try to explain it heuristically. In view of Theorem 5.1.4, it will be convenient for us to work with a sphere of radius $\sqrt{n}$ rather the unit sphere.

Lemma 5.1.7 (Blow-up) *Let A be a subset of the sphere $\sqrt{n}S^{n-1}$, and let σ denote the normalized area on that sphere. If $\sigma(A) \ge 1/2$ then,*[4] *for every $t \ge 0$,*

$$\sigma(A_t) \ge 1 - 2\exp(-ct^2).$$

Proof Consider the hemisphere defined by the first coordinate:

$$H := \left\{x \in \sqrt{n}S^{n-1} : x_1 \le 0\right\}.$$

By assumption, $\sigma(A) \ge 1/2 = \sigma(H)$, so the isoperimetric inequality (Theorem 5.1.6) implies that

$$\sigma(A_t) \ge \sigma(H_t). \tag{5.1}$$

The neighborhood H_t of the hemisphere H is a spherical cap, and we could compute its area by a direct calculation. It is, however, easier to use Theorem 3.4.6 instead, which states that a random vector

$$X \sim \mathrm{Unif}(\sqrt{n}S^{n-1})$$

is sub-gaussian, and $\|X\|_{\psi_2} \le C$. Since σ is the uniform probability measure on the sphere, it follows that

$$\sigma(H_t) = \mathbb{P}\left\{X \in H_t\right\}.$$

Now, the definition of the neighborhood implies that

$$H_t \supset \left\{x \in \sqrt{n}S^{n-1} : x_1 \le t/\sqrt{2}\right\}. \tag{5.2}$$

(Check this – see Exercise 5.1.8.) Thus

$$\sigma(H_t) \ge \mathbb{P}\left\{X_1 \le t/\sqrt{2}\right\} \ge 1 - 2\exp(-ct^2).$$

The last inequality holds because $\|X_1\|_{\psi_2} \le \|X\|_{\psi_2} \le C$. In view of (5.1), the lemma is proved. ■

[4] Here the neighborhood A_t of a set A is defined in the same way as before, that is, $A_t := \left\{x \in \sqrt{n}S^{n-1} : \exists y \in A \text{ such that } \|x - y\|_2 \le t\right\}$.

Exercise 5.1.8 Prove inclusion (5.2).

The number 1/2 for the area bound in Lemma 5.1.7 was rather arbitrary. As the next exercise shows, it can be changed to any constant and even to an exponentially small quantity.

Exercise 5.1.9 (Blow-up of exponentially small sets) Let A be a subset of the sphere $\sqrt{n}S^{n-1}$ such that

$$\sigma(A) > 2\exp(-cs^2) \quad \text{for some } s > 0.$$

(a) Prove that $\sigma(A_s) > 1/2$.
(b) Deduce from this that, for any $t \geq s$,

$$\sigma(A_{2t}) \geq 1 - 2\exp(-ct^2).$$

Here $c > 0$ is the absolute constant from Lemma 5.1.7. ☞

Remark 5.1.10 (Zero–one law) The blow-up phenomenon we just saw may be quite counterintuitive at first sight. How can the exponentially small set A in Exercise 5.1.9 undergo such a dramatic transition to an exponentially large set A_{2t} under such a small perturbation $2t$? (Remember that t can be much smaller than the radius $\sqrt{n}$ of the sphere.) However perplexing this may seem, this is a typical phenomenon in high dimensions. It is reminiscent of *zero–one laws* in probability theory, which basically state that events that are determined by many random variables tend to have probabilities either zero or one.

5.1.4 Proof of Theorem 5.1.4

Without loss of generality, we can assume that $\|f\|_{\text{Lip}} = 1$. (Why?) Let M denote a median of $f(X)$, which by definition is a number satisfying[5]

$$\mathbb{P}\left\{f(X) \leq M\right\} \geq \frac{1}{2} \quad \text{and} \quad \mathbb{P}\left\{f(X) \geq M\right\} \geq \frac{1}{2}.$$

Consider the sub-level set

$$A := \left\{x \in \sqrt{n}S^{n-1} : f(x) \leq M\right\}.$$

Since $\mathbb{P}\left\{X \in A\right\} \geq 1/2$, Lemma 5.1.7 implies that on the one hand

$$\mathbb{P}\left\{X \in A_t\right\} \geq 1 - 2\exp(-ct^2). \tag{5.3}$$

On the other hand, we claim that

$$\mathbb{P}\left\{X \in A_t\right\} \leq \mathbb{P}\left\{f(X) \leq M + t\right\}. \tag{5.4}$$

[5] The median may not be unique. However, for continuous and one-to-one functions f, the median is unique. (Check!)

Indeed, if $X \in A_t$ then $\|X - y\|_2 \le t$ for some point $y \in A$. By definition, $f(y) \le M$. Since f is Lipschitz with $\|f\|_{\mathrm{Lip}} = 1$, it follows that

$$f(X) \le f(y) + \|X - y\|_2 \le M + t.$$

This proves our claim (5.4).

Combining (5.3) and (5.4), we conclude that

$$\mathbb{P}\left\{f(X) \le M + t\right\} \ge 1 - 2\exp(-ct^2).$$

Repeating the argument for $-f$, we obtain a similar bound for the probability that $f(X) \ge M - t$. (Do this!) Combining the two, we obtain a similar bound for the probability that $|f(X) - M| \le t$, and conclude that

$$\|f(X) - M\|_{\psi_2} \le C.$$

It remains to replace the median M by the expectation $\mathbb{E} f$. This can be done easily by applying the centering lemma 2.6.8. (How?) The proof of Theorem 5.1.4 is now complete. ■

Exercise 5.1.11 (Geodesic metric)☕☕☕ We proved Theorem 5.1.4 for functions f that are Lipschitz with respect to the Euclidean metric $\|x - y\|_2$ on the sphere. Argue that the same result holds for the geodesic metric, which is the length of the shortest arc connecting x and y.

Exercise 5.1.12 (Concentration for the unit sphere)☕ We stated Theorem 5.1.4 for the scaled sphere $\sqrt{n}S^{n-1}$. Deduce that a Lipschitz function f on the *unit* sphere S^{n-1} satisfies

$$\|f(X) - \mathbb{E} f(X)\|_{\psi_2} \le \frac{C\|f\|_{\mathrm{Lip}}}{\sqrt{n}}, \tag{5.5}$$

where $X \sim \mathrm{Unif}(S^{n-1})$. Equivalently, for every $t \ge 0$, we have

$$\mathbb{P}\left\{|f(X) - \mathbb{E} f(X)| \ge t\right\} \le 2\exp\left(-\frac{cnt^2}{\|f\|_{\mathrm{Lip}}^2}\right). \tag{5.6}$$

In the geometric approach to concentration that we have just presented, we first (a) proved a blow-up inequality (Lemma 5.1.7), then (b) deduced the concentration about the median, and (c) replaced the median by the expectation. The next exercise shows that these steps can be reversed.

Exercise 5.1.13 (Concentration about the expectation and concentration about the median are equivalent)☕☕ Consider a random variable Z with median M. Show that

$$c\|Z - \mathbb{E} Z\|_{\psi_2} \le \|Z - M\|_{\psi_2} \le C\|Z - \mathbb{E} Z\|_{\psi_2},$$

where $c, C > 0$ are some absolute constants. ☞

Exercise 5.1.14 (Concentration and blow-up are equivalent) Consider a random vector X taking values in some metric space (T, d). Assume that there exists a $K > 0$ such that

$$\|f(X) - \mathbb{E} f(X)\|_{\psi_2} \leq K \|f\|_{\text{Lip}}$$

for every Lipschitz function f: $T \to \mathbb{R}$. For a subset $A \subset T$, define $\sigma(A) := \mathbb{P}(X \in A)$. (Then σ is a probability measure on T.) Show that if $\sigma(A) \geq 1/2$ then,[6] for every $t \geq 0$,

$$\sigma(A_t) \geq 1 - 2\exp(-ct^2/K^2),$$

where $c > 0$ is an absolute constant. ☞

Exercise 5.1.15 (Exponential set of mutually almost orthogonal points) From linear algebra, we know that any set of orthonormal vectors in $\mathbb{R}^n$ must contain at most n vectors. However, if we allow the vectors to be almost orthogonal, there can be *exponentially many* of them! Prove this counterintuitive fact as follows. Fix $\varepsilon \in (0, 1)$. Show that there exists a set $\{x_1, \ldots, x_N\}$ of unit vectors in $\mathbb{R}^n$ which are mutually almost orthogonal,

$$|\langle x_i, x_j \rangle| \leq \varepsilon \quad \text{for all } i \neq j,$$

and that the set is *exponentially large* in n:

$$N \geq \exp\left(c(\varepsilon)n\right).$$

☞

5.2 Concentration for Other Metric Measure Spaces

In this section, we extend the concentration for the sphere to other spaces. To do this, note that our proof of Theorem 5.1.4 was based on two main ingredients:

(i) an isoperimetric inequality;
(ii) a blow-up of the minimizers for the isoperimetric inequality.

These two ingredients are not special to the sphere. Many other metric measure spaces satisfy (i) and (ii) as well, and thus concentration can be proved in such spaces as well. We will discuss two such examples, which lead to Gaussian concentration in $\mathbb{R}^n$ and concentration on the Hamming cube, and then we will mention a few other situations where concentration can be shown.

5.2.1 Gaussian Concentration

The classical isoperimetric inequality in $\mathbb{R}^n$, Theorem 5.1.5, holds not only with respect to the volume but also with respect to the *Gaussian measure* on $\mathbb{R}^n$. The Gaussian measure of a (Borel) set $A \subset \mathbb{R}^n$ is defined as[7]

$$\gamma_n(A) := \mathbb{P}\left\{X \in A\right\} = \frac{1}{(2\pi)^{n/2}} \int_A e^{-\|x\|_2^2/2}\, dx$$

where $X \sim N(0, I_n)$ is the standard normal random vector in $\mathbb{R}^n$.

[6] Here the neighborhood A_t of a set A is defined in the same way as before, that is, $A_t := \{x \in T : \exists y \in A \text{ such that } d(x, y) \leq t\}$.

[7] Recall the definition of the standard normal distribution in $\mathbb{R}^n$ from Section 3.3.2.

Theorem 5.2.1 (Gaussian isoperimetric inequality) *Let $\varepsilon > 0$. Then, among all sets $A \subset \mathbb{R}^n$ with fixed Gaussian measure $\gamma_n(A)$, the half-spaces minimize the Gaussian measure of the neighborhood $\gamma_n(A_\varepsilon)$.*

Using the method we developed for the sphere, we can deduce from Theorem 5.2.1 the following Gaussian concentration inequality.

Theorem 5.2.2 (Gaussian concentration) *Consider a random vector $X \sim N(0, I_n)$ and a Lipschitz function $f: \mathbb{R}^n \to \mathbb{R}$ (with respect to the Euclidean metric). Then*

$$\|f(X) - \mathbb{E} f(X)\|_{\psi_2} \le C\|f\|_{\mathrm{Lip}}. \tag{5.7}$$

Exercise 5.2.3 ☕☕☕ Deduce the Gaussian concentration inequality (Theorem 5.2.2) from the Gaussian isoperimetric inequality (Theorem 5.2.1). ☞

Two partial cases of Theorem 5.2.2 should already be familiar:

(i) For *linear functions* f, Theorem 5.2.2 follows easily since the normal distribution $N(0, I_n)$ is sub-gaussian.
(ii) For the *Euclidean norm* $f(x) = \|x\|_2$, Theorem 5.2.2 follows from Theorem 3.1.1.

Exercise 5.2.4 (Replacing expectation by L^p norm) ☕☕☕ Prove that in the concentration results for the sphere and for Gaussian space (Theorems 5.1.4 and 5.2.2), the expectation $\mathbb{E} f(X)$ can be replaced by the L^p norm $(\mathbb{E} f^p)^{1/p}$ for any $p \ge 1$ and for any non-negative function f. The constants may depend on p.

5.2.2 Hamming Cube

We saw how isoperimetry leads to concentration in two metric measure spaces, namely (a) the sphere S^{n-1} equipped with the Euclidean (or geodesic) metric and the uniform measure, and (b) $\mathbb{R}^n$ equipped with the Euclidean metric and the Gaussian measure. A similar method yields the concentration for many other metric measure spaces. One of them is the Hamming cube

$$\left(\{0, 1\}^n, d, \mathbb{P}\right),$$

which we introduced in Definition 4.2.14. It will be convenient here to assume that $d(x, y)$ is the *normalized* Hamming distance, which is the fraction of the digits on which the binary strings x and y disagree, thus

$$d(x, y) = \frac{1}{n} \left|\{i: x_i \ne y_i\}\right|.$$

The measure $\mathbb{P}$ is the uniform probability measure on the Hamming cube, i.e.,

$$\mathbb{P}(A) = \frac{|A|}{2^n} \quad \text{for any } A \subset \{0, 1\}^n.$$

Theorem 5.2.5 (Concentration for the Hamming cube) *Consider a random vector* $X \sim \mathrm{Unif}\{0,1\}^n$. *(Thus, the coordinates of* X *are independent* $\mathrm{Ber}(1/2)$ *random variables.) Consider a function* $f : \{0,1\}^n \to \mathbb{R}$. *Then*

$$\|f(X) - \mathbb{E}\, f(X)\|_{\psi_2} \le \frac{C\|f\|_{\mathrm{Lip}}}{\sqrt{n}}. \tag{5.8}$$

This result can be deduced from the isoperimetric inequality on the Hamming cube, whose minimizers are known to be the *Hamming balls* – the neighborhoods of single points with respect to the Hamming distance.

5.2.3 Symmetric Group

The symmetric group S_n consists of all $n!$ permutations of n symbols, which we choose to be $\{1, \dots, n\}$ to be specific. We can view the symmetric group as a metric measure space

$$(S_n, d, \mathbb{P}).$$

Here $d(\pi, \rho)$ is the normalized Hamming distance – the fraction of the symbols on which the permutations π and ρ disagree:

$$d(\pi, \rho) = \frac{1}{n} \left|\{i : \pi(i) \ne \rho(i)\}\right|.$$

The measure $\mathbb{P}$ is the uniform probability measure on S_n, i.e.,

$$\mathbb{P}(A) = \frac{|A|}{n!} \quad \text{for any } A \subset S_n.$$

Theorem 5.2.6 (Concentration for the symmetric group) *Consider a random permutation* $X \sim \mathrm{Unif}(S_n)$ *and a function* $f : S_n \to \mathbb{R}$. *Then the concentration inequality* (5.8) *holds.*

5.2.4 Riemannian Manifolds with Strictly Positive Curvature

A wide general class of examples with nice concentration properties is covered by the notion of a *Riemannian manifold.* Since we do not assume that the reader has necessary background in differential geometry, the rest of this section is optional.

Let (M, g) be a compact connected smooth Riemannian manifold. The canonical distance $d(x, y)$ on M is defined as the arclength (with respect to the Riemannian tensor g) of a minimizing geodesic connecting x and y. The Riemannian manifold can be viewed as a metric measure space

$$(M, d, \mathbb{P})$$

where $\mathbb{P} = dv/V$ is the probability measure on M obtained from the Riemann volume element dv by dividing by V, the total volume of M.

Let $c(M)$ denote the infimum of the Ricci curvature tensor over all tangent vectors. Assuming that $c(M) > 0$, it can be proved using semigroup tools that

$$\|f(X) - \mathbb{E}\, f(X)\|_{\psi_2} \le \frac{C\|f\|_{\mathrm{Lip}}}{\sqrt{c(M)}} \tag{5.9}$$

for any Lipschitz function $f : M \to \mathbb{R}$.

To give an example, it is known that $c(S^{n-1}) = n - 1$. Thus (5.9) gives an alternative approach to concentration inequality (5.5) for the sphere S^{n-1}. We give several other examples next.

5.2.5 Special Orthogonal Group

The special orthogonal group $SO(n)$ consists of all distance-preserving linear transformations on $\mathbb{R}^n$. Equivalently, the elements of $SO(n)$ are $n \times n$ orthogonal matrices whose determinant equals 1. We can view the special orthogonal group as a metric measure space

$$(SO(n), \|\cdot\|_F, \mathbb{P}),$$

where the distance is the Frobenius norm[8] $\|A - B\|_F$ and $\mathbb{P}$ is the uniform probability measure on $SO(n)$.

Theorem 5.2.7 (Concentration for the special orthogonal group) *Consider a random orthogonal matrix* $X \sim \text{Unif}(SO(n))$ *and a function* $f: SO(n) \to \mathbb{R}$. *Then the concentration inequality* (5.8) *holds.*

This result can be deduced from the result for concentration on general Riemannian manifolds which we discussed in Section 5.2.4.

Remark 5.2.8 (Haar measure) Here we do not go into detail about the formal definition of the uniform probability measure $\mathbb{P}$ on $SO(n)$. Let us just mention for an interested reader that $\mathbb{P}$ is the *Haar measure* on $SO(n)$ – the unique probability measure that is invariant under the actions of the group.[9]

One can explicitly construct a random orthogonal matrix $X \sim \text{Unif}(SO(n))$ in several ways. For example, we can make it from an $n \times n$ Gaussian random matrix G with $N(0, 1)$ independent entries. Indeed, consider the singular value decomposition

$$G = U \Sigma V^\mathsf{T}.$$

Then the matrix of the left singular vectors $X := U$ is uniformly distributed in $SO(n)$. One can then define the Haar measure μ on $SO(n)$ by setting

$$\mu(A) := \mathbb{P}\{X \in A\} \quad \text{for } A \subset SO(n).$$

(The rotation invariance should be straightforward – check it!)

5.2.6 Grassmannian

The Grassmannian, or Grassmann manifold $G_{n,m}$, consists of all m-dimensional subspaces of $\mathbb{R}^n$. In the special case where $m = 1$, the Grassmann manifold can be identified with the sphere S^{n-1} (how?), so the concentration result that we are about to state will include the concentration for the sphere as a special case.

[8] The definition of the Frobenius norm was given in Section 4.1.3.

[9] A measure μ on $\mathrm{SO}(n)$ is rotation invariant if, for any measurable set $E \subset \mathrm{SO}(n)$ and any $T \in \mathrm{SO}(n)$, one has $\mu(E) = \mu(T(E))$.

We can view the Grassmann manifold as a metric measure space

$$(G_{n,m}, d, \mathbb{P}).$$

The distance between subspaces E and F can be defined as the operator norm[10]

$$d(E, F) = \|P_E - P_F\|,$$

where P_E and P_F are the orthogonal projections onto E and F, respectively.

The probability $\mathbb{P}$ is, as before, the uniform (Haar) probability measure on $G_{n,m}$. This measure allows us to talk about a *random m-dimensional subspace of* $\mathbb{R}^n$

$$E \sim \mathrm{Unif}(G_{n,m}).$$

Alternatively, a random subspace E (and thus the Haar measure on the Grassmannian) can be constructed by computing the column span (i.e., the image) of a random $n \times m$ Gaussian random matrix G with i.i.d. $N(0, 1)$ entries. (Again the rotation invariance should be straightforward – check it!)

Theorem 5.2.9 (Concentration for the Grassmannian) *Consider a random subspace* $X \sim \mathrm{Unif}(G_{n,m})$ *and a function* $f : G_{n,m} \to \mathbb{R}$. *Then the concentration inequality* (5.8) *holds.*

This result can be deduced from that for concentration on the special orthogonal group from Section 5.2.5. (For the interested reader let us mention how this is done: one can express that Grassmannian as the quotient $G_{n,k} = SO(n)/(SO_m \times SO_{n-m})$ and use the fact that concentration is passed on to quotients.)

5.2.7 Continuous Cube and Euclidean Ball

Similar concentration inequalities can be proved for the unit Euclidean cube $[0, 1]^n$ and the Euclidean ball[11] $\sqrt{n}B_2^n$, both equipped with Euclidean distance and uniform probability measures. They can be deduced from Gaussian concentration by *pushing forward* the Gaussian measure to uniform measures on the ball and the cube, respectively. We state these two theorems and prove them in a few exercises.

Theorem 5.2.10 (Concentration for the continuous cube) *Consider a random vector* $X \sim \mathrm{Unif}([0, 1]^n)$. *(Thus, the coordinates of* X *are independent random variables uniformly distributed on* $[0, 1]$.*) Consider a Lipschitz function* $f : [0, 1]^n \to \mathbb{R}$. *(The Lipschitz norm is with respect to the Euclidean distance.) Then the concentration inequality* (5.7) *holds.*

[10] The operator norm was introduced in Section 4.1.2.

[11] Recall that B_2^n denotes the unit Euclidean ball, i.e. $B_2^n = \{x \in \mathbb{R}^n : \|x\|_2 \le 1\}$, and $\sqrt{n}B_2^n$ is the Euclidean ball of radius $\sqrt{n}$.

Exercise 5.2.11 (Pushing forward the Gaussian to the uniform distribution) Let $\Phi(x)$ denote the cumulative distribution function of the standard normal distribution $N(0, 1)$. Consider a random vector $Z = (Z_1, \ldots, Z_n) \sim N(0, I_n)$. Check that

$$\phi(Z) := \big(\Phi(Z_1), \ldots, \Phi(Z_n)\big) \sim \mathrm{Unif}([0, 1]^n).$$

Exercise 5.2.12 (Proving concentration for the continuous cube) Expressing $X = \phi(Z)$ by means of the previous exercise, use the Gaussian concentration to control the deviation of $f \circ \phi(Z) = f(\phi(Z))$ in terms of $\|f \circ \phi\|_{\mathrm{Lip}} \le \|f\|_{\mathrm{Lip}} \|\phi\|_{\mathrm{Lip}}$. Show that $\|\phi\|_{\mathrm{Lip}}$ is bounded by an absolute constant and complete the proof of Theorem 5.2.10.

Theorem 5.2.13 (Concentration for the Euclidean ball) *Consider a random vector $X \sim \mathrm{Unif}(\sqrt{n}B_2^n)$. Consider a Lipschitz function $f\colon \sqrt{n}B_2^n \to \mathbb{R}$. (The Lipschitz norm is with respect to the Euclidean distance.) Then the concentration inequality* (5.7) *holds.*

Exercise 5.2.14 (Proving concentration for the Euclidean ball) Use a similar method as in the previous exercise to prove Theorem 5.2.13. Define a function $\phi\colon \mathbb{R}^n \to \sqrt{n}B_2^n$ that pushes forward the Gaussian measure on $\mathbb{R}^n$ into the uniform measure on $\sqrt{n}B_2^n$, and check that ϕ has bounded Lipschitz norm.

5.2.8 Densities $e^{-U(x)}$

The push-forward approach from last section can be used to obtain the concentration for many other distributions in $\mathbb{R}^n$. In particular, suppose that a random vector X has density of the form

$$f(x) = e^{-U(x)}$$

for some function $U\colon \mathbb{R}^n \to \mathbb{R}$. As an example, if $X \sim N(0, I_n)$ then the normal density (3.4) gives $U(x) = \|x\|_2^2 + c$, where c is a constant (that depends on n but not x), and Gaussian concentration holds for X.

Now, if U is a general function whose curvature goes at least as $\|x\|_2^2$ then we should expect at least Gaussian concentration. This is exactly what the next theorem states. The curvature of U is measured with the help of the *Hessian* $\mathrm{Hess}\, U(x)$, which by definition is the $n \times n$ symmetric matrix whose (i, j)th entry equals $\partial^2 U/\partial x_i \partial x_j$.

Theorem 5.2.15 *Consider a random vector X in $\mathbb{R}^n$ whose density has the form $f(x) = e^{-U(x)}$ for some function $U\colon \mathbb{R}^n \to \mathbb{R}$. Assume there exists $\kappa > 0$ such that*[12]

$$\mathrm{Hess}\, U(x) \succeq \kappa I_n \quad \textit{for all } x \in \mathbb{R}^n.$$

Then any Lipschitz function $f\colon \mathbb{R}^n \to \mathbb{R}$ satisfies

$$\|f(X) - \mathbb{E} f(X)\|_{\psi_2} \le \frac{C\|f\|_{\mathrm{Lip}}}{\sqrt{\kappa}}.$$

[12] The matrix inequality here means that $\mathrm{Hess}\, U(x) - \kappa I_n$ is a positive-semidefinite matrix.

Note the similarity of this theorem to the concentration inequality (5.9) for Riemannian manifolds. Both can be proved using semigroup tools, which we do not present in this book.

5.2.9 Random Vectors with Independent Bounded Coordinates

There is a remarkable partial generalization of Theorem 5.2.10 for random vectors $X = (X_1, \ldots, X_n)$ whose coordinates are independent and have *arbitrary* bounded distributions. By scaling, there is no loss of generality in assuming that $|X_i| \leq 1$, but we no longer require that the X_i be *uniformly* distributed.

Theorem 5.2.16 (Talagrand's concentration inequality) *Consider a random vector $X = (X_1, \ldots, X_n)$ whose coordinates are independent and satisfy*

$$|X_i| \leq 1 \quad \textit{almost surely.}$$

Then concentration inequality (5.7) *holds for any* convex *Lipschitz function $f\colon [0,1]^n \to \mathbb{R}$.*

In particular, Talagrand's concentration ineqiuality holds for any *norm* on $\mathbb{R}^n$. We do not prove this result here.

5.3 Application: Johnson–Lindenstrauss Lemma

Suppose that we have N data points in $\mathbb{R}^n$, where n is very large. We would like to reduce the dimension of the data without sacrificing too much of its geometry. The simplest form of dimension reduction is to project the data points onto a low-dimensional subspace

$$E \subset \mathbb{R}^n, \quad \dim(E) := m \ll n;$$

see Figure 5.2 for an illustration. How shall we choose the subspace E, and how small can its dimension m be?

The Johnson–Lindenstrauss lemma, given below, states that the geometry of the data is well preserved if we choose E to be a *random subspace* of dimension

$$m \asymp \log N.$$

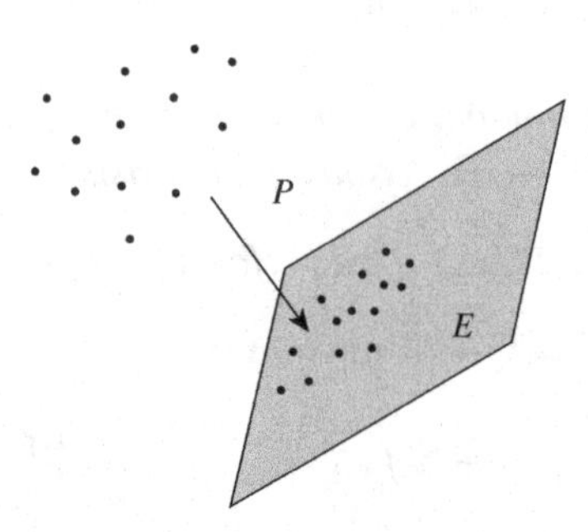

Figure 5.2 In the Johnson–Lindenstrauss lemma, the dimension of the data is reduced by projection P onto a random low-dimensional subspace E.

We have already come across the notion of a random subspace in Section 5.2.6; let us recall it here. We say that E is a random m-dimensional subspace in $\mathbb{R}^n$ uniformly distributed in the Grassmannian $G_{n,m}$, i.e.,

$$E \sim \mathrm{Unif}(G_{n,m}),$$

if E is a random m-dimensional subspace of $\mathbb{R}^n$ whose distribution is rotation invariant, i.e.,

$$\mathbb{P}\left\{E \in \mathcal{E}\right\} = \mathbb{P}\left\{U(E) \in \mathcal{E}\right\}$$

for any fixed subset $\mathcal{E} \subset G_{n,m}$ and $n \times n$ orthogonal matrix U.

Theorem 5.3.1 (Johnson–Lindenstrauss lemma) *Let $\mathcal{X}$ be a set of N points in $\mathbb{R}^n$ and $\varepsilon > 0$. Assume that*

$$m \geq (C/\varepsilon^2) \log N.$$

Consider a random m-dimensional subspace E in $\mathbb{R}^n$ uniformly distributed in $G_{n,m}$. Denote the orthogonal projection onto E by P. Then, with probability at least $1 - 2\exp(-c\varepsilon^2 m)$, the scaled projection

$$Q := \sqrt{\frac{n}{m}}\, P$$

is an approximate isometry on $\mathcal{X}$:

$$(1 - \varepsilon)\|x - y\|_2 \leq \|Qx - Qy\|_2 \leq (1 + \varepsilon)\|x - y\|_2 \quad \text{for all } x, y \in \mathcal{X}. \tag{5.10}$$

The proof of the Johnson–Lindenstrauss lemma will be based on the concentration of Lipschitz functions for the sphere, which we studied in Section 5.1. We use this to first examine how the random projection P acts on a *fixed* vector $x - y$, and then we take the *union bound* over all N^2 differences $x - y$.

Lemma 5.3.2 (Random projection) *Let P be a projection from $\mathbb{R}^n$ onto a random m-dimensional subspace uniformly distributed in $G_{n,m}$. Let $z \in \mathbb{R}^n$ be a (fixed) point and $\varepsilon > 0$. Then:*

(i) $\left(\mathbb{E}\, \|Pz\|_2^2\right)^{1/2} = \sqrt{\frac{m}{n}}\, \|z\|_2.$

(ii) With probability at least $1 - 2\exp(-c\varepsilon^2 m)$, we have

$$(1 - \varepsilon)\sqrt{\frac{m}{n}}\, \|z\|_2 \leq \|Pz\|_2 \leq (1 + \varepsilon)\sqrt{\frac{m}{n}}\, \|z\|_2.$$

Proof Without loss of generality, we may assume that $\|z\|_2 = 1$. (Why?) Next, we consider an equivalent model: instead of a random projection P acting on a fixed vector z, we consider a fixed projection P acting on a random vector z. Specifically, the distribution of $\|Pz\|_2$ will not change if we let P be fixed and let

$$z \sim \mathrm{Unif}(S^{n-1}).$$

(Check this using rotation invariance!)

Using rotation invariance again, we may assume without loss of generality that P is the *coordinate projection* onto the first m coordinates in $\mathbb{R}^n$. Thus

$$\mathbb{E}\,\|Pz\|_2^2 = \mathbb{E}\sum_{i=1}^m z_i^2 = \sum_{i=1}^m \mathbb{E}\,z_i^2 = m\,\mathbb{E}\,z_1^2, \tag{5.11}$$

since the coordinates z_i of the random vector $z \sim \mathrm{Unif}(S^{n-1})$ are identically distributed. To compute $\mathbb{E}\,z_1^2$, note that

$$1 = \|z\|_2^2 = \sum_{i=1}^n z_i^2.$$

Taking expectations of both sides, we obtain

$$1 = \sum_{i=1}^n \mathbb{E}\,z_i^2 = n\,\mathbb{E}\,z_1^2,$$

which yields

$$\mathbb{E}\,z_1^2 = \frac{1}{n}.$$

Putting this into (5.11), we get

$$\mathbb{E}\,\|Pz\|_2^2 = \frac{m}{n}.$$

This proves the first part of the lemma.

The second part follows from the concentration of Lipschitz functions for the sphere. Indeed,

$$f(x) := \|Px\|_2$$

is a Lipschitz function on S^{n-1}, and $\|f\|_{\mathrm{Lip}} = 1$. (Check!) Then the concentration inequality (5.6) yields

$$\mathbb{P}\left\{\left|\|Px\|_2 - \sqrt{\frac{m}{n}}\right| \ge t\right\} \le 2\exp(-cnt^2).$$

(Here we have also used Exercise 5.2.4 to replace $\mathbb{E}\,\|x\|_2$ by the quantity $(\mathbb{E}\,\|x\|_2^2)^{1/2}$ in the concentration inequality.) Choosing $t := \varepsilon\sqrt{m/n}$ completes the proof of the lemma. ■

Proof of Johnson–Lindenstrauss lemma Consider the difference set

$$\mathcal{X} - \mathcal{X} := \{x - y : x, y \in \mathcal{X}\}.$$

We would like to show that, with the required probability, the inequality

$$(1-\varepsilon)\|z\|_2 \le \|Qz\|_2 \le (1+\varepsilon)\|z\|_2$$

holds for all $z \in \mathcal{X} - \mathcal{X}$. Since $Q = \sqrt{n/m}\,P$, this inequality is equivalent to

$$(1-\varepsilon)\sqrt{\frac{m}{n}}\,\|z\|_2 \le \|Pz\|_2 \le (1+\varepsilon)\sqrt{\frac{m}{n}}\,\|z\|_2. \tag{5.12}$$

For any fixed z, Lemma 5.3.2 states that (5.12) holds with probability at least $1 - 2\exp(-c\varepsilon^2 m)$. It remains to take a union bound over $z \in \mathcal{X} - \mathcal{X}$. It follows that inequality (5.12) holds simultaneously for all $z \in \mathcal{X} - \mathcal{X}$, with probability at least

$$1 - |\mathcal{X} - \mathcal{X}|\, 2\exp(-c\varepsilon^2 m) \geq 1 - N^2\, 2\exp(-c\varepsilon^2 m).$$

If $m \geq (C/\varepsilon^2)\log N$ then this probability is at least $1 - 2\exp(-c\varepsilon^2 m/2)$, as claimed. The Johnson–Lindenstrauss lemma is proved. ■

A remarkable feature of the Johnson–Lindenstrauss lemma is that the dimension reduction map A is *non-adaptive*: it does not depend on the data. Note also that the ambient dimension n of the data plays no role in this result.

Exercise 5.3.3 (Johnson–Lindenstrauss with sub-gaussian matrices) Let A be an $m \times n$ random matrix whose rows are independent mean-zero sub-gaussian isotropic random vectors in $\mathbb{R}^n$. Show that the conclusion of the Johnson–Lindenstrauss lemma holds for $Q = (1/\sqrt{n})A$.

Exercise 5.3.4 (Optimality of Johnson–Lindenstrauss) Give an example of a set $\mathcal{X}$ of N points for which no scaled projection onto a subspace of dimension $m \ll \log N$ is an approximate isometry. ☞

5.4 Matrix Bernstein Inequality

In this section, we show how to generalize concentration inequalities for sums of independent random variables $\sum X_i$ to sums of independent *random matrices*.

We will prove a matrix version of Bernstein's inequality (Theorem 2.8.4), where the random variables X_i are replaced by random matrices, and the absolute value $|\cdot|$ is replaced by the operator norm $\|\cdot\|$. Remarkably, we will not require independence of the entries, rows, or columns within each random matrix X_i.

Theorem 5.4.1 (Matrix Bernstein inequality) *Let $X_1, \ldots, X_N$ be independent mean-zero $n \times n$ symmetric random matrices, such that $\|X_i\| \leq K$ almost surely for all i. Then, for every $t \geq 0$, we have*

$$\mathbb{P}\left\{\left\|\sum_{i=1}^N X_i\right\| \geq t\right\} \leq 2n\exp\left(-\frac{t^2/2}{\sigma^2 + Kt/3}\right).$$

Here $\sigma^2 = \left\|\sum_{i=1}^N \mathbb{E}\, X_i^2\right\|$ is the norm of the matrix variance of the sum.

In particular, we can express this bound as a mixture of sub-gaussian and sub-exponential tails, just as for the scalar Bernstein inequality:

$$\mathbb{P}\left\{\left\|\sum_{i=1}^N X_i\right\| \geq t\right\} \leq 2n\exp\left(-c\min\left(\frac{t^2}{\sigma^2}, \frac{t}{K}\right)\right).$$

The proof of the matrix Bernstein inequality is based on the following naïve idea. We will try to repeat the classical argument based on moment generating functions (see Section 2.8), replacing scalars by matrices at each occurrence. In most of our argument this idea will work, except for one step that is nontrivial to generalize. Before we dive into this argument, let us develop some *matrix calculus* which will allow us to treat matrices as scalars.

5.4.1 Matrix Calculus

Throughout this section, we work with symmetric $n \times n$ matrices. As we know, the operation of addition, $A + B$, generalizes painlessly from scalars to matrices. We need to be more careful with multiplication, since it is not commutative for matrices: in general, $AB \neq BA$. For this reason, the matrix Bernstein inequality is sometimes called the *non-commutative* Bernstein inequality. Functions of matrices are defined as follows.

Definition 5.4.2 (Functions of matrices) Consider a function $f : \mathbb{R} \to \mathbb{R}$ and an $n \times n$ symmetric matrix X with eigenvalues λ_i and corresponding eigenvectors u_i. Recall that X can be represented as a spectral decomposition:

$$X = \sum_{i=1}^{n} \lambda_i u_i u_i^\mathsf{T}.$$

Then define

$$f(X) := \sum_{i=1}^{n} f(\lambda_i) u_i u_i^\mathsf{T}.$$

In other words, to obtain the matrix $f(X)$ from X, we do not change the eigenvectors but we apply f to the eigenvalues.

In the following exercise, we will check that the definition of a function of matrices agrees with the basic rules of matrix addition and multiplication.

Exercise 5.4.3 (Matrix polynomials and power series)☕☕

(a) Consider a polynomial

$$f(x) = a_0 + a_1 x + \cdots + a_p x^p.$$

Check that for a matrix X, we have

$$f(X) = a_0 I + a_1 X + \cdots + a_p X^p.$$

On the right-hand side we use the standard rules for matrix addition and multiplication, so in particular $X^p = X \cdots X$ (p times) there.

(b) Consider a convergent power series expansion of f about x_0:

$$f(x) = \sum_{k=1}^{\infty} a_k (x - x_0)^k.$$

Check that the series of matrix terms converges and that

$$f(X) = \sum_{k=1}^{\infty} a_k (X - X_0)^k.$$

As an example, for each $n \times n$ symmetric matrix X we have

$$e^X = I + X + \frac{X^2}{2!} + \frac{X^3}{3!} + \cdots$$

Just like scalars, matrices can be compared with each other. To do this, we define a *partial order* on the set of $n \times n$ symmetric matrices, as follows.

Definition 5.4.4 (positive-semidefinite order) We say that

$$X \succeq 0$$

if X is a positive-semidefinite matrix. Equivalently, $X \succeq 0$ if X is symmetric and all eigenvalues of X satisfy $\lambda_i(X) \geq 0$. Next, we set

$$X \succeq Y \quad \text{and} \quad Y \preceq X$$

if $X - Y \succeq 0$.

Note that $\succeq$ is a partial, as opposed to total, order, since there are matrices for which neither $X \succeq Y$ nor $Y \succeq X$ hold. (Give an example!)

Exercise 5.4.5☕☕☕ Prove the following properties.

(a) $\|X\| \leq t$ if and only if $-tI \preceq X \preceq tI$.
(b) Let $f, g \colon \mathbb{R} \to \mathbb{R}$ be two functions. If $f(x) \leq g(x)$ for all $x \in \mathbb{R}$ satisfying $|x| \leq K$, then $f(X) \preceq g(X)$ for all X satisfying $\|X\| \leq K$.
(c) Let $f \colon \mathbb{R} \to \mathbb{R}$ be an increasing function and X, Y be commuting matrices. Then $X \preceq Y$ implies $f(X) \preceq f(Y)$.
(d) Give an example showing that property (c) may fail for non-commuting matrices. ☞
(e) In the following parts of the exercise, we develop weaker versions of property (c) that hold for arbitrary, not necessarily commuting, matrices. First, show that $X \preceq Y$ always implies $\operatorname{tr} f(X) \leq \operatorname{tr} f(Y)$ for any increasing function $f \colon \mathbb{R} \to \mathbb{R}$. ☞
(f) Show that $0 \preceq X \preceq Y$ implies $X^{-1} \succeq Y^{-1}$ if X is invertible. ☞
(g) Show that $0 \preceq X \preceq Y$ implies $\log X \preceq \log Y$. ☞

5.4.2 Trace Inequalities

So far, our attempts to extend scalar concepts for matrices have not met any resistance. But this does not always go so smoothly. The non-commutativity of the matrix product ($AB \neq BA$) causes some important scalar identities to fail for matrices. One such identity is $e^{x+y} = e^x e^y$, which holds for scalars but fails for matrices:

Exercise 5.4.6 ☕☕☕ Let X and Y be $n \times n$ symmetric matrices.

(a) Show that if the matrices commute, i.e. $XY = YX$, then
$$e^{X+Y} = e^X e^Y.$$

(b) Find an example of matrices X and Y such that
$$e^{X+Y} \neq e^X e^Y.$$

This is unfortunate for us, because we used the identity $e^{x+y} = e^x e^y$ in a crucial way in our approach to the concentration of sums of random variables. Indeed, this identity allowed us to break the moment generating function $\mathbb{E}\exp(\lambda S)$ of the sum into the product of exponentials: see (2.6).

Nevertheless, there exist useful substitutes for the missing identity $e^{X+Y} = e^X e^Y$. We state two of them here without proof; they belong to the rich family of *trace inequalities*.

Theorem 5.4.7 (Golden–Thompson inequality) *For any $n \times n$ symmetric matrices A and B, we have*
$$\operatorname{tr}(e^{A+B}) \leq \operatorname{tr}(e^A e^B).$$

Theorem 5.4.8 (Lieb's inequality) *Let H be an $n \times n$ symmetric matrix. Define the function on matrices*
$$f(X) := \operatorname{tr}\exp(H + \log X).$$
Then f is concave on the space of positive-definite $n \times n$ symmetric matrices.[13]

Note that in the scalar case, where $n = 1$, the function f is linear and Lieb's inequality holds trivially.

A proof of the matrix Bernstein inequality can be based on either the Golden–Thompson inequality or Lieb's inequality. We will use Lieb's inequality, which we will now restate for random matrices. If X is a random matrix then Lieb's and Jensen's inequalities imply that
$$\mathbb{E} f(X) \leq f(\mathbb{E} X).$$
(Why does Jensen's inequality hold for random matrices?) Applying this with $X = e^Z$, we obtain the following.

Lemma 5.4.9 (Lieb's inequality for random matrices) *Let H be a fixed $n \times n$ symmetric matrix and Z be a random $n \times n$ symmetric matrix. Then*
$$\mathbb{E}\operatorname{tr}\exp(H + Z) \leq \operatorname{tr}\exp(H + \log \mathbb{E} e^Z).$$

[13] Concavity means that the inequality $f(\lambda X + (1-\lambda)Y) \geq \lambda f(X) + (1-\lambda) f(Y)$ holds for matrices X and Y, and for $\lambda \in [0, 1]$.

5.4.3 Proof of Matrix Bernstein Inequality

We are now ready to prove the matrix Bernstein inequality, Theorem 5.4.1, using Lieb's inequality.

Step 1: Reduction to MGF. To bound the norm of the sum

$$S := \sum_{i=1}^{N} X_i,$$

we need to control the largest and smallest eigenvalues of S. We can do this separately. To put it formally, consider the largest eigenvalue,

$$\lambda_{\max}(S) := \max_i \lambda_i(S),$$

and note that

$$\|S\| = \max_i |\lambda_i(S)| = \max\left(\lambda_{\max}(S),\ \lambda_{\max}(-S)\right). \tag{5.13}$$

To bound $\lambda_{\max}(S)$, we proceed with the method based on computing the moment generating function that we used in the scalar case, e.g. in Section 2.2. To this end, fix $\lambda \geq 0$ and use Markov's inequality to obtain

$$\mathbb{P}\left\{\lambda_{\max}(S) \geq t\right\} = \mathbb{P}\left\{e^{\lambda \cdot \lambda_{\max}(S)} \geq e^{\lambda t}\right\} \leq e^{-\lambda t}\, \mathbb{E}\, e^{\lambda \lambda_{\max}(S)}. \tag{5.14}$$

Since by Definition 5.4.2 the eigenvalues of $e^{\lambda S}$ are $e^{\lambda \lambda_i(S)}$, we have

$$E := \mathbb{E}\, e^{\lambda \lambda_{\max}(S)} = \mathbb{E}\, \lambda_{\max}(e^{\lambda S}).$$

Since the eigenvalues of $e^{\lambda S}$ are all positive, the maximal eigenvalue of $e^{\lambda S}$ is bounded by the sum of all the eigenvalues, the trace of $e^{\lambda S}$, which leads to

$$E \leq \mathbb{E} \operatorname{tr} e^{\lambda S}.$$

Step 2: Application of Lieb's inequality. To prepare for an application of Lieb's inequality (Lemma 5.4.9), let us separate the last term from the sum S:

$$E \leq \mathbb{E} \operatorname{tr} \exp\left(\sum_{i=1}^{N-1} \lambda X_i + \lambda X_N\right).$$

Condition on $(X_i)_{i=1}^{N-1}$ and apply Lemma 5.4.9 for the fixed matrix $H := \sum_{i=1}^{N-1} \lambda X_i$ and the random matrix $Z := \lambda X_N$. We obtain

$$E \leq \mathbb{E} \operatorname{tr} \exp\left(\sum_{i=1}^{N-1} \lambda X_i + \log \mathbb{E}\, e^{\lambda X_N}\right).$$

(To be more specific here, we first apply Lemma 5.4.9 for the conditional expectation, and then take the expectation of both sides using the law of total expectation.)

We continue in a similar way: separate the next term λX_{N-1} from the sum $\sum_{i=1}^{N-1} \lambda X_i$ and apply Lemma 5.4.9 again for $Z = \lambda X_{N-1}$. Repeating N times, we obtain

$$E \leq \operatorname{tr} \exp\left(\sum_{i=1}^{N} \log \mathbb{E}\, e^{\lambda X_i}\right). \tag{5.15}$$

Step 3: MGF of the individual terms. It remains to bound the matrix-valued moment generating function $\mathbb{E}\, e^{\lambda X_i}$ for each term X_i. This is a standard task, and the argument will be similar to the scalar case.

Lemma 5.4.10 (Moment generating function) *Let X be an $n \times n$ symmetric mean-zero random matrix such that $\|X\| \le K$ almost surely. Then*

$$\mathbb{E} \exp(\lambda X) \preceq \exp\left(g(\lambda)\, \mathbb{E}\, X^2\right) \quad \textit{where} \quad g(\lambda) = \frac{\lambda^2/2}{1 - |\lambda| K/3},$$

provided that $|\lambda| < 3/K$.

Proof First, note that we can bound the (scalar) exponential function by the first few terms of its Taylor's expansion as follows:

$$e^z \le 1 + z + \frac{1}{1 - |z|/3} \frac{z^2}{2}, \quad \text{if } |z| < 3.$$

(To obtain this inequality, write $e^z = 1 + z + z^2 \sum_{p=2}^{\infty} z^{p-2}/p!$ and use the bound $p! \ge 2 \times 3^{p-2}$.) Next, apply this inequality for $z = \lambda x$. If $|x| \le K$ and $|\lambda| < 3/K$ then we obtain

$$e^{\lambda x} \le 1 + \lambda x + g(\lambda) x^2,$$

where $g(\lambda)$ is the function in the statement of the lemma.

Finally, we can transfer this inequality from scalars to matrices using part (b) of Exercise 5.4.5. We obtain that if $\|X\| \le K$ and $|\lambda| < 3/K$ then

$$e^{\lambda X} \preceq I + \lambda X + g(\lambda) X^2.$$

Take expectations of both sides and use the assumption that $\mathbb{E}\, X = 0$ to obtain

$$\mathbb{E}\, e^{\lambda X} \preceq I + g(\lambda)\, \mathbb{E}\, X^2.$$

To bound the right-hand side, we can use the inequality $1 + z \le e^z$, which holds for all scalars z. Thus the inequality $I + Z \preceq e^Z$ holds for all matrices Z, and in particular for $Z = g(\lambda)\, \mathbb{E}\, X^2$. (Here we again refer to part (b) of Exercise 5.4.5.) This yields the conclusion of the lemma. ∎

Step 4: Completion of the proof. Let us return to bounding the quantity in (5.15). Using Lemma 5.4.10, we obtain

$$E \le \operatorname{tr} \exp\left(\sum_{i=1}^N \log \mathbb{E}\, e^{\lambda X_i} \right) \le \operatorname{tr} \exp\left(g(\lambda) Z\right), \quad \text{where} \quad Z := \sum_{i=1}^N \mathbb{E}\, X_i^2.$$

(Here we used Exercise 5.4.5 again: part (g) to take logarithms on both sides, and then part (e) to take traces of the exponential of both sides.)

Since the trace of $\exp\left(g(\lambda) Z\right)$ is a sum of n positive eigenvalues, it is bounded by n times the maximum eigenvalue, so

$$\begin{aligned} E &\le n\lambda_{\max}\left(\exp(g(\lambda)Z)\right) = n\exp\left[g(\lambda)\lambda_{\max}(Z)\right] \qquad \text{(why?)} \\ &= n\exp\left(g(\lambda)\|Z\|\right) \qquad \text{(since } Z \succeq 0\text{)} \\ &= n\exp(g(\lambda)\sigma^2) \qquad \text{(by definition of } \sigma \text{ in the theorem).} \end{aligned}$$

Substituting this bound for $E = \mathbb{E}\, e^{\lambda\lambda_{\max}(S)}$ into (5.14), we obtain

$$\mathbb{P}\left\{\lambda_{\max}(S) \ge t\right\} \le n\, \exp\left(- \lambda t + g(\lambda)\sigma^2\right).$$

We have obtained a bound that holds for any $\lambda > 0$ such that $|\lambda| < 3/K$, so we can minimize it in λ. The minimum is attained for $\lambda = t/(\sigma^2 + Kt/3)$ (check!), which gives

$$\mathbb{P}\left\{\lambda_{\max}(S) \ge t\right\} \le n\, \exp\left(- \frac{t^2/2}{\sigma^2 + Kt/3}\right).$$

Repeating the argument for $-S$ and combining the two bounds via (5.13), we complete the proof of Theorem 5.4.1. (Do this!) ■

5.4.4 Matrix Khintchine Inequality

The matrix Bernstein inequality gives a good *tail bound* on $\|\sum_{i=1}^N X_i\|$, and this in particular implies a nontrivial bound on the *expectation*.

Exercise 5.4.11 (Matrix Bernstein inequality: expectation) ☕☕☕ Let $X_1, \dots, X_N$ be independent mean-zero $n \times n$ symmetric random matrices such that $\|X_i\| \le K$ almost surely for all i. Deduce from Bernstein's inequality that

$$\mathbb{E}\left\|\sum_{i=1}^N X_i\right\| \lesssim \left\|\sum_{i=1}^N \mathbb{E}\, X_i^2\right\|^{1/2} \sqrt{1+\log n} + K(1+\log n).$$ ☞

Note that in the scalar case, where $n = 1$, a bound on the expectation is trivial. Indeed, in this case we have

$$\mathbb{E}\left|\sum_{i=1}^N X_i\right| \le \left(\mathbb{E}\left|\sum_{i=1}^N X_i\right|^2\right)^{1/2} = \left(\sum_{i=1}^N \mathbb{E}\, X_i^2\right)^{1/2},$$

where we have used that the variance of a sum of independent random variables equals the sum of the variances.

The techniques developed in the proof of the matrix Bernstein inequality can be used to give matrix versions of other classical concentration inequalities. In the next two exercises, the reader can prove matrix versions of Hoeffding's inequality (Theorem 2.2.2) and Khintchine's inequality (Exercise 2.6.6).

Exercise 5.4.12 (Matrix Hoeffding inequality) ☕☕☕ Let $\varepsilon_1, \dots, \varepsilon_n$ be independent symmetric Bernoulli random variables and let $A_1, \dots, A_N$ be symmetric $n \times n$ matrices (deterministic). Prove that, for any $t \ge 0$, we have

$$\mathbb{P}\left\{\left\|\sum_{i=1}^N \varepsilon_i A_i\right\| \geq t\right\} \leq 2n\exp(-t^2/2\sigma^2),$$

where $\sigma^2 = \|\sum_{i=1}^N A_i^2\|$. ☞

From this, one can deduce a matrix version of Khintchine's inequality:

Exercise 5.4.13 (Matrix Khintchine inequality) Let $\varepsilon_1, \ldots, \varepsilon_N$ be independent symmetric Bernoulli random variables and let $A_1, \ldots, A_N$ be symmetric $n \times n$ matrices (deterministic).

(a) Prove that

$$\mathbb{E}\left\|\sum_{i=1}^N \varepsilon_i A_i\right\| \leq C\sqrt{\log n}\left\|\sum_{i=1}^N A_i^2\right\|^{1/2}.$$

(b) More generally, prove that for every $p \in [1, \infty)$ we have

$$\left(\mathbb{E}\left\|\sum_{i=1}^N \varepsilon_i A_i\right\|^p\right)^{1/p} \leq C\sqrt{p+\log n}\left\|\sum_{i=1}^N A_i^2\right\|^{1/2}.$$

The price of going from scalars to matrices is the pre-factor n in the probability bound in Theorem 5.4.1. This is a small price, considering that this factor becomes *logarithmic* in the dimension n in the expectation bounds of Exercises 5.4.11–5.4.13. The following example demonstrates that the logarithmic factor is needed in general.

Exercise 5.4.14 (Sharpness of matrix Bernstein inequality) Let X be an $n \times n$ random matrix that takes values $e_k e_k^\mathsf{T}$, $k = 1, \ldots, n$, with probability $1/n$ each. (Here (e_k) denotes the standard basis in $\mathbb{R}^n$.) Let $X_1, \ldots, X_N$ be independent copies of X. Consider the sum

$$S := \sum_{i=1}^N X_i,$$

which is a diagonal matrix.

(a) Show that the entry S_{ii} has the same distribution as the number of balls in the ith bin when N balls are thrown into n bins independently.
(b) Relating this to the classical coupon collector's problem, show that if $N \asymp n$ then[14]

$$\mathbb{E}\|S\| \asymp \frac{\log n}{\log\log n}.$$

Deduce that the bound in Exercise 5.4.11 would fail if the logarithmic factors were removed from it.

[14] Here we write $a_n \asymp b_n$ if there exist constants $c, C > 0$ such that $ca_n < b_n \leq Ca_n$ for all n.

The following exercise extends the matrix Bernstein inequality by dropping both the symmetry and square assumptions on the matrices X_i.

Exercise 5.4.15 (Matrix Bernstein inequality for rectangular matrices) Let $X_1, \dots, X_N$ be independent mean-zero $m \times n$ random matrices, such that $\|X_i\| \le K$ almost surely for all i. Prove that, for $t \ge 0$, we have

$$\mathbb{P}\left\{\left\|\sum_{i=1}^N X_i\right\| \ge t\right\} \le 2(m+n)\exp\left(-\frac{t^2/2}{\sigma^2 + Kt/3}\right),$$

where

$$\sigma^2 = \max\left(\left\|\sum_{i=1}^N \mathbb{E}\, X_i^\mathsf{T} X_i\right\|, \left\|\sum_{i=1}^N \mathbb{E}\, X_i X_i^\mathsf{T}\right\|\right).$$

☞

5.5 Application: Community Detection in Sparse Networks

In Section 4.5 we analyzed a basic method for community detection in networks – the spectral clustering algorithm. We examined the performance of spectral clustering for the stochastic block model $G(n, p, q)$ with two communities, and we found how the communities can be identified with high accuracy and high probability (Theorem 4.5.6).

We now re-examine the performance of spectral clustering using the matrix Bernstein's inequality. In the following two exercises, we find that spectral clustering actually works for *much sparser* networks than are implied by Theorem 4.5.6.

Just as in Section 4.5, we denote by A the adjacency matrix of a random graph from $G(n, p, q)$, and we express A as

$$A = D + R,$$

where $D = \mathbb{E}\, A$ is a deterministic matrix (the "signal") and R is random (the "noise"). As we know, the success of spectral clustering method hinges on the fact that the noise $\|R\|$ is small with high probability (recall (4.18)). In the following exercise, the matrix Bernstein inequality is used to derive a better bound on $\|R\|$.

Exercise 5.5.1 (Controlling the noise)

(a) Represent the adjacency matrix A as a sum of independent random matrices,

$$A = \sum_{1 \le i < j \le n}^{n} Z_{ij},$$

in such a way that each Z_{ij} encodes the contribution of the edge between vertices i and j. Thus, the only nonzero entries of Z_{ij} should be (ij) and (ji), and they should be the same as in A.

(b) Apply the matrix Bernstein inequality to find that

$$\mathbb{E}\,\|R\| \lesssim \sqrt{d \log n} + \log n,$$

where $d = \frac{1}{2}(p+q)n$ is the *expected average degree* of the graph.

Exercise 5.5.2 (Spectral clustering for sparse networks) Use the bound from Exercise 5.5.1 to give better guarantees for the performance of spectral clustering than we had in Section 4.5. In particular, argue that spectral clustering works for *sparse networks* as long as the average expected degrees satisfy

$$d \gg \log n.$$

5.6 Application: Covariance Estimation for General Distributions

In Section 3.2, we saw how the covariance matrix of a sub-gaussian distribution in $\mathbb{R}^n$ can be estimated accurately using a sample of size $O(n)$. In this section, we remove the sub-gaussian requirement, and thus make covariance estimation possible for very general, in particular, discrete, distributions. The price we pay is very small – just a logarithmic oversampling factor. Indeed, the following theorem shows that $O(n \log n)$ samples suffice for the covariance estimation of general distributions in $\mathbb{R}^n$.

As in Section 4.7, we estimate the second-moment matrix $\Sigma = \mathbb{E}\, XX^\mathsf{T}$ by its sample version

$$\Sigma_m = \frac{1}{m} \sum_{i=1}^{m} X_i X_i^\mathsf{T}.$$

If we assume that X has zero mean (which we often do for simplicity), Σ is the covariance matrix of X and Σ_m is the sample covariance matrix of X.

Theorem 5.6.1 (General covariance estimation) *Let X be a random vector in $\mathbb{R}^n$, $n \ge 2$. Assume that, for some $K \ge 1$,*

$$\|X\|_2 \le K\, (\mathbb{E}\, \|X\|_2^2)^{1/2} \quad \textit{almost surely.} \tag{5.16}$$

Then, for every positive integer m, we have

$$\mathbb{E}\, \|\Sigma_m - \Sigma\| \le C\Big(\sqrt{\frac{K^2 n \log n}{m}} + \frac{K^2 n \log n}{m}\Big)\|\Sigma\|.$$

Proof Before we start proving the bound, let us pause to note that $\mathbb{E}\, \|X\|_2^2 = \operatorname{tr}(\Sigma)$. (Check this in the same way as in the proof of Lemma 3.2.4.) So, the assumption (5.16) becomes

$$\|X\|_2^2 \le K^2 \operatorname{tr}(\Sigma) \quad \text{almost surely.} \tag{5.17}$$

Apply the expectation version of the matrix Bernstein inequality (Exercise 5.4.11) for the sum of i.i.d. mean-zero random matrices $X_i X_i^\mathsf{T} - \Sigma$ and get[15]

$$\mathbb{E}\, \|\Sigma_m - \Sigma\| = \frac{1}{m}\, \mathbb{E}\, \Big\| \sum_{i=1}^{m} (X_i X_i^\mathsf{T} - \Sigma) \Big\| \lesssim \frac{1}{m}\Big(\sigma \sqrt{\log n} + M \log n\Big), \tag{5.18}$$

[15] As usual, the notation $a \lesssim b$ hides absolute constant factors, i.e., it means that $a \le Cb$ where C is an absolute constant.

where

$$\sigma^2 = \left\| \sum_{i=1}^{m} \mathbb{E}(X_i X_i^\mathsf{T} - \Sigma)^2 \right\| = m \| \mathbb{E}(XX^\mathsf{T} - \Sigma)^2 \|$$

and M is any number chosen so that

$$\|XX^\mathsf{T} - \Sigma\| \le M \quad \text{almost surely.}$$

To complete the proof, it remains to bound σ^2 and M.

Let us start with σ^2. Expanding the square, we find that[16]

$$\mathbb{E}(XX^\mathsf{T} - \Sigma)^2 = \mathbb{E}(XX^\mathsf{T})^2 - \Sigma^2 \preceq \mathbb{E}(XX^\mathsf{T})^2. \tag{5.19}$$

Further, the assumption (5.17) gives

$$(XX^\mathsf{T})^2 = \|X\|^2 XX^\mathsf{T} \preceq K^2 \operatorname{tr}(\Sigma)\, XX^\mathsf{T}.$$

Taking expectations and recalling that $\mathbb{E}\, XX^\mathsf{T} = \Sigma$, we obtain

$$\mathbb{E}(XX^\mathsf{T})^2 \preceq K^2 \operatorname{tr}(\Sigma)\, \Sigma.$$

Substituting this bound into (5.19), we obtain a good bound on σ, namely

$$\sigma^2 \le K^2 m \operatorname{tr}(\Sigma)\, \|\Sigma\|.$$

Bounding M is simple: indeed,

$$\begin{aligned}
\|XX^\mathsf{T} - \Sigma\| &\le \|X\|_2^2 + \|\Sigma\| \quad \text{(by the triangle inequality)} \\
&\le K^2 \operatorname{tr}(\Sigma) + \|\Sigma\| \quad \text{(by assumption (5.17))} \\
&\le 2K^2 \operatorname{tr}(\Sigma) =: M \quad \text{(since } \|\Sigma\| \le \operatorname{tr}(\Sigma) \text{ and } K \ge 1).
\end{aligned}$$

Substituting our bounds for σ and M into (5.18), we get

$$\mathbb{E}\, \|\Sigma_m - \Sigma\| \le \frac{1}{m}\Big(\sqrt{K^2 m \operatorname{tr}(\Sigma)\, \|\Sigma\|}\, \sqrt{\log n} + 2K^2 \operatorname{tr}(\Sigma)\, \log n\Big).$$

To complete the proof, we use the inequality $\operatorname{tr}(\Sigma) \le n\|\Sigma\|$ and simplify the bound. ∎

Remark 5.6.2 (Sample complexity) Theorem 5.6.1 implies that, for any $\varepsilon \in (0, 1)$, we are guaranteed to have covariance estimation with a good relative error,

$$\mathbb{E}\, \|\Sigma_m - \Sigma\| \le \varepsilon \|\Sigma\|, \tag{5.20}$$

if we take a sample of size

$$m \asymp \varepsilon^{-2} n \log n.$$

Compare this with the sample complexity $m \asymp \varepsilon^{-2} n$ for sub-gaussian distributions (recall Remark 4.7.2). We see that the price of dropping the sub-gaussian requirement turns out to be very small – it is just a logarithmic oversampling factor.

[16] Recall Definition 5.4.4 for the positive-semidefinite order $\preceq$ used here.

Remark 5.6.3 (Lower-dimensional distributions) At the end of the proof of Theorem 5.6.1 we used the crude bound $\operatorname{tr}(\Sigma) \le n\|\Sigma\|$. But we may choose not to do that, and instead get a bound in terms of the *intrinsic dimension*

$$r = \frac{\operatorname{tr}(\Sigma)}{\|\Sigma\|},$$

namely

$$\mathbb{E}\,\|\Sigma_m - \Sigma\| \le C\Big(\sqrt{\frac{K^2 r \log n}{m}} + \frac{K^2 r \log n}{m}\Big)\|\Sigma\|.$$

In particular, this stronger bound implies that a sample of size

$$m \asymp \varepsilon^{-2} r \log n$$

is sufficient to estimate the covariance matrix as in (5.20). Note that we always have $r \le n$ (why?), so the new bound is always as good as the bound in Theorem 5.6.1. But for *approximately low-dimensional* distributions – those that tend to concentrate near low-dimensional subspaces – we may have $r \ll n$, and in this case estimate covariance using a much smaller sample. We will return to this discussion in Section 7.6, where we introduce the notions of stable dimension and stable rank.

Exercise 5.6.4 (Tail bound) Our argument also implies the following high-probability guarantee. Check that, for any $u \ge 0$, we have

$$\|\Sigma_m - \Sigma\| \le C\Big(\sqrt{\frac{K^2 r(\log n + u)}{m}} + \frac{K^2 r(\log n + u)}{m}\Big)\|\Sigma\|$$

with probability at least $1 - 2e^{-u}$. Here $r = \operatorname{tr}(\Sigma)/\|\Sigma\| \le n$ as before.

Exercise 5.6.5 (Necessity of boundedness assumption) Show that if the boundedness assumption (5.16) is removed from Theorem 5.6.1, the conclusion may fail in general.

Exercise 5.6.6 (Sampling from frames) Consider an equal-norm tight frame[17] $(u_i)_{i=1}^N$ in $\mathbb{R}^n$. State and prove a result that shows that a random sample of

$$m \gtrsim n \log n$$

elements of (u_i) forms a frame with good frame bounds (i.e., as close to tight as one wants). The quality of the result should not depend on the frame size N.

Exercise 5.6.7 (Necessity of logarithmic oversampling) Show that, in general, logarithmic oversampling is necessary for covariance estimation. More precisely, give an example of a distribution in $\mathbb{R}^n$ for which the bound (5.20) must fail for every $\varepsilon < 1$ unless $m \gtrsim n \log n$. ☞

[17] The concept of frames was introduced in Section 3.3.4. By equal-norm frame we mean that $\|u_i\|_2 = \|u_j\|_2$ for all i and j.

Exercise 5.6.8 (Random matrices with general independent rows) Prove a version of Theorem 4.6.1 which holds for random matrices with arbitrary, not necessarily sub-gaussian, distributions of rows.

Let A be an $m \times n$ matrix whose rows A_i are independent isotropic random vectors in $\mathbb{R}^n$. Assume that, for some $L \geq 0$,

$$\|A_i\|_2 \leq K\sqrt{n} \quad \text{almost surely for every } i. \tag{5.21}$$

Prove that, for every $t \geq 1$, one has

$$\sqrt{m} - Kt\sqrt{n \log n} \leq s_n(A) \leq s_1(A) \leq \sqrt{m} + Kt\sqrt{n \log n}, \tag{5.22}$$

with probability at least $1 - 2n^{-ct^2}$.

5.7 Notes

There are several introductory texts about concentration, such as [11, Chapter 3], [146, 126, 125, 29], and an elementary tutorial [13].

The approach to concentration via isoperimetric inequalities that we presented in Section 5.1 was first discovered by P. Lévy, to whom Theorems 5.1.5 and 5.1.4 are due (see [89]).

When V. Milman realized the power and generality of Lévy's approach in the 1970s, this led to far-reaching extensions of the *concentration of measure* principle, some of which we surveyed in Section 5.2. To keep this book concise, we have left out some important approaches to concentration, including the bounded differences inequality, martingale, semigroup, and transportation methods, the Poincaré inequality, the log-Sobolev inequality, hypercontractivity, Stein's method, and Talagrand's concentration inequalities; see [206, 125, 29]. Most of the material we covered in Sections 5.1 and 5.2 can be found in [11, Chapter 3], [146, 125].

The Gaussian isoperimetric inequality (Theorem 5.2.1) was first proved by V. N. Sudakov and B. S. Cirelson (Tsirelson) and independently by C. Borell [27]. There are several other proofs of the Gaussian isoperimetric inequality; see [23, 12, 15]. There is also an elementary derivation of Gaussian concentration (Theorem 5.2.2) from Gaussian interpolation instead of from isoperimetry; see [163].

The concentration result for the Hamming cube (Theorem 5.2.5) is a consequence of Harper's theorem, which is an isoperimetric inequality for the Hamming cube [96]; see [24]. Concentration for the symmetric group (Theorem 5.2.6) is due to B. Maurey [135]. Both Theorems 5.2.5 and 5.2.6 can be also proved using martingale methods; see [146, Chapter 7].

The proof of concentration for Riemannian manifolds with positive curvature can be found e.g. in [125, Proposition 2.17]. Many interesting special cases follow from this general result, including Theorem 5.2.7 for the special orthogonal group [146, Section 6.5.1] and, consequently, Theorem 5.2.9 for the Grassmannian [146, Section 6.7.2]. The construction of the Haar measure that we mentioned in Remark 5.2.8 can be found e.g. in [146, Chapter 1] and [74, Chapter 2]; the survey [143] discusses numerically stable ways to generate random unitary matrices.

Concentration for the continuous cube (Theorem 5.2.10) can be found in [125, Proposition 2.8], and concentration for the Euclidean ball (Theorem 5.2.13) in [125, Proposition 2.9]. Theorem 5.2.15 on concentration for exponential densities is borrowed from [125, Proposition 2.18]. The original proof of Talagrand's concentration inequality (Theorem 5.2.16) can be found in [192, Theorem 6.6] and [125, Corollary 4.10].

The original formulation of the Johnson–Lindenstrauss lemma is from [107]. For various versions of this lemma, related results, applications, and bibliographic notes, see [134, Section 15.2]. The condition $m \gtrsim \varepsilon^{-2} \log N$ is known to be optimal [120].

The approach to matrix concentration inequalities followed in Section 5.4 originates in the work of R. Ahlswede and A. Winter [4]. A short proof of the Golden–Thompson inequality (Theorem 5.4.7), a result on which the Ahlswede–Winter approach rests, can be found in e.g. [20, Theorem 9.3.7] and [215]. While the work of R. Ahlswede and A. Winter was motivated by problems of quantum information theory, the usefulness of their approach was gradually understood in other areas as well; the early work includes [219, 214, 90, 155].

The original argument of Ahlswede and Winter yields a version of the matrix Bernstein inequality that is somewhat weaker than Theorem 5.4.1, namely with $\sum_{i=1}^{N} \| \mathbb{E} X_i^2 \|$ instead of σ. This quantity was later tightened by R. Oliveira [156] by a modification of the Ahlswede–Winter method and independently by J. Tropp [200] using Lieb's inequality (Theorem 5.4.8) instead of that of Golden and Thompson. In this book, we mainly follow Tropp's proof of Theorem 5.4.1. The book [201] presents a self-contained proof of Lieb's inequality (Theorem 5.4.8), the matrix Hoeffding inequality from Exercise 5.4.12, the matrix Chernoff inequality, and much more. Owing to Tropp's contributions, there now exist matrix analogs of almost all the classical scalar concentration results [201]. The survey [161] discusses several other useful trace inequalities and outlines proofs of the Golden–Thompson inequality (in Section 3) and Lieb's inequality (embedded in the proof of Proposition 7). The book [76] also contains a detailed exposition of the matrix Bernstein inequality and some of its variants (Section 8.5) and a proof of Lieb's inequality (Appendix B.6).

Instead of using the matrix Bernstein inequality one can deduce the result of Exercise 5.4.11 from Gaussian integration by parts and a trace inequality [203]. The matrix Khintchine inequality from Exercise 5.4.13 can be deduced alternatively from the *non-commutative Khintchine inequality*, which is due to F. Lust-Piquard [130]; see also [131, 39, 40, 168]. This derivation was first observed and used by M. Rudelson [171], who proved a version of Exercise 5.4.13.

For the problem of community detection in networks discussed in Section 5.5, see the notes at the end of Chapter 4. The approach to concentration in random graphs using the matrix Bernstein inequality outlined in Section 5.5 was first proposed by R. Oliveira [156].

In Section 5.6 we discussed covariance estimation for general high-dimensional distributions, following [216]. An alternative and earlier approach to covariance estimation, which gives similar results, relies on the matrix Khintchine inequalities (also known as the non-commutative Khintchine inequalities); it was developed earlier by M. Rudelson [171]. For more references on covariance estimation problem, see the notes at the end of Chapter 4. The result of Exercise 5.6.8 is from [216, Section 5.4.2].

6

Quadratic Forms, Symmetrization, and Contraction

In this chapter we introduce a number of basic tools of high-dimensional probability: decoupling in Section 6.1, the concentration of quadratic forms (the Hanson–Wright inequality) in Section 6.2, symmetrization in Section 6.4, and contraction in Section 6.7.

We illustrate these tools in a number of applications. In Section 6.3, we use the Hanson–Wright inequality to establish concentration for anisotropic random vectors (thus extending Theorem 3.1.1) and for the distances between random vectors and subspaces. In Section 6.5 we combine the matrix Bernstein inequality with symmetrization arguments to analyze the operator norm of a random matrix; we show that it is almost equivalent to the largest Euclidean norm of the rows and columns. This result is used in Section 6.6 for the problem of matrix completion, where one is shown a few randomly chosen entries of a given matrix and is asked to fill in the missing entries.

6.1 Decoupling

At the beginning of this book, we made a thorough study of independent random variables of the type

$$\sum_{i=1}^{n} a_i X_i, \tag{6.1}$$

where $X_1, \dots, X_n$ are independent random variables and the a_i are fixed coefficients. In this section we study *quadratic forms* of the type

$$\sum_{i,j=1}^{n} a_{ij} X_i X_j = X^\mathsf{T} A X = \langle X, AX \rangle, \tag{6.2}$$

where $A = (a_{ij})$ is an $n \times n$ matrix of coefficients and $X = (X_1, \dots, X_n)$ is a random vector with independent coordinates. Such a quadratic form is called a *chaos* in probability theory.

Computing the expectation of a chaos is easy. For simplicity, let us assume that the X_i have zero means and unit variances. Then

$$\mathbb{E}\, X^\mathsf{T} A X = \sum_{i,j=1}^{n} a_{ij}\, \mathbb{E}\, X_i X_j = \sum_{i=1}^{n} a_{ii} = \operatorname{tr} A.$$

It is harder to establish the concentration of a chaos. The main difficulty is that the terms of the sum in (6.2) are not independent. This difficulty can be overcome by the *decoupling* technique, which we will study now.

The purpose of decoupling is to replace the quadratic form (6.2) with the *bilinear* form

$$\sum_{i,j=1}^{n} a_{ij} X_i X'_j = X^{\mathsf{T}} A X' = \langle AX, X' \rangle,$$

where $X' = (X'_1, \ldots, X'_n)$ is a random vector which is independent of X yet has the same distribution as X. Such an X' is called an *independent copy* of X. The point here is that the bilinear form is easier to analyze than the quadratic form, since it is linear rather than quadratic in X. Indeed, if we condition on X' we may treat the bilinear form as a sum of independent random variables

$$\sum_{i=1}^{n} \Big(\sum_{j=1}^{n} a_{ij} X'_j \Big) X_i = \sum_{i=1}^{n} c_i X_i$$

with fixed coefficients c_i, much as we have treated the sums (6.1).

Theorem 6.1.1 (Decoupling) *Let A be an $n \times n$ diagonal-free matrix (i.e., the diagonal entries of A equal zero). Let $X = (X_1, \ldots, X_n)$ be a random vector with independent mean-zero coordinates X_i. Then, for every convex function $F \colon \mathbb{R} \to \mathbb{R}$, one has*

$$\mathbb{E}\, F(X^{\mathsf{T}} A X) \le \mathbb{E}\, F(4 X^{\mathsf{T}} A X') \tag{6.3}$$

where X' is an independent copy of X.

The proof will be based on the following observation.

Lemma 6.1.2 *Let Y and Z be independent random variables such that $\mathbb{E}\, Z = 0$. Then, for every convex function F, one has*

$$\mathbb{E}\, F(Y) \le \mathbb{E}\, F(Y + Z).$$

Proof This is a simple consequence of Jensen's inequality. First let us fix an arbitrary $y \in \mathbb{R}$ and use $\mathbb{E}\, Z = 0$ to get

$$F(y) = F(y + \mathbb{E}\, Z) = F(\mathbb{E}(y + Z)) \le \mathbb{E}\, F(y + Z).$$

Now choose $y = Y$ and take expectations of both sides to complete the proof. (To check whether you have understood this argument, find where the independence of Y and Z was used!) ∎

Proof of Theorem 6.1.1 Here is what our proof will look like in a nutshell. First, we replace the chaos $X^{\mathsf{T}} A X = \sum_{i,j} a_{ij} X_i X_j$ by the "partial chaos"

$$\sum_{(i,j) \in I \times I^c} a_{ij} X_i X_j,$$

where the subset of indices $I \subset \{1, \ldots, n\}$ is chosen by random sampling. The advantage of partial chaos is that the summation is done over disjoint sets for i and j. Thus one can automatically replace X_j by X'_j without changing the distribution. Finally, we complete the partial chaos to the full sum $X^{\mathsf{T}} A X' = \sum_{i,j} a_{ij} X_i X'_j$ using Lemma 6.1.2.

Now we pass to a detailed proof. To randomly select a subset of indices I, let us consider *selectors* $\delta_1, \dots, \delta_n \in \{0, 1\}$, which are independent Bernoulli random variables with $\mathbb{P}\{\delta_i = 0\} = \mathbb{P}\{\delta_i = 1\} = 1/2$. Define

$$I := \{i : \delta_i = 1\}.$$

Next we condition on X. Since by assumption $a_{ii} = 0$ and

$$\mathbb{E}\, \delta_i(1 - \delta_j) = \frac{1}{2} \times \frac{1}{2} = \frac{1}{4} \quad \text{for all } i \neq j,$$

we may express the chaos as

$$X^{\mathsf{T}} A X = \sum_{i \neq j} a_{ij} X_i X_j = 4\, \mathbb{E}_\delta \sum_{i \neq j} \delta_i (1 - \delta_j) a_{ij} X_i X_j = 4\, \mathbb{E}_I \sum_{(i,j) \in I \times I^c} a_{ij} X_i X_j.$$

(The subscripts δ and I are intended to remind us about the sources of randomness used in taking these conditional expectations. Since we have fixed X, the conditional expectations are over the random selectors $\delta = (\delta_1, \dots, \delta_n)$ or, equivalently, over the random set of indices I. We will continue to use similar notation later.)

Apply the function F to both sides and take expectations over X. Using Jensen's inequality and the Fubini theorem, we obtain

$$\mathbb{E}_X\, F(X^{\mathsf{T}} A X) \le \mathbb{E}_I\, \mathbb{E}_X\, F\Big(4 \sum_{(i,j) \in I \times I^c} a_{ij} X_i X_j\Big).$$

It follows that there exists a realization of a random subset I such that

$$\mathbb{E}_X\, F(X^{\mathsf{T}} A X) \le \mathbb{E}_X\, F\Big(4 \sum_{(i,j) \in I \times I^c} a_{ij} X_i X_j\Big).$$

Fix such a realization of I until the end of the proof (and drop the subscript X on the expectation for convenience). Since the random variables $(X_i)_{i \in I}$ are independent of $(X_j)_{j \in I^c}$, the distribution of the sum on the right-hand side will not change if we replace X_j by X'_j. So we get

$$\mathbb{E}\, F(X^{\mathsf{T}} A X) \le \mathbb{E}\, F\Big(4 \sum_{(i,j) \in I \times I^c} a_{ij} X_i X'_j\Big).$$

It remains to complete the sum on the right-hand side to a sum over all pairs of indices. In other words, we want to show that

$$\mathbb{E}\, F\Big(4 \sum_{(i,j) \in I \times I^c} a_{ij} X_i X'_j\Big) \le \mathbb{E}\, F\Big(4 \sum_{(i,j) \in [n] \times [n]} a_{ij} X_i X'_j\Big), \tag{6.4}$$

where we use the notation $[n] = \{1, \dots, n\}$. To do this, we decompose the sum on the right-hand side as follows:

$$\sum_{(i,j) \in [n] \times [n]} a_{ij} X_i X'_j = Y + Z_1 + Z_2,$$

where

$$Y = \sum_{(i,j)\in I\times I^c} a_{ij} X_i X'_j, \quad Z_1 = \sum_{(i,j)\in I\times I} a_{ij} X_i X'_j, \quad Z_2 = \sum_{(i,j)\in I^c\times [n]} a_{ij} X_i X'_j.$$

Now condition on all random variables except $(X'_j)_{j\in I}$ and $(X_i)_{i\in I^c}$. This fixes Y, while Z_1 and Z_2 are random variables with zero conditional expectations (check!). Use Lemma 6.1.2 to conclude that the conditional expectation, which we denote $\mathbb{E}'$, satisfies

$$F(4Y) \le \mathbb{E}' F(4Y + 4Z_1 + 4Z_2).$$

Finally, taking the expectation of both sides over all other random variables, we conclude that

$$\mathbb{E} F(4Y) \le \mathbb{E} F(4Y + 4Z_1 + 4Z_2).$$

This proves (6.4) and finishes the argument. ■

Remark 6.1.3 We have actually proved a slightly stronger version of the decoupling inequality, in which A need not be diagonal-free. Thus, for any square matrix $A = (a_{ij})$ we showed that

$$\mathbb{E} F\Big(\sum_{i,j:\, i\ne j} a_{ij} X_i X_j \Big) \le \mathbb{E} F\Big(4 \sum_{i,j} a_{ij} X_i X'_j \Big).$$

Exercise 6.1.4 (Decoupling in Hilbert spaces) Prove the following generalization of Theorem 6.1.1. Let $A = (a_{ij})$ be an $n\times n$ matrix. Let $X_1, \dots, X_n$ be independent mean-zero random vectors in some Hilbert space. Show that, for every convex function $F: \mathbb{R} \to \mathbb{R}$, one has

$$\mathbb{E} F\Big(\sum_{i,j:\, i\ne j} a_{ij} \langle X_i, X_j \rangle \Big) \le \mathbb{E} F\Big(4 \sum_{i,j} a_{ij} \langle X_i, X'_j \rangle \Big),$$

where (X'_i) is an independent copy of (X_i).

Exercise 6.1.5 (Decoupling in normed spaces) Prove the following alternative generalization of Theorem 6.1.1. Let $(u_{ij})_{i,j=1}^n$ be fixed vectors in some normed space. Let $X_1, \dots, X_n$ be independent mean-zero random variables. Show that, for every convex and increasing function F, one has

$$\mathbb{E} F\Big(\Big\| \sum_{i,j:\, i\ne j} X_i X_j u_{ij} \Big\| \Big) \le \mathbb{E} F\Big(4 \Big\| \sum_{i,j} X_i X'_j u_{ij} \Big\| \Big).$$

where (X'_i) is an independent copy of (X_i).

6.2 Hanson–Wright Inequality

We now prove a general concentration inequality for a chaos. It can be viewed as a chaos version of Bernstein's inequality.

Theorem 6.2.1 (Hanson–Wright inequality) *Let $X = (X_1, \dots, X_n) \in \mathbb{R}^n$ be a random vector with independent mean zero sub-gaussian coordinates. Let A be an $n \times n$ matrix. Then, for every $t \ge 0$, we have*

$$\mathbb{P}\left\{|X^\mathsf{T} AX - \mathbb{E}\, X^\mathsf{T} AX| \ge t\right\} \le 2\exp\Big(- c\min\Big(\frac{t^2}{K^4\|A\|_F^2}, \frac{t}{K^2\|A\|}\Big)\Big),$$

where $K = \max_i \|X_i\|_{\psi_2}$.

As many times before, our proof of the Hanson–Wright inequality will be based on bounding the moment generating function of $X^\mathsf{T} AX$. We use decoupling to replace this chaos by $X^\mathsf{T} AX'$. Next, we bound the MGF of the decoupled chaos in the easier, Gaussian, case where $X \sim N(0, I_n)$. Finally, we extend the bound to general sub-gaussian distributions using a replacement trick.

Lemma 6.2.2 (MGF of Gaussian chaos) *Let $X, X' \sim N(0, I_n)$ be independent and let $A = (a_{ij})$ be an $n \times n$ matrix. Then*

$$\mathbb{E}\exp(\lambda X^\mathsf{T} AX') \le \exp(C\lambda^2\|A\|_F^2)$$

for all λ satisfying $|\lambda| \le c/\|A\|$.

Proof First let us use rotation invariance to reduce to the case where the matrix A is diagonal. Expressing A through its singular value decomposition

$$A = \sum_i s_i u_i v_i^\mathsf{T},$$

we can write

$$X^\mathsf{T} AX' = \sum_i s_i \langle u_i, X\rangle \langle v_i, X'\rangle.$$

By the rotation invariance of the normal distribution, $g := (\langle u_i, X\rangle)_{i=1}^n$ and $g' := (\langle v_i, X'\rangle)_{i=1}^n$ are independent standard normal random vectors in $\mathbb{R}^n$ (recall Exercise 3.3.3). In other words, we represented the chaos as

$$X^\mathsf{T} AX' = \sum_i s_i g_i g_i',$$

where $g, g' \sim N(0, I_n)$ are independent and the s_i are the singular values of A. This is a sum of independent random variables, which is easy to handle. Indeed, independence gives

$$\mathbb{E}\exp(\lambda X^\mathsf{T} AX') = \prod_i \mathbb{E}\exp(\lambda s_i g_i g_i'). \tag{6.5}$$

Now, for each i, we have

$$\mathbb{E}\exp(\lambda s_i g_i g_i') = \mathbb{E}\exp(\lambda^2 s_i^2 g_i^2/2) \le \exp(C\lambda^2 s_i^2) \quad \text{provided that } \lambda^2 s_i^2 \le c.$$

To get the first identity here, condition on g_i and use the formula (2.12) for the MGF of the normal random variable g_i'. At the second step, we used part (iii) of Proposition 2.7.1 for the sub-exponential random variable g_i^2.

Substituting this bound into (6.5), we obtain

$$\mathbb{E}\exp(\lambda X^\mathsf{T} A X') \le \exp\Big(C\lambda^2 \sum_i s_i^2\Big) \quad \text{provided that } \lambda^2 \le \frac{c}{\max_i s_i^2}.$$

It remains to recall that the s_i are the singular values of A, so $\sum_i s_i^2 = \|A\|_F^2$ and $\max_i s_i = \|A\|$. The lemma is proved. ■

To extend Lemma 6.2.2 to general sub-gaussian distributions, we use a replacement trick to compare the MGFs of general and Gaussian types of chaos.

Lemma 6.2.3 (Comparison) *Consider independent mean-zero sub-gaussian random vectors X, X' in $\mathbb{R}^n$ with $\|X\|_{\psi_2} \le K$ and $\|X'\|_{\psi_2} \le K$. Consider also independent random vectors $g, g' \sim N(0, I_n)$. Let A be an $n \times n$ matrix. Then*

$$\mathbb{E}\exp(\lambda X^\mathsf{T} A X') \le \mathbb{E}\exp(CK^2 \lambda g^\mathsf{T} A g')$$

for any $\lambda \in \mathbb{R}$.

Proof Condition on X' and take the expectation over X, which we denote $\mathbb{E}_X$. Then the random variable $X^\mathsf{T} A X' = \langle X, AX' \rangle$ is (conditionally) sub-gaussian, and its sub-gaussian norm[1] is bounded by $K\|AX'\|_2$. Then the bound (2.16) on the MGF of sub-gaussian random variables gives

$$\mathbb{E}_X \exp(\lambda X^\mathsf{T} A X') \le \exp(C\lambda^2 K^2 \|AX'\|_2^2), \quad \lambda \in \mathbb{R}. \tag{6.6}$$

Compare this with the formula (2.12) for the MGF of the normal distribution. Applied to the normal random variable $g^\mathsf{T} A X' = \langle g, AX' \rangle$ (still conditioning on X'), it gives

$$\mathbb{E}_g \exp(\mu g^\mathsf{T} A X') = \exp(\mu^2 \|AX'\|_2^2 / 2), \quad \mu \in \mathbb{R}. \tag{6.7}$$

Choosing $\mu = \sqrt{2C}K\lambda$, we can match the right-hand sides of (6.6) and (6.7) and thus get

$$\mathbb{E}_X \exp(\lambda X^\mathsf{T} A X') \le \mathbb{E}_g \exp(\sqrt{2C}K\lambda g^\mathsf{T} A X').$$

Taking the expectation over X' of both sides, we see that we have successfully replaced X by g in the chaos, and we have paid a factor $\sqrt{2C}K$. Going through a similar argument again, this time for X', we can now replace X' with g' and pay an extra factor $\sqrt{2C}K$. (Exercise 6.2.4 below asks you to carefully write down the details of this step.) The proof of the lemma is complete. ■

Exercise 6.2.4 (Comparison) Complete the proof of Lemma 6.2.3. Replace X' by g'; write down all the details carefully.

Proof of Theorem 6.2.1 Without loss of generality, we may assume that $K = 1$. (Why?) As usual, it is enough to bound the one-sided tail:

$$p := \mathbb{P}\left\{X^\mathsf{T} A X - \mathbb{E}\, X^\mathsf{T} A X \ge t\right\}.$$

[1] Recall Definition 3.4.1.

Indeed, once we have a bound on this upper tail, a similar bound will hold for the lower tail as well (since one can replace A with $-A$). By combining the two tails, we can then complete the proof.

In terms of the entries of $A = (a_{ij})_{i,j=1}^n$, we have

$$X^\mathsf{T} A X = \sum_{i,j} a_{ij} X_i X_j \quad \text{and} \quad \mathbb{E}\, X^\mathsf{T} A X = \sum_i a_{ii}\, \mathbb{E}\, X_i^2,$$

where we have used the mean-zero assumption and independence. So we can express the deviation as

$$X^\mathsf{T} A X - \mathbb{E}\, X^\mathsf{T} A X = \sum_i a_{ii}(X_i^2 - \mathbb{E}\, X_i^2) + \sum_{i,j:\, i \neq j} a_{ij} X_i X_j.$$

The problem reduces to estimating the diagonal and off-diagonal sums:

$$p \le \mathbb{P}\Big\{ \sum_i a_{ii}(X_i^2 - \mathbb{E}\, X_i^2) \ge t/2 \Big\} + \mathbb{P}\Big\{ \sum_{i,j:\, i \neq j} a_{ij} X_i X_j \ge t/2 \Big\} =: p_1 + p_2.$$

Step 1: diagonal sum. Since the X_i are independent sub-gaussian random variables, the random variables $X_i^2 - \mathbb{E}\, X_i^2$ are independent, mean-zero, and sub-exponential. Thus

$$\|X_i^2 - \mathbb{E}\, X_i^2\|_{\psi_1} \lesssim \|X_i^2\|_{\psi_1} \lesssim \|X_i\|_{\psi_2}^2 \lesssim 1.$$

(This follows from the centering exercise 2.7.10 and Lemma 2.7.6.) Then Bernstein's inequality (Theorem 2.8.2) gives

$$p_1 \le \exp\Big(-c \min\Big(\frac{t^2}{\sum_i a_{ii}^2}, \frac{t}{\max_i |a_{ii}|} \Big) \Big) \le \exp\Big(-c \min\Big(\frac{t^2}{\|A\|_F^2}, \frac{t}{\|A\|} \Big) \Big).$$

Step 2: off-diagonal sum. It remains to bound the off-diagonal sum

$$S := \sum_{i,j:\, i \neq j} a_{ij} X_i X_j.$$

Let $\lambda > 0$ be a parameter whose value we will determine later. By Markov's inequality, we have

$$p_2 = \mathbb{P}\left\{ S \ge t/2 \right\} = \mathbb{P}\left\{ \lambda S \ge \lambda t/2 \right\} \le \exp(-\lambda t/2)\, \mathbb{E} \exp(\lambda S). \tag{6.8}$$

Now,

$$\begin{aligned} \mathbb{E} \exp(\lambda S) &\le \mathbb{E} \exp(4\lambda X^\mathsf{T} A X') \quad \text{(by decoupling – see Remark 6.1.3)} \\ &\le \mathbb{E} \exp(C_1 \lambda g^\mathsf{T} A g') \quad \text{(by the comparison lemma 6.2.3)} \\ &\le \exp(C\lambda^2 \|A\|_F^2) \quad \text{(by Lemma 6.2.2 for Gaussian chaos),} \end{aligned}$$

provided that $|\lambda| \le c/\|A\|$. Substituting this bound into (6.8), we obtain

$$p_2 \le \exp\big(-\lambda t/2 + C\lambda^2 \|A\|_F^2 \big).$$

Optimizing over $0 \le \lambda \le c/\|A\|$, we conclude that

$$p_2 \le \exp\Big(-c \min\Big(\frac{t^2}{\|A\|_F^2}, \frac{t}{\|A\|} \Big) \Big).$$

(Check!)

Summarizing, we have obtained the desired bounds for the probabilities of diagonal deviation p_1 and off-diagonal deviation p_2. Putting them together, we complete the proof of Theorem 6.2.1. ■

Exercise 6.2.5 Give an alternative proof of the Hanson–Wright inequality for normal distributions, without separating the diagonal part or decoupling. ☞

Exercise 6.2.6 Consider a mean-zero sub-gaussian random vector X in $\mathbb{R}^n$ with $\|X\|_{\psi_2} \le K$. Let B be an $m \times n$ matrix. Show that

$$\mathbb{E}\exp(\lambda^2\|BX\|_2^2) \le \exp(CK^2\lambda^2\|B\|_F^2) \quad \text{provided that } |\lambda| \le \frac{c}{K\|B\|}.$$

To prove this bound, replace X with a Gaussian random vector $g \sim N(0, I_n)$ along the following lines:

(a) Prove the comparison inequality

$$\mathbb{E}\exp(\lambda^2\|BX\|_2^2) \le \mathbb{E}\exp(CK^2\lambda^2\|Bg\|_2^2)$$

for every $\lambda \in \mathbb{R}$. ☞

(b) Check that

$$\mathbb{E}\exp(\lambda^2\|Bg\|_2^2) \le \exp(C\lambda^2\|B\|_F^2),$$

provided that $|\lambda| \le c/\|B\|$. ☞

Exercise 6.2.7 (Higher-dimensional Hanson–Wright inequality) Let $X_1, \dots, X_n$ be independent mean-zero sub-gaussian random vectors in $\mathbb{R}^d$. Let $A = (a_{ij})$ be an $n \times n$ matrix. Prove that, for every $t \ge 0$, we have

$$\mathbb{P}\left\{\left|\sum_{i,j:\, i\ne j}^{n} a_{ij}\langle X_i, X_j\rangle\right| \ge t\right\} \le 2\exp\left(-c\min\left(\frac{t^2}{K^4 d\|A\|_F^2}, \frac{t}{K^2\|A\|}\right)\right)$$

where $K = \max_i \|X_i\|_{\psi_2}$. ☞

6.3 Concentration for Anisotropic Random Vectors

As a consequence of the Hanson–Wright inequality, we can now obtain the concentration for *anisotropic* random vectors, which have the form BX where B is a fixed matrix and X is an isotropic random vector.

Exercise 6.3.1 Let B be an $m \times n$ matrix and X be an isotropic random vector in $\mathbb{R}^n$. Check that

$$\mathbb{E}\,\|BX\|_2^2 = \|B\|_F^2.$$

Theorem 6.3.2 (Concentration for random vectors) *Let B be an $m \times n$ matrix, and let $X = (X_1, \dots, X_n) \in \mathbb{R}^n$ be a random vector with independent mean-zero unit-variance sub-gaussian coordinates. Then*

$$\Big\| \|BX\|_2 - \|B\|_F \Big\|_{\psi_2} \le CK^2 \|B\|,$$

where $K = \max_i \|X_i\|_{\psi_2}$.

There is an important partial case of this theorem when $B = I_n$. In this case, the inequality we obtain is

$$\Big\| \|X\|_2 - \sqrt{n} \Big\|_{\psi_2} \le CK^2,$$

which we proved in Theorem 3.1.1.

Proof of Theorem 6.3.2. For simplicity we assume that $K \ge 1$. (Argue that we can make this assumption.) We apply the Hanson–Wright inequality (Theorem 6.2.1) for the matrix $A := B^\mathsf{T} B$. Let us express the main terms appearing in the Hanson–Wright inequality in terms of B. We have

$$X^\mathsf{T} A X = \|BX\|_2^2, \qquad \mathbb{E}\, X^\mathsf{T} A X = \|B\|_F^2,$$

and

$$\|A\| = \|B\|^2, \quad \|A\|_F = \|B^\mathsf{T} B\|_F \le \|B^\mathsf{T}\| \|B\|_F = \|B\| \|B\|_F.$$

(You will be asked to check the inequality in Exercise 6.3.3.) Thus, we have for every $u \ge 0$ that

$$\mathbb{P}\left\{ \big| \|BX\|_2^2 - \|B\|_F^2 \big| \ge u \right\} \le 2 \exp\Big(- \frac{c}{K^4} \min\Big(\frac{u^2}{\|B\|^2 \|B\|_F^2}, \frac{u}{\|B\|^2} \Big)\Big).$$

(Here we have used $K^4 \ge K^2$ since we assume that $K \ge 1$.)

Substitute the value $u = \varepsilon \|B\|_F^2$ for $\varepsilon \ge 0$ and obtain

$$\mathbb{P}\left\{ \big| \|BX\|_2^2 - \|B\|_F^2 \big| \ge \varepsilon \|B\|_F^2 \right\} \le 2 \exp\Big(- c \min(\varepsilon^2, \varepsilon) \frac{\|B\|_F^2}{K^4 \|B\|^2} \Big).$$

This is a good concentration inequality for $\|BX\|_2^2$, from which we are going to deduce a concentration inequality for $\|BX\|_2$. Denote $\delta^2 = \min(\varepsilon^2, \varepsilon)$, or equivalently set $\varepsilon = \max(\delta, \delta^2)$. Observe that the following implication holds:

$$\text{If } \big| \|BX\|_2 - \|B\|_F \big| \ge \delta \|B\|_F \text{ then } \big| \|BX\|_2^2 - \|B\|_F^2 \big| \ge \varepsilon \|B\|_F^2.$$

(Check it! This is the same elementary inequality as (3.2), once we divide through by $\|B\|_F^2$.) Thus we get

$$\mathbb{P}\left\{ \big| \|BX\|_2 - \|B\|_F \big| \ge \delta \|B\|_F \right\} \le 2 \exp\Big(- c\delta^2 \frac{\|B\|_F^2}{K^4 \|B\|^2} \Big).$$

Changing variables to $t = \delta \|B\|_F$, we obtain

$$\mathbb{P}\left\{ \big| \|BX\|_2 - \|B\|_F \big| > t \right\} \le 2 \exp\Big(- \frac{ct^2}{K^4 \|B\|^2} \Big).$$

Since this inequality holds for all $t \ge 0$, the conclusion of the theorem follows from the definition of sub-gaussian distributions. ∎

Exercise 6.3.3 Let D be a $k \times m$ matrix and B be an $m \times n$ matrix. Prove that

$$\|DB\|_F \le \|D\| \|B\|_F.$$

Exercise 6.3.4 (Distance to a subspace) Let E be a subspace of $\mathbb{R}^n$ of dimension d. Consider a random vector $X = (X_1, \dots, X_n) \in \mathbb{R}^n$ with independent mean-zero unit-variance sub-gaussian coordinates.

(a) Check that

$$\left(\mathbb{E}\,\mathrm{dist}(X, E)^2\right)^{1/2} = \sqrt{n-d}.$$

(b) Prove that, for any $t \ge 0$, the distance concentrates nicely:

$$\mathbb{P}\left\{\left|d(X, E) - \sqrt{n-d}\right| > t\right\} \le 2\exp(-ct^2/K^4),$$

where $K = \max_i \|X_i\|_{\psi_2}$.

Let us prove a weaker version of Theorem 6.3.2 without assuming independence of the coordinates of X:

Exercise 6.3.5 (Tails of sub-gaussian random vectors) Let B be an $m \times n$ matrix, and let X be a mean-zero sub-gaussian random vector in $\mathbb{R}^n$ with $\|X\|_{\psi_2} \le K$. Prove that, for any $t \ge 0$, we have

$$\mathbb{P}\left\{\|BX\|_2 \ge CK\|B\|_F + t\right\} \le \exp\left(-\frac{ct^2}{K^2\|B\|^2}\right).$$

☞

The following exercise explains why the concentration inequality *must* be weaker than in Theorem 3.1.1 if we do not assume independence of the coordinates of X.

Exercise 6.3.6 Show that there exists a mean-zero isotropic sub-gaussian random vector X in $\mathbb{R}^n$ such that

$$\mathbb{P}\left\{\|X\|_2 = 0\right\} = \mathbb{P}\left\{\|X\|_2 \ge 1.4\sqrt{n}\right\} = \frac{1}{2}.$$

In other words, $\|X\|_2$ does not concentrate near $\sqrt{n}$.

6.4 Symmetrization

A random variable X is *symmetric* if X and $-X$ have the same distribution. A simple example of a symmetric random variable is the *symmetric Bernoulli*, which takes values -1 and 1 with probabilities $1/2$ each:

$$\mathbb{P}\left\{\xi = 1\right\} = \mathbb{P}\left\{\xi = -1\right\} = \frac{1}{2}.$$

A normal mean-zero random variable $X \sim N(0, \sigma^2)$ is also symmetric, while Poisson or exponential random variables are not.

In this section we develop the simple and useful technique of *symmetrization*. It allows one to reduce arbitrary distributions to symmetric distributions and in some cases even to the symmetric Bernoulli distribution.

Exercise 6.4.1 (Constructing symmetric distributions) Let X be a random variable and ξ be an independent symmetric Bernoulli random variable.

(a) Check that ξX and $\xi |X|$ are symmetric random variables and that they have the same distribution.
(b) If X is symmetric, show that the distributions of ξX and $\xi |X|$ are the same as that of X.
(c) Let X' be an independent copy of X. Check that $X - X'$ is symmetric.

Throughout this section, we denote by

$$\varepsilon_1, \varepsilon_2, \varepsilon_3, \ldots$$

a sequence of independent symmetric Bernoulli random variables. We assume that they are (jointly) independent not only of each other but also of any other random variables in question.

Lemma 6.4.2 (Symmetrization) *Let $X_1, \ldots, X_N$ be independent mean-zero random vectors in a normed space. Then*

$$\frac{1}{2}\mathbb{E}\left\|\sum_{i=1}^N \varepsilon_i X_i\right\| \le \mathbb{E}\left\|\sum_{i=1}^N X_i\right\| \le 2\,\mathbb{E}\left\|\sum_{i=1}^N \varepsilon_i X_i\right\|.$$

The purpose of this lemma is to let us replace general random variables X_i by the symmetric random variables $\varepsilon_i X_i$.

Proof Upper bound. Let (X_i') be an independent copy of the random vectors (X_i). Since $\sum_i X_i'$ has zero mean, we have

$$p := \mathbb{E}\left\|\sum_i X_i\right\| \le \mathbb{E}\left\|\sum_i X_i - \sum_i X_i'\right\| = \mathbb{E}\left\|\sum_i (X_i - X_i')\right\|.$$

The inequality here is an application of the following version of Lemma 6.1.2 for independent random vectors Y and Z:

$$\text{if } \mathbb{E}\,Z = 0 \text{ then } \mathbb{E}\,\|Y\| \le \mathbb{E}\,\|Y + Z\|. \tag{6.9}$$

(Check it!)

Next, since $X_i - X_i'$ are symmetric random vectors, they have the same distribution as $\varepsilon_i (X_i - X_i')$ (see Exercise 6.4.1). Then

$$\begin{aligned} p &\le \mathbb{E}\left\|\sum_i \varepsilon_i (X_i - X_i')\right\| \\ &\le \mathbb{E}\left\|\sum_i \varepsilon_i X_i\right\| + \mathbb{E}\left\|\sum_i \varepsilon_i X_i'\right\| \quad \text{(by the triangle inequality)} \\ &= 2\,\mathbb{E}\left\|\sum_i \varepsilon_i X_i\right\| \quad \text{(since the two terms are identically distributed).} \end{aligned}$$

Lower bound. The argument here is similar:

$$\begin{aligned}
\mathbb{E}\Big\|\sum_i \varepsilon_i X_i\Big\| &\le \mathbb{E}\Big\|\sum_i \varepsilon_i (X_i - X_i')\Big\| \quad \text{(condition on } \varepsilon_i \text{ and use (6.9))}\\
&= \mathbb{E}\Big\|\sum_i (X_i - X_i')\Big\| \quad \text{(the distribution is the same)}\\
&\le \mathbb{E}\Big\|\sum_i X_i\Big\| + \mathbb{E}\Big\|\sum_i X_i'\Big\| \quad \text{(by the triangle inequality)}\\
&\le 2\,\mathbb{E}\Big\|\sum_i X_i\Big\| \quad \text{(the distributions are identical).}
\end{aligned}$$

This completes the proof of the symmetrization lemma. ■

Exercise 6.4.3 Where in this argument did we use the independence of the random variables X_i? Is the mean-zero assumption needed for both the upper and lower bounds?

Exercise 6.4.4 (Removing the mean-zero assumption)

(a) Prove the following generalization of the symmetrization lemma 6.4.2 for random vectors X_i that do not necessarily have zero means:

$$\mathbb{E}\Big\|\sum_{i=1}^N X_i - \sum_{i=1}^N \mathbb{E}\, X_i\Big\| \le 2\,\mathbb{E}\Big\|\sum_{i=1}^N \varepsilon_i X_i\Big\|.$$

(b) Argue that there cannot be any nontrivial reverse inequality.

Exercise 6.4.5 Prove the following generalization of the symmetrization lemma 6.4.2. Let $F\colon \mathbb{R}_+ \to \mathbb{R}$ be an increasing, convex, function. Show that the same inequalities as in Lemma 6.4.2 hold if the norm $\|\cdot\|$ is replaced with $F(\|\cdot\|)$, namely

$$\mathbb{E}\, F\Big(\frac{1}{2}\Big\|\sum_{i=1}^N \varepsilon_i X_i\Big\|\Big) \le \mathbb{E}\, F\Big(\Big\|\sum_{i=1}^N X_i\Big\|\Big) \le \mathbb{E}\, F\Big(2\Big\|\sum_{i=1}^N \varepsilon_i X_i\Big\|\Big).$$

Exercise 6.4.6 Let $X_1, \dots, X_N$ be independent, mean-zero random variables. Show that their sum $\sum_i X_i$ is sub-gaussian if and only if $\sum_i \varepsilon_i X_i$ is sub-gaussian, and

$$c\Big\|\sum_{i=1}^N \varepsilon_i X_i\Big\|_{\psi_2} \le \Big\|\sum_{i=1}^N X_i\Big\|_{\psi_2} \le C\Big\|\sum_{i=1}^N \varepsilon_i X_i\Big\|_{\psi_2}.$$

6.5 Random Matrices With Non-I.I.D. Entries

A typical use of symmetrization technique consists of two steps. First, general random variables X_i are replaced by symmetric random variables $\varepsilon_i X_i$. Next, we condition on X_i, which leaves the entire randomness with ε_i. This reduces the problem to one involving *symmetric*

Bernoulli random variables ε_i, which are often simpler to deal with. We illustrate this technique by proving a general bound on the norms of random matrices with independent but not identically distributed entries.

Theorem 6.5.1 (Norms of random matrices with non-i.i.d. entries) *Let A be an $n \times n$ symmetric random matrix whose entries on and above the diagonal are independent mean-zero random variables. Then*

$$\mathbb{E}\,\|A\| \le C\sqrt{\log n}\;\mathbb{E}\max_i \|A_i\|_2,$$

where the A_i denote the rows of A.

Before we pass to the proof of this theorem, let us note that it is sharp up to the logarithmic factor. Indeed, since the operator norm of any matrix is bounded below by the Euclidean norms of the rows (why?), we have trivially

$$\mathbb{E}\,\|A\| \ge \mathbb{E}\max_i \|A_i\|_2.$$

Note also that, unlike the results we have seen before, Theorem 6.5.1 does not require *any moment assumptions* on the entries of A.

Proof of Theorem 6.5.1 Our argument will be based on a combination of symmetrization with the matrix Khintchine inequality (Exercise 5.4.13).

First decompose A into a sum of independent mean-zero symmetric random matrices Z_{ij}, each of which contains a pair of symmetric entries of A (or one diagonal entry). Precisely, we have

$$A = \sum_{i \le j} Z_{ij}, \quad \text{where} \quad Z_{ij} := \begin{cases} A_{ij}(e_i e_j^{\mathsf T} + e_j e_i^{\mathsf T}), & i < j, \\ A_{ii} e_i e_i^{\mathsf T}, & i = j, \end{cases}$$

and where (e_i) denotes the canonical basis of $\mathbb{R}^n$.

Apply the symmetrization lemma 6.4.2, which gives

$$\mathbb{E}\,\|A\| = \mathbb{E}\Big\|\sum\nolimits_{i\le j} Z_{ij}\Big\| \le 2\,\mathbb{E}\Big\|\sum\nolimits_{i\le j} \varepsilon_{ij} Z_{ij}\Big\|, \tag{6.10}$$

where (ε_{ij}) are independent symmetric Bernoulli random variables.

Condition on (Z_{ij}), apply the matrix Khintchine inequality (Exercise 5.4.13), and then take the expectation with respect to (Z_{ij}). This gives

$$\mathbb{E}\Big\|\sum\nolimits_{i\le j} \varepsilon_{ij} Z_{ij}\Big\| \le C\sqrt{\log n}\;\mathbb{E}\Big(\Big\|\sum\nolimits_{i\le j} Z_{ij}^2\Big\|\Big)^{1/2}. \tag{6.11}$$

Now, a quick check verifies that each Z_{ij}^2 is a diagonal matrix; more precisely,

$$Z_{ij}^2 = \begin{cases} A_{ij}^2(e_i e_i^{\mathsf T} + e_j e_j^{\mathsf T}), & i < j \\ A_{ii}^2 e_i e_i^{\mathsf T}, & i = j. \end{cases}$$

Summing up, we get

$$\sum_{i\le j} Z_{ij}^2 = \sum_{i=1}^n \Big(\sum_{j=1}^n A_{ij}^2\Big) e_i e_i^{\mathsf T} = \sum_{i=1}^n \|A_i\|_2^2 e_i e_i^{\mathsf T}.$$

(Check the first identity carefully!) In other words, $\sum_{i \le j} Z_{ij}^2$ is a diagonal matrix, and its diagonal entries equal $\|A_i\|_2^2$. The operator norm of a diagonal matrix is the maximal absolute value of its entries (why?); thus

$$\Big\| \sum_{i \le j} Z_{ij}^2 \Big\| = \max_i \|A_i\|_2^2.$$

Substitute this into (6.11) and then into (6.10) and so complete the proof. ∎

In the following exercise we derive a version of Theorem 6.5.1 for non-symmetric rectangular matrices using the so-called "Hermitization trick".

Exercise 6.5.2 (Rectangular matrices) Let A be an $m \times n$ random matrix whose entries are independent mean-zero random variables. Show that

$$\mathbb{E}\|A\| \le C\sqrt{\log(m+n)} \Big(\mathbb{E} \max_i \|A_i\|_2 + \mathbb{E} \max_j \|A^j\|_2 \Big)$$

where A_i and A^j denote the rows and columns of A, respectively. ☞

Exercise 6.5.3 (Sharpness) Show that the result of Exercise 6.5.2 is sharp up to the logarithmic factor, i.e., one always has

$$\mathbb{E}\|A\| \ge c\Big(\mathbb{E} \max_i \|A_i\|_2 + \mathbb{E} \max_j \|A^j\|_2 \Big).$$

Exercise 6.5.4 (Sharpness) Show that the logarithmic factor in Theorem 6.5.1 is needed: construct a random matrix A satisfying the assumptions of the theorem and for which

$$\mathbb{E}\|A\| \ge c\sqrt{\log n}\, \mathbb{E} \max_i \|A_i\|_2.$$

6.6 Application: Matrix Completion

A remarkable application of the methods we have studied is to the problem of matrix completion. Suppose we are shown a few entries of a matrix; can we guess the other entries? We obviously cannot unless we know something else about the matrix. In this section we show that if the matrix has *low rank* then matrix completion is possible.

To describe the problem mathematically, consider a fixed $n \times n$ matrix X with

$$\operatorname{rank}(X) = r$$

where $r \ll n$. Suppose we are shown a few *randomly chosen entries* of X. Each entry X_{ij} is revealed to us independently with some probability $p \in (0,1)$ or is hidden from us with probability $1-p$. In other words, assume that we are shown the $n \times n$ matrix Y whose entries are

$$Y_{ij} := \delta_{ij} X_{ij} \quad \text{where the} \quad \delta_{ij} \sim \operatorname{Ber}(p) \text{ are independent.}$$

These δ_{ij} are *selectors* – Bernoulli random variables that indicate whether an entry is revealed to us or not (in the latter case the entry is replaced with zero). If

$$p = \frac{m}{n^2} \tag{6.12}$$

then we are *shown m entries of X on average*.

How can we infer X from Y? Although X has small rank r by assumption, Y may not have small rank. (Why?) It is thus natural to enforce small rank by choosing a *best rank-r approximation* to Y. (Recall the notion of the best rank-k approximation in Section 4.1.4.) The result, properly scaled, will be a good approximation to X.

Theorem 6.6.1 (Matrix completion) *Let $\hat{X}$ be a best rank-r approximation to $p^{-1}Y$. Then*

$$\mathbb{E}\frac{1}{n}\|\hat{X} - X\|_F \le C\sqrt{\frac{rn\log n}{m}}\,\|X\|_\infty,$$

as long as $m \ge n\log n$. Here $\|X\|_\infty = \max_{i,j}|X_{ij}|$ is the maximum magnitude of the entries of X.

Before we pass to the proof, let us pause quickly to note that Theorem 6.6.1 bounds the recovery error

$$\frac{1}{n}\|\hat{X} - X\|_F = \Big(\frac{1}{n^2}\sum_{i,j=1}^{n}|\hat{X}_{ij} - X_{ij}|^2\Big)^{1/2}.$$

This is simply the average error per entry (in the L^2 sense). If we choose the average number of observed entries m so that

$$m \ge C' rn\log n$$

with a large constant C' then Theorem 6.6.1 guarantees that the average error is much smaller than $\|X\|_\infty$.

To summarize, *matrix completion is possible if the number of observed entries exceeds rn by a logarithmic margin*. In this case, the expected average error per entry is much smaller than the maximal magnitude of an entry. Thus, for low-rank matrices, matrix completion is possible with few observed entries.

Proof We first bound the recovery error in the operator norm and then pass to the Frobenius norm using the low-rank assumption.

Step 1: Bounding the error in the operator norm. Using the triangle inequality, let us split the error as follows:

$$\|\hat{X} - X\| \le \|\hat{X} - p^{-1}Y\| + \|p^{-1}Y - X\|.$$

Since we have chosen $\hat{X}$ as a best rank-r approximation to $p^{-1}Y$, the second summand dominates, i.e., $\|\hat{X} - p^{-1}Y\| \le \|p^{-1}Y - X\|$, so we have

$$\|\hat{X} - X\| \le 2\|p^{-1}Y - X\| = \frac{2}{p}\|Y - pX\|. \tag{6.13}$$

Note that the matrix $\hat{X}$, which would be hard to handle, has now disappeared from the bound. Instead, $Y - pX$ is a matrix that is easy to understand. Its entries

$$(Y - pX)_{ij} = (\delta_{ij} - p)X_{ij}$$

are independent mean-zero random variables. So we can apply the result of Exercise 6.5.2, which gives

$$\mathbb{E}\,\|Y - pX\| \leq C\sqrt{\log n}\,\Big(\mathbb{E}\max_{i\in[n]}\|(Y - pX)_i\|_2 + \mathbb{E}\max_{j\in[n]}\|(Y - pX)^j\|_2\Big). \tag{6.14}$$

To bound the norms of the rows of $Y - pX$, we can express them as

$$\|(Y - pX)_i\|_2^2 = \sum_{j=1}^n (\delta_{ij} - p)^2 X_{ij}^2 \leq \sum_{j=1}^n (\delta_{ij} - p)^2\,\|X\|_\infty^2,$$

and similarly for the columns. These sums of independent random variables can be easily bounded using Bernstein's (or Chernoff's) inequality, which yields

$$\mathbb{E}\max_{i\in[n]}\sum_{j=1}^n (\delta_{ij} - p)^2 \leq Cpn.$$

(We do this calculation in Exercise 6.6.2.) Combining with a similar bound for the columns and substituting into (6.14), we obtain

$$\mathbb{E}\,\|Y - pX\| \lesssim \sqrt{pn\log n}\,\|X\|_\infty.$$

Then, by (6.13), we get

$$\mathbb{E}\,\|\hat{X} - X\| \lesssim \sqrt{\frac{n\log n}{p}}\,\|X\|_\infty. \tag{6.15}$$

Step 2: Passing to the Frobenius norm. We have not used the low-rank assumption yet, and will do this now. Since $\operatorname{rank}(X) \leq r$ by assumption and $\operatorname{rank}(\hat{X}) \leq r$ by construction, we have $\operatorname{rank}(\hat{X} - X) \leq 2r$. The relationship (4.4) between the operator and Frobenius norms thus gives

$$\|\hat{X} - X\|_F \leq \sqrt{2r}\,\|\hat{X} - X\|.$$

Taking expectations and using the bound on the error in the operator norm (6.15), we get

$$\mathbb{E}\,\|\hat{X} - X\|_F \leq \sqrt{2r}\,\mathbb{E}\,\|\hat{X} - X\| \lesssim \sqrt{\frac{rn\log n}{p}}\,\|X\|_\infty.$$

Dividing both sides by n, we can rewrite this bound as

$$\mathbb{E}\,\frac{1}{n}\|\hat{X} - X\|_F \lesssim \sqrt{\frac{rn\log n}{pn^2}}\,\|X\|_\infty.$$

To finish the proof, recall that $pn^2 = m$ by the definition (6.12) of p. ■

Exercise 6.6.2 (Bounding rows of random matrices)☕☕☕ Consider i.i.d. random variables $\delta_{ij} \sim \operatorname{Ber}(p)$, where $i, j = 1, \ldots, n$. Assuming that $pn \geq \log n$, show that

$$\mathbb{E}\max_{i\in[n]}\sum_{j=1}^n (\delta_{ij} - p)^2 \leq Cpn.$$

☞

Exercise 6.6.3 (Rectangular matrices) State and prove a version of the matrix completion theorem 6.6.1 for general rectangular $n_1 \times n_2$ matrices X.

Exercise 6.6.4 (Noisy observations) Extend the matrix completion theorem 6.6.1 to noisy observations, where we are shown noisy versions $X_{ij} + \nu_{ij}$ of some entries of X. Here the ν_{ij} are independent mean-zero sub-gaussian random variables representing noise.

Remark 6.6.5 (Improvements) The logarithmic factor can be removed from the bound of Theorem 6.6.1, and in some cases matrix completion can be *exact*, i.e. with zero error. See the notes at the end of this chapter for details.

6.7 Contraction Principle

We conclude the main part of this chapter with one more useful inequality. We still denote by $\varepsilon_1, \varepsilon_2, \varepsilon_3, \ldots$ a sequence of independent symmetric Bernoulli random variables (which are also independent of any other random variables in question).

Theorem 6.7.1 (Contraction principle) *Let $x_1, \ldots, x_N$ be (deterministic) vectors in some normed space, and let $a = (a_1, \ldots, a_n) \in \mathbb{R}^n$. Then*

$$\mathbb{E}\left\| \sum_{i=1}^N a_i \varepsilon_i x_i \right\| \le \|a\|_\infty \, \mathbb{E}\left\| \sum_{i=1}^N \varepsilon_i x_i \right\|.$$

Proof Without loss of generality, we may assume that $\|a\|_\infty \le 1$. (Why?) Define the function

$$f(a) := \mathbb{E}\left\| \sum_{i=1}^N a_i \varepsilon_i x_i \right\|. \tag{6.16}$$

Then $f : \mathbb{R}^N \to \mathbb{R}$ is a convex function. (See Exercise 6.7.2.)

Our goal is to find a bound for f on the set of points a satisfying $\|a\|_\infty \le 1$, i.e., on the unit cube $[-1, 1]^n$. By the elementary maximum principle for convex functions, the maximum of a convex function on a compact set in $\mathbb{R}^n$ is attained at an extreme point of the set. Thus f attains its maximum at one of the vertices of the cube, i.e. at a point a whose coefficients are all $a_i = \pm 1$.

For this point a, the random variables $(\varepsilon_i a_i)$ have the same distribution as (ε_i) owing to symmetry. Thus

$$\mathbb{E}\left\| \sum_{i=1}^N a_i \varepsilon_i x_i \right\| = \mathbb{E}\left\| \sum_{i=1}^N \varepsilon_i x_i \right\|.$$

Summarizing, we have shown that $f(a) \le \mathbb{E}\left\| \sum_{i=1}^N \varepsilon_i x_i \right\|$ whenever $\|a\|_\infty \le 1$. This completes the proof. ■

Exercise 6.7.2 Check that the function f defined in (6.16) is convex.

Exercise 6.7.3 (Contraction principle for general distributions) Prove the following generalization of Theorem 6.7.1. Let $X_1, \ldots, X_N$ be independent mean-zero random vectors in a normed space, and let $a = (a_1, \ldots, a_n) \in \mathbb{R}^n$. Then

$$\mathbb{E}\left\|\sum_{i=1}^{N} a_i X_i\right\| \le 4\|a\|_\infty \,\mathbb{E}\left\|\sum_{i=1}^{N} X_i\right\|.$$

☞

As an application, let us show how symmetrization can be done using *Gaussian* random variables $g_i \sim N(0,1)$ instead of symmetric Bernoulli random variables ε_i.

Lemma 6.7.4 (Symmetrization with Gaussians) *Let $X_1, \dots, X_N$ be independent mean-zero random vectors in a normed space. Let $g_1, \dots, g_N \sim N(0,1)$ be independent Gaussian random variables which are also independent of X_i. Then*

$$\frac{c}{\sqrt{\log N}}\,\mathbb{E}\left\|\sum_{i=1}^{N} g_i X_i\right\| \le \mathbb{E}\left\|\sum_{i=1}^{N} X_i\right\| \le 3\,\mathbb{E}\left\|\sum_{i=1}^{N} g_i X_i\right\|.$$

Proof Upper bound. By symmetrization (Lemma 6.4.2), we have

$$E := \mathbb{E}\left\|\sum_{i=1}^{N} X_i\right\| \le 2\,\mathbb{E}\left\|\sum_{i=1}^{N} \varepsilon_i X_i\right\|.$$

To interject Gaussian random variables, recall that $\mathbb{E}\,|g_i| = \sqrt{2/\pi}$. Thus we can continue our bound as follows:[2]

$$\begin{aligned}
E &\le 2\sqrt{\frac{\pi}{2}}\,\mathbb{E}_X\left\|\sum_{i=1}^{N} \varepsilon_i\,\mathbb{E}_g\,|g_i| X_i\right\| \\
&\le 2\sqrt{\frac{\pi}{2}}\,\mathbb{E}\left\|\sum_{i=1}^{N} \varepsilon_i |g_i| X_i\right\| \quad \text{(by Jensen's inequality)} \\
&= 2\sqrt{\frac{\pi}{2}}\,\mathbb{E}\left\|\sum_{i=1}^{N} g_i X_i\right\|.
\end{aligned}$$

The last equality follows by the symmetry of the Gaussian distribution, which implies that the random variables $\varepsilon_i |g_i|$ have the same distribution as the g_i (recall Exercise 6.4.1).

Lower bound. This can be proved by using the contraction principle (Theorem 6.7.1) and symmetrization (Lemma 6.4.2). We have

$$\begin{aligned}
\mathbb{E}\left\|\sum_{i=1}^{N} g_i X_i\right\| &= \mathbb{E}\left\|\sum_{i=1}^{N} \varepsilon_i g_i X_i\right\| \quad \text{(by the symmetry of the } g_i\text{)} \\
&\le \mathbb{E}_g\,\mathbb{E}_X\left(\|g\|_\infty\,\mathbb{E}_\varepsilon\left\|\sum_{i=1}^{N} \varepsilon_i X_i\right\|\right) \quad \text{(by Theorem 6.7.1)} \\
&= \mathbb{E}_g\left(\|g\|_\infty\,\mathbb{E}_\varepsilon\,\mathbb{E}_X\left\|\sum_{i=1}^{N} \varepsilon_i X_i\right\|\right) \quad \text{(by independence)} \\
&\le 2\,\mathbb{E}_g\left(\|g\|_\infty\,\mathbb{E}_X\left\|\sum_{i=1}^{N} X_i\right\|\right) \quad \text{(by Lemma 6.4.2)} \\
&= 2\Big(\mathbb{E}\,\|g\|_\infty\Big)\left(\mathbb{E}\left\|\sum_{i=1}^{N} X_i\right\|\right) \quad \text{(by independence).}
\end{aligned}$$

[2] Here we use the index g in $\mathbb{E}_g$ to indicate that this is an expectation "over (g_i)", i.e., conditional on (X_i). Similarly, E_X denotes the expectation over (X_i).

It remains to recall from Exercise 2.5.10 that

$$\mathbb{E}\,\|g\|_\infty \le C\sqrt{\log N}.$$

The proof is complete. ■

Exercise 6.7.5 Show that the factor $\sqrt{\log N}$ in Lemma 6.7.4 is needed in general and is optimal. Thus, symmetrization with Gaussian random variables is generally weaker than symmetrization with symmetric Bernoullis.

Exercise 6.7.6 (Symmetrization and contraction for functions of norms) Let $F: \mathbb{R}_+ \to \mathbb{R}$ be a convex increasing function. Generalize the symmetrization and contraction results of this and the previous section by replacing the norm $\|\cdot\|$ with $F(\|\cdot\|)$ throughout.

In the following exercise we set foot in the study of random processes, on which we focus fully in the next chapter.

Exercise 6.7.7 (Talagrand's contraction principle) Consider a bounded subset $T \subset \mathbb{R}^n$, and let $\varepsilon_1, \ldots, \varepsilon_n$ be independent symmetric Bernoulli random variables. Let $\phi_i : \mathbb{R} \to \mathbb{R}$ be contractions, i.e., Lipschitz functions with $\|\phi_i\|_{\mathrm{Lip}} \le 1$. Then

$$\mathbb{E}\sup_{t\in T}\sum_{i=1}^n \varepsilon_i\phi_i(t_i) \le \mathbb{E}\sup_{t\in T}\sum_{i=1}^n \varepsilon_i t_i. \tag{6.17}$$

To prove this result, do the following steps:

(a) First let $n = 2$. Consider a subset $T \subset \mathbb{R}^2$ and contraction $\phi: \mathbb{R} \to \mathbb{R}$, and check that

$$\sup_{t\in T}(t_1 + \phi(t_2)) + \sup_{t\in T}(t_1 - \phi(t_2)) \le \sup_{t\in T}(t_1 + t_2) + \sup_{t\in T}(t_1 - t_2).$$

(b) Use induction on n to complete the proof. ☞

Exercise 6.7.8 Generalize Talagrand's contraction principle for arbitrary Lipschitz functions $\phi_i : \mathbb{R} \to \mathbb{R}$ without restriction on their Lipschitz norms. ☞

6.8 Notes

The version of the decoupling inequality that we stated in Theorem 6.1.1 and Exercise 6.1.5 was originally proved by J. Bourgain and L. Tzafriri [31]. We refer the reader to the papers [60] and the books [59] and [76, Section 8.4] for related results and extensions.

The original form of the Hanson–Wright inequality, which is somewhat weaker than Theorem 6.2.1, goes back to [95, 220]. The version of Theorem 6.2.1 and its proof that we gave in Section 6.2 are from [175]. Several special cases of the Hanson–Wright inequality appeared earlier in [76, Proposition 8.13] for Bernoulli random variables, in [193,

Lemma 2.5.1] for Gaussian random variables, and in [16] for diagonal-free matrices. Concentration for anisotropic random vectors (Theorem 6.3.2) and the bound on the distance between a random vector and a subspace (Exercise 6.3.4) were taken from [175].

The symmetrization lemma 6.4.2 and its proof can be found in, e.g., [126, Lemma 6.3] and [76, Section 8.2].

Although a precise statement of Theorem 6.5.1 is difficult to locate in the existing literature, the result is essentially known. It can be deduced, for example, from the inequalities in [200, 201]. We refer the reader to [207, Section 4] and [121] for more elaborate results, which attempt to describe the operator norm of the random matrix A in terms of the variances of its entries.

Theorem 6.6.1 on matrix completion and its proof are from [166, Section 2.5]. Earlier, E. Candes and B. Recht [44] had demonstrated that, under some additional incoherence assumptions, *exact* matrix completion is possible with $m \asymp rn \log^2(n)$ randomly sampled entries. We refer the reader to the papers [46, 169, 90, 56] for many further developments on matrix completion.

The contraction principle (Theorem 6.7.1) is taken from [126, Section 4.2]; see also [126, Corollary 3.17 and Theorem 4.12] for different versions of this principle for random processes. Lemma 6.7.4 can be found in [126, inequality (4.9)]. While the logarithmic factor is in general needed there, it can be removed if the normed space has a nontrivial cotype; see [126, Proposition 9.14]. Talagrand's contraction principle (Exercise 6.7.7) can be found in [126, Corollary 3.17], where one can also find a more general result (with a convex and increasing function of the supremum). Exercise 6.7.7 was adapted from [206, Exercise 7.4]. A Gaussian version of Talagrand's contraction principle will be given in Exercise 7.2.13.

7

Random Processes

In this chapter we begin to study random processes – collections of random variables $(X_t)_{t\in T}$ that are not necessarily independent. In many classical examples in probability theory such as Brownian motion, t stands for time and thus T is a subset of $\mathbb{R}$. But in high-dimensional probability it is important to go beyond this case and allow T to be a general abstract set. An important example is the so-called canonical Gaussian process

$$X_t = \langle g, t \rangle , \quad t \in T,$$

where T is an arbitrary subset of $\mathbb{R}^n$ and g is a standard normal random vector in $\mathbb{R}^n$. We discuss this in Section 7.1.

In Section 7.2 we prove remarkably sharp comparison inequalities for Gaussian processes – Slepian's inequality, the Sudakov–Fernique inequality, and Gordon's inequality. Our argument introduces the useful technique of Gaussian interpolation. In Section 7.3 we illustrate the comparison inequalities by proving a sharp bound $\mathbb{E}\,\|A\| \le \sqrt{m} + \sqrt{n}$ on the operator norm of an $m \times n$ Gaussian random matrix A.

It is important to understand how the probabilistic properties of random processes, and in particular canonical Gaussian process, are related to the geometry of the underlying set T. In Section 7.4 we prove Sudakov's minoration inequality, which gives a lower bound on the magnitude of the canonical Gaussian process

$$w(T) = \mathbb{E} \sup_{t\in T} \langle g, t\rangle$$

in terms of the covering numbers of T; the upper bound will be studied in Chapter 8. The quantity $w(T)$ is called the Gaussian width of the set $T \subset \mathbb{R}^n$. We study this key geometric parameter in detail in Section 7.5, where we relate it to other notions including those of stable dimension, stable rank, and Gaussian complexity.

In Section 7.7 we give an example that highlights the importance of the Gaussian width in high-dimensional geometric problems. We examine how random projections affect a given set $T \subset \mathbb{R}^n$, and we find that the Gaussian width of T plays a key role in determining the sizes of random projections of T.

7.1 Basic Concepts and Examples

Definition 7.1.1 (Random process) A *random process* is a collection of random variables $(X_t)_{t\in T}$ on the same probability space, which are indexed by the elements t of some set T.

In some classical examples, t stands for *time*, in which case T is a subset of $\mathbb{R}$. But we primarily study processes in high-dimensional settings, where T is a subset of $\mathbb{R}^n$ and where the analogy with time will be lost.

Example 7.1.2 (Discrete time) If $T = \{1, \ldots, n\}$ then the random process

$$(X_1, \ldots, X_n)$$

can be identified with a *random vector* in $\mathbb{R}^n$.

Example 7.1.3 (Random walks) If $T = \mathbb{N}$, a discrete-time random process $(X_n)_{n\in\mathbb{N}}$ is simply a *sequence* of random variables. An important example is a *random walk*, defined as

$$X_n := \sum_{i=1}^{n} Z_i,$$

where the increments Z_i are independent mean-zero random variables. See Figure 7.1 for an illustration.

Example 7.1.4 (Brownian motion) The most classical continuous-time random process is the standard *Brownian motion* $(X_t)_{t\geq 0}$, also called the *Wiener process*. It can be characterized as follows:

(i) The process has continuous sample paths, i.e., the random function $f(t) := X_t$ is continuous almost surely;
(ii) The increments are independent and satisfy $X_t - X_s \sim N(0, t - s)$ for all $t \geq s$.

Figure 7.1 (right) illustrates a few trials of the standard Brownian motion.

Example 7.1.5 (Random fields) When the index set T is a subset of $\mathbb{R}^n$, a random process $(X_t)_{t\in T}$ is sometimes called a spatial random process, or a *random field*. For example, the

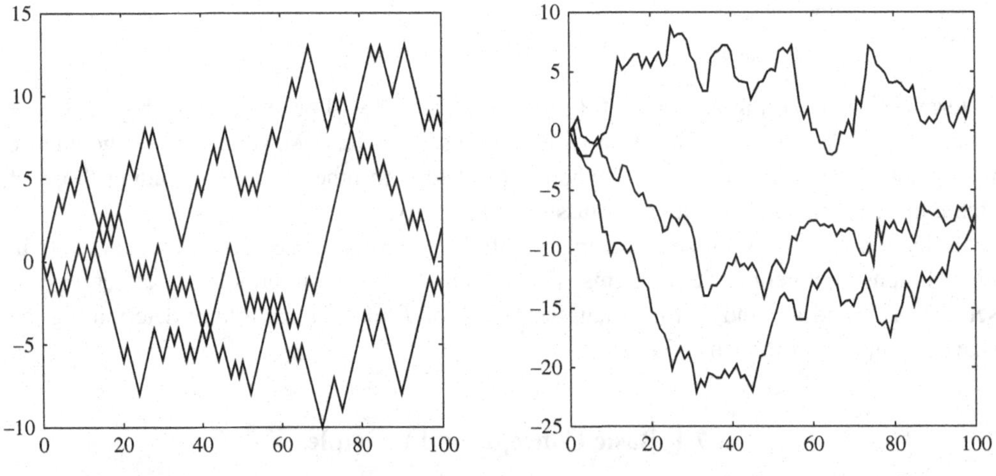

Figure 7.1 A few trials of a random walk with symmetric Bernoulli steps Z_i (left) and a few trials of the standard Brownian motion in $\mathbb{R}$ (right).

water temperature X_t at a location on Earth that is parametrized by t can be modeled as a spatial random process.

7.1.1 Covariance and Increments

In Section 3.2 we introduced the notion of the covariance matrix of a random vector. We now define the *covariance function* of a random process $(X_t)_{t\in T}$ in a similar manner. For simplicity, let us assume in this section that the random process has zero mean, i.e.

$$\mathbb{E}\, X_t = 0 \quad \text{for all } t \in T.$$

(The adjustments for the general case will be obvious.) The covariance function of the process is defined as

$$\Sigma(t,s) := \operatorname{cov}(X_t, X_s) = \mathbb{E}\, X_t X_s, \quad t,s \in T.$$

Similarly, the *increments* of the random process are defined as

$$d(t,s) := \|X_t - X_s\|_{L^2} = \left(\mathbb{E}(X_t - X_s)^2\right)^{1/2}, \quad t,s \in T.$$

Example 7.1.6 The increments of the standard Brownian motion satisfy

$$d(t,s) = \sqrt{t-s}, \quad t \geq s,$$

by definition. The increments of the random walk of Example 7.1.3 with $\mathbb{E}\, Z_i^2 = 1$ behave similarly:

$$d(n,m) = \sqrt{n-m}, \quad n \geq m.$$

(Check!)

Remark 7.1.7 (Canonical metric) As we emphasized at the beginning of the chapter, the index set T of a general random process may be an abstract set without any geometric structure. But even in this case, the increments $d(t,s)$ always define a *metric* on T, thus automatically turning T into a *metric space*.[1] However, Example 7.1.6 shows that this metric may not agree with the standard metric on $\mathbb{R}$, where the distance between t and s is $|t-s|$.

Exercise 7.1.8 (Covariance vs. increments) ☕☕ Consider a random process $(X_t)_{t\in T}$.

(a) Express the increments $\|X_t - X_s\|_2$ in terms of the covariance function $\Sigma(t,s)$.
(b) Assuming that the zero random variable 0 belongs to the process, express the covariance function $\Sigma(t,s)$ in terms of the increments $\|X_t - X_s\|_2$.

Exercise 7.1.9 (Symmetrization for random processes) ☕☕☕ Let $X_1(t), \ldots, X_N(t)$ be N independent mean-zero random processes indexed by points $t \in T$. Let $\varepsilon_1, \ldots, \varepsilon_N$ be independent symmetric Bernoulli random variables. Prove that

$$\frac{1}{2}\,\mathbb{E} \sup_{t\in T} \sum_{i=1}^N \varepsilon_i X_i(t) \leq \mathbb{E} \sup_{t\in T} \sum_{i=1}^N X_i(t) \leq 2\,\mathbb{E} \sup_{t\in T} \sum_{i=1}^N \varepsilon_i X_i(t).$$ ☞

[1] More precisely, $d(t,s)$ is a *pseudometric* on T since the distance between two distinct points can be zero, i.e., $d(t,s) = 0$ does not necessarily imply $t = s$.

7.1.2 Gaussian Processes

Definition 7.1.10 (Gaussian process) A random process $(X_t)_{t\in T}$ is called a *Gaussian process* if, for any finite subset $T_0 \subset T$, the random vector $(X_t)_{t\in T_0}$ has a normal distribution. Equivalently, $(X_t)_{t\in T}$ is Gaussian if every finite linear combination $\sum_{t\in T_0} a_t X_t$ is a normal random variable. (This equivalence is due to the characterization of the normal distribution in Exercise 3.3.4.)

The notion of Gaussian processes generalizes that of Gaussian random vectors in $\mathbb{R}^n$. A classical example of a Gaussian process is the standard Brownian motion.

Remark 7.1.11 (Distribution is determined by covariance or increments) From the formula (3.5) for the multivariate normal density we may recall that the distribution of a mean-zero Gaussian random vector X in $\mathbb{R}^n$ is completely determined by its covariance matrix. Then, by definition, the distribution of a mean-zero Gaussian process $(X_t)_{t\in T}$ is also completely determined[2] by its covariance function $\Sigma(t, s)$. Equivalently (owing to Exercise 7.1.8), the distribution of the process is determined by the increments $d(t, s)$.

We now consider a wide class of examples of a Gaussian processes indexed by higher-dimensional sets $T \subset \mathbb{R}^n$. Consider the standard normal random vector $g \sim N(0, I_n)$ and define the random process

$$X_t := \langle g, t\rangle, \quad t \in T. \tag{7.1}$$

Then $(X_t)_{t\in T}$ is clearly a Gaussian process, and we call it a *canonical Gaussian process*. The increments of this process define the Euclidean distance

$$\|X_t - X_s\|_2 = \|t - s\|_2, \quad t, s \in T.$$

(Check!)

Actually, one can realize any Gaussian process as the canonical process (7.1). This follows from a simple observation about Gaussian vectors.

Lemma 7.1.12 (Gaussian random vectors) *Let Y be a mean-zero Gaussian random vector in $\mathbb{R}^n$. Then there exist points $t_1, \ldots, t_n \in \mathbb{R}^n$ such that*

$$Y \equiv (\langle g, t_i\rangle)_{i=1}^n, \text{ where } g \sim N(0, I_n).$$

Here the equivalence symbol means that the distributions of the two random vectors are the same.

Proof Let Σ denote the covariance matrix of Y. Then we may realize

$$Y \equiv \Sigma^{1/2} g \text{ where } g \sim N(0, I_n)$$

(recall Section 3.3.2). Next, the coordinates of the vector $\Sigma^{1/2} g$ are $\langle t_i, g\rangle$, where the t_i denote the rows of the matrix $\Sigma^{1/2}$. This completes the proof. ■

[2] To avoid measurability issues, we do not formally define the distribution of a random process here. So, the statement above should be understood as the fact that the covariance function determines the distribution of all marginals $(X_t)_{t\in T_0}$ with finite $T_0 \subset T$.

It follows that, for any Gaussian process $(Y_s)_{s\in S}$, all finite-dimensional marginals $(Y_s)_{s\in S_0}$, $|S_0| = n$, can be represented as the canonical Gaussian process (7.1) indexed in a certain subset $T_0 \subset \mathbb{R}^n$.

Exercise 7.1.13 ☕☕ Realize an N-step random walk of Example 7.1.3 with $Z_i \sim N(0,1)$ as a canonical Gaussian process (7.1) with $T \subset \mathbb{R}^N$. ☞

7.2 Slepian's Inequality

In many applications, it is useful to have a uniform control on a random process $(X_t)_{t\in T}$, i.e., to have a bound on[3]

$$\mathbb{E} \sup_{t\in T} X_t.$$

For some processes, this quantity can be computed exactly. For example, if (X_t) is a standard Brownian motion, the so-called *reflection principle* yields

$$\mathbb{E} \sup_{t\le t_0} X_t = \sqrt{\frac{2t_0}{\pi}} \quad \text{for every } t_0 \ge 0.$$

For general random processes, even if they are Gaussian, the problem is emphatically nontrivial.

The first general bound we will prove is Slepian's comparison inequality for Gaussian processes. It basically states that the faster the process grows (in terms of the magnitude of the increments), the farther it gets.

Theorem 7.2.1 (Slepian's inequality) *Let $(X_t)_{t\in T}$ and $(Y_t)_{t\in T}$ be two mean-zero Gaussian processes. Assume that, for all $t, s \in T$, we have*

$$\mathbb{E} X_t^2 = \mathbb{E} Y_t^2 \quad \textit{and} \quad \mathbb{E}(X_t - X_s)^2 \le \mathbb{E}(Y_t - Y_s)^2. \tag{7.2}$$

Then for every $\tau \in \mathbb{R}$ we have

$$\mathbb{P}\left\{\sup_{t\in T} X_t \ge \tau\right\} \le \mathbb{P}\left\{\sup_{t\in T} Y_t \ge \tau\right\}. \tag{7.3}$$

Consequently,

$$\mathbb{E} \sup_{t\in T} X_t \le \mathbb{E} \sup_{t\in T} Y_t. \tag{7.4}$$

Whenever the tail comparison inequality (7.3) holds, we say that the random variable X is *stochastically dominated* by the random variable Y.

We now prepare for the proof of Slepian's inequality.

[3] To avoid measurability issues, we study random processes through their finite-dimensional marginals as before. Thus we interpret $\mathbb{E}\sup_{t\in T} X_t$ more formally as $\sup_{T_0\subset T} \mathbb{E}\max_{t\in T_0} X_t$, where the supremum is over all finite subsets $T_0 \subset T$.

7.2.1 Gaussian Interpolation

The proof of Slepian's inequality that we are about to give will be based on the technique of *Gaussian interpolation*. Let us describe it briefly. Assume that T is finite; then $X = (X_t)_{t \in T}$ and $Y = (Y_t)_{t \in T}$ are Gaussian random vectors in $\mathbb{R}^n$ where $n = |T|$. We may also assume that X and Y are independent. (Why?)

Define a Gaussian random vector $Z(u)$ in $\mathbb{R}^n$ that continuously interpolates between $Z(0) = Y$ and $Z(1) = X$:

$$Z(u) := \sqrt{u}\, X + \sqrt{1-u}\, Y, \quad u \in [0, 1].$$

Exercise 7.2.2 Check that the covariance matrix of $Z(u)$ interpolates linearly between the covariance matrices of Y and X:

$$\Sigma(Z(u)) = u\, \Sigma(X) + (1-u)\, \Sigma(Y).$$

For a given function $f \colon \mathbb{R}^n \to \mathbb{R}$, we now study how the quantity $\mathbb{E}\, f(Z(u))$ changes as u increases from 0 to 1. Of specific interest to us is the function

$$f(x) = \mathbf{1}_{\{\max_i x_i < \tau\}}.$$

We will be able to show that, in this case, $\mathbb{E}\, f(Z(u))$ *increases* in u. This would imply the conclusion of Slepian's inequality at once, since then

$$\mathbb{E}\, f(Z(1)) \geq \mathbb{E}\, f(Z(0)), \quad \text{thus} \quad \mathbb{P}\Big\{ \max_i X_i < \tau \Big\} \geq \mathbb{P}\Big\{ \max_i Y_i < \tau \Big\},$$

as claimed.

Now let us pass to a detailed argument. To develop Gaussian interpolation, let us start with the following useful identity.

Lemma 7.2.3 (Gaussian integration by parts) *Let $X \sim N(0, 1)$. Then for any differentiable function $f \colon \mathbb{R} \to \mathbb{R}$ we have*

$$\mathbb{E}\, f'(X) = \mathbb{E}\, X f(X).$$

Proof Assume first that f has bounded support. Denoting the Gaussian density of X by

$$p(x) = \frac{1}{\sqrt{2\pi}} e^{-x^2/2},$$

we can express the expectation as an integral, and integrate it by parts:

$$\mathbb{E}\, f'(X) = \int_{\mathbb{R}} f'(x) p(x)\, dx = -\int_{\mathbb{R}} f(x) p'(x)\, dx. \tag{7.5}$$

Now, a direct check gives

$$p'(x) = -x p(x),$$

so the integral in (7.5) equals

$$\int_{\mathbb{R}} f(x) p(x) x\, dx = \mathbb{E}\, X f(X),$$

as claimed. The identity can be extended to general functions by an approximation argument. The lemma is proved. ■

Exercise 7.2.4 If $X \sim N(0, \sigma^2)$, show that

$$\mathbb{E}\, X f(X) = \sigma^2\, \mathbb{E}\, f'(X).$$

☞

Gaussian integration by parts generalizes nicely to high dimenions.

Lemma 7.2.5 (Multivariate Gaussian integration by parts) *Let $X \sim N(0, \Sigma)$. Then for any differentiable function $f : \mathbb{R}^n \to \mathbb{R}$ we have*

$$\mathbb{E}\, X f(X) = \Sigma\; \mathbb{E}\, \nabla f(X).$$

Exercise 7.2.6 Prove Lemma 7.2.5. According to matrix-by-vector multiplication, note that the conclusion of the lemma is equivalent to

$$\mathbb{E}\, X_i f(X) = \sum_{j=1}^{n} \Sigma_{ij}\, \mathbb{E}\, \frac{\partial f}{\partial x_j}(X), \quad i = 1, \ldots, n. \tag{7.6}$$

☞

Lemma 7.2.7 (Gaussian interpolation) *Consider two independent Gaussian random vectors $X \sim N(0, \Sigma^X)$ and $Y \sim N(0, \Sigma^Y)$. Define the interpolation Gaussian vector*

$$Z(u) := \sqrt{u}\, X + \sqrt{1-u}\, Y, \quad u \in [0, 1]. \tag{7.7}$$

Then for any twice-differentiable function $f : \mathbb{R}^n \to \mathbb{R}$, we have

$$\frac{d}{du}\, \mathbb{E}\, f(Z(u)) = \frac{1}{2} \sum_{i,j=1}^{n} (\Sigma^X_{ij} - \Sigma^Y_{ij})\, \mathbb{E}\Big(\frac{\partial^2 f}{\partial x_i\, \partial x_j}(Z(u))\Big). \tag{7.8}$$

Proof Using the chain rule,[4] we have

$$\begin{aligned} \frac{d}{du}\, \mathbb{E}\, f(Z(u)) &= \sum_{i=1}^{n} \mathbb{E}\, \frac{\partial f}{\partial x_i}(Z(u)) \frac{dZ_i}{du} \\ &= \frac{1}{2} \sum_{i=1}^{n} \mathbb{E}\, \frac{\partial f}{\partial x_i}(Z(u)) \Big(\frac{X_i}{\sqrt{u}} - \frac{Y_i}{\sqrt{1-u}}\Big) \quad \text{(by (7.7))}. \end{aligned} \tag{7.9}$$

Let us break this sum into two, and first compute the contribution of the terms containing X_i. To this end, we condition on Y and express

$$\sum_{i=1}^{n} \frac{1}{\sqrt{u}}\, \mathbb{E}\, X_i \frac{\partial f}{\partial x_i}(Z(u)) = \sum_{i=1}^{n} \frac{1}{\sqrt{u}}\, \mathbb{E}\, X_i g_i(X), \tag{7.10}$$

[4] Here we use the multivariate chain rule to differentiate a function $f(g_1(u), \ldots, g_n(u))$, where $g_i : \mathbb{R} \to \mathbb{R}$ and $f : \mathbb{R}^n \to \mathbb{R}$, as follows: $df/du = \sum_{i=1}^n (\partial f/\partial x_i)(dg_i/du)$.

where

$$g_i(X) = \frac{\partial f}{\partial x_i}(\sqrt{u}\, X + \sqrt{1-u}\, Y).$$

Apply the multivariate Gaussian integration by parts (Lemma 7.2.5). According to (7.6), we have

$$\begin{aligned}
\mathbb{E}\, X_i g_i(X) &= \sum_{j=1}^{n} \Sigma_{ij}^X \, \mathbb{E}\, \frac{\partial g_i}{\partial x_j}(X) \\
&= \sum_{j=1}^{n} \Sigma_{ij}^X \, \mathbb{E}\, \frac{\partial^2 f}{\partial x_i\, \partial x_j}(\sqrt{u}\, X + \sqrt{1-u}\, Y)\sqrt{u}.
\end{aligned}$$

Substitute this into (7.10) to get

$$\sum_{i=1}^{n} \frac{1}{\sqrt{u}}\, \mathbb{E}\, X_i \frac{\partial f}{\partial x_i}(Z(u)) = \sum_{i,j=1}^{n} \Sigma_{ij}^X \, \mathbb{E}\, \frac{\partial^2 f}{\partial x_i\, \partial x_j}(Z(u)).$$

Taking expectation of both sides with respect to Y, we lift the conditioning on Y.

We can simiarly evaluate the other sum in (7.9), the one containing the terms Y_i. Combining the two sums we complete the proof. ■

7.2.2 *Proof of Slepian's Inequality*

We are ready to establish a preliminary functional form of Slepian's inequality.

Lemma 7.2.8 (Slepian's inequality, functional form) *Consider two mean-zero Gaussian random vectors X and Y in $\mathbb{R}^n$. Assume that for all $i, j = 1, \dots, n$, we have*

$$\mathbb{E}\, X_i^2 = \mathbb{E}\, Y_i^2 \quad \textit{and} \quad \mathbb{E}(X_i - X_j)^2 \le \mathbb{E}(Y_i - Y_j)^2.$$

Consider a twice-differentiable function $f : \mathbb{R}^n \to \mathbb{R}$ such that

$$\frac{\partial^2 f}{\partial x_i\, \partial x_j} \ge 0 \quad \textit{for all } i \ne j.$$

Then

$$\mathbb{E}\, f(X) \ge \mathbb{E}\, f(Y).$$

Proof The assumptions imply that the entries of the covariance matrices Σ^X and Σ^Y of X and Y satisfy

$$\Sigma_{ii}^X = \Sigma_{ii}^Y \quad \text{and} \quad \Sigma_{ij}^X \ge \Sigma_{ij}^Y.$$

for all $i, j = 1, \dots, n$. We can assume that X and Y are independent. (Why?) Applying Lemma 7.2.7 and using our assumptions, we conclude that

$$\frac{d}{du}\, \mathbb{E}\, f(Z(u)) \ge 0,$$

so that $\mathbb{E}\, f(Z(u))$ increases in u. Then $\mathbb{E}\, f(Z(1)) = \mathbb{E}\, f(X)$ is at least as large as $\mathbb{E}\, f(Z(0)) = \mathbb{E}\, f(Y)$. This completes the proof. ■

Now we are ready to prove Slepian's inequality, Theorem 7.2.1. Let us state and prove it in the equivalent form for Gaussian random vectors.

Theorem 7.2.9 (Slepian's inequality) *Let X and Y be Gaussian random vectors as in Lemma 7.2.8. Then for every $\tau \in \mathbb{R}$ we have*

$$\mathbb{P}\Big\{\max_{i\le n} X_i \ge \tau\Big\} \le \mathbb{P}\Big\{\max_{i\le n} Y_i \ge \tau\Big\}.$$

Consequently,

$$\mathbb{E}\max_{i\le n} X_i \le \mathbb{E}\max_{i\le n} Y_i.$$

Proof Let $h\colon \mathbb{R} \to [0,1]$ be a twice-differentiable non-increasing approximation to the indicator function of the interval $(-\infty,\tau)$:

$$h(x) \approx \mathbf{1}_{(-\infty,\tau)};$$

see Figure 7.2. Define the function $f\colon \mathbb{R}^n \to \mathbb{R}$ by $f(x) = h(x_1)\cdots h(x_n)$.

Then $f(x)$ is an approximation to the indicator function

$$f(x) \approx \mathbf{1}_{\{\max_i x_i < \tau\}}.$$

We are aiming to apply the functional form of Slepian's inequality, Lemma 7.2.8, for $f(x)$. To check the assumptions of this result, note that for $i \ne j$ we have

$$\frac{\partial^2 f}{\partial x_i\,\partial x_j} = h'(x_i)h'(x_j) \prod_{k\notin\{i,j\}} h(x_k).$$

The first two factors are non-positive and the others are non-negative by assumption. Thus the second derivative is non-negative, as required.

It follows that

$$\mathbb{E} f(X) \ge \mathbb{E} f(Y).$$

By approximation, this implies that

$$\mathbb{P}\Big\{\max_{i\le n} X_i < \tau\Big\} \ge \mathbb{P}\Big\{\max_{i\le n} Y_i < \tau\Big\}.$$

This proves the first part of the conclusion. The second part follows using the integral identity in Lemma 1.2.1; see the following exercise. ■

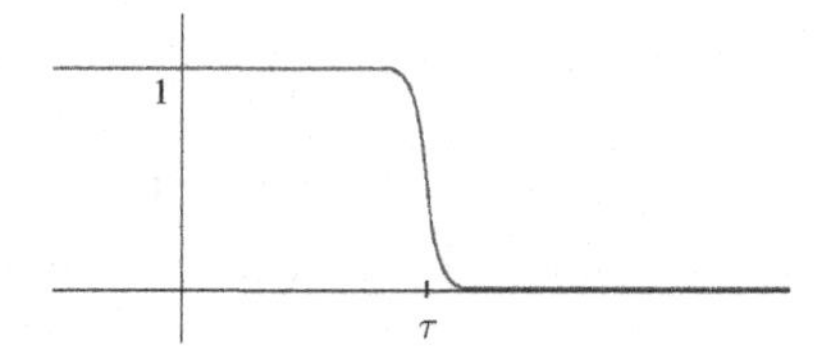

Figure 7.2 The function $h(x)$ is a smooth non-increasing approximation to the indicator function $\mathbf{1}_{(-\infty,\tau)}$.

Exercise 7.2.10 Using the integral identity in Exercise 1.2.2, deduce the second part of Slepian's inequality (the comparison of expectations).

7.2.3 The Sudakov–Fernique Inequality and Gordon's Inequality

Slepian's inequality has two assumptions on the processes (X_t) and (Y_t) in (7.2): the equality of variances and the dominance of increments. If we now remove the assumption on the equality of variances then we will still be able to obtain (7.4). This more practically useful result is due to Sudakov and Fernique.

Theorem 7.2.11 (Sudakov–Fernique inequality) *Let $(X_t)_{t\in T}$ and $(Y_t)_{t\in T}$ be two mean-zero Gaussian processes. Assume that, for all $t, s \in T$, we have*

$$\mathbb{E}(X_t - X_s)^2 \le \mathbb{E}(Y_t - Y_s)^2.$$

Then

$$\mathbb{E} \sup_{t\in T} X_t \le \mathbb{E} \sup_{t\in T} Y_t.$$

Proof It is enough to prove this theorem for Gaussian random vectors X and Y in $\mathbb{R}^n$, just as we did for Slepian's inequality in Theorem 7.2.9. We again deduce the result from the Gaussian interpolation lemma 7.2.7. But this time, instead of choosing a function $f(x)$ that approximates the indicator function of $\{\max_i x_i < \tau\}$, we want $f(x)$ to approximate $\max_i x_i$.

To this end, let $\beta > 0$ be a parameter and define the function[5]

$$f(x) := \frac{1}{\beta} \log \sum_{i=1}^{n} e^{\beta x_i}. \tag{7.11}$$

A quick check shows that

$$f(x) \to \max_{i\le n} x_i \quad \text{as } \beta \to \infty.$$

(Do this!) Substituting $f(x)$ into the Gaussian interpolation formula (7.8) and simplifying the expression shows that $(d/du)\,\mathbb{E} f(Z(u)) \le 0$ for all u (see Exercise 7.2.12 below). The proof can then be completed in just the same way as the proof of Slepian's inequality. ∎

Exercise 7.2.12 Show that $(d/du)\,\mathbb{E} f(Z(u)) \le 0$ in the Sudakov–Fernique theorem 7.2.11. ☞

Exercise 7.2.13 (Gaussian contraction inequality) The following is a Gaussian version of Talagrand's contraction principle, which we proved in Exercise 6.7.7. Consider a

[5] The motivation for considering this form of $f(x)$ comes from statistical mechanics, where the right-hand side of (7.11) can be interpreted as the log of a *partition function* and β as the *inverse temperature*.

bounded subset $T \subset \mathbb{R}^n$, and let $g_1, \ldots, g_n$ be independent $N(0,1)$ random variables. Let $\phi_i : \mathbb{R} \to \mathbb{R}$ be contractions, i.e., Lipschitz functions with $\|\phi_i\|_{\mathrm{Lip}} \le 1$. Prove that

$$\mathbb{E} \sup_{t \in T} \sum_{i=1}^n g_i \phi_i(t_i) \le \mathbb{E} \sup_{t \in T} \sum_{i=1}^n g_i t_i.$$ ☞

Exercise 7.2.14 (Gordon's inequality) Prove the following extension of Slepian's inequality due to Y. Gordon. Let $(X_{ut})_{u \in U, t \in T}$ and $Y = (Y_{ut})_{u \in U, t \in T}$ be two mean-zero Gaussian processes indexed by pairs of points (u, t) in a product set $U \times T$. Assume that we have

$$\mathbb{E} X_{ut}^2 = \mathbb{E} Y_{ut}^2, \quad \mathbb{E}(X_{ut} - X_{us})^2 \le \mathbb{E}(Y_{ut} - Y_{us})^2 \quad \text{for all } u, t, s;$$
$$\mathbb{E}(X_{ut} - X_{vs})^2 \ge \mathbb{E}(Y_{ut} - Y_{vs})^2 \quad \text{for all } u \ne v \text{ and all } t, s.$$

Then for every $\tau \ge 0$ we have

$$\mathbb{P}\left\{ \inf_{u \in U} \sup_{t \in T} X_{ut} \ge \tau \right\} \le \mathbb{P}\left\{ \inf_{u \in U} \sup_{t \in T} Y_{ut} \ge \tau \right\}.$$

Consequently,

$$\mathbb{E} \inf_{u \in U} \sup_{t \in T} X_{ut} \le \mathbb{E} \inf_{u \in U} \sup_{t \in T} Y_{ut}. \tag{7.12}$$

☞

As for the Sudakov–Fernique inequality, it is possible to remove the assumption of equal variances from Gordon's theorem and still be able to derive (7.12). We do not prove this result.

7.3 Sharp Bounds on Gaussian Matrices

We now illustrate the Gaussian comparison inequalities that we have just proved with an application to random matrices. In Section 4.6 we studied $m \times n$ random matrices A with independent sub-gaussian rows. We used the ε-net argument to control the norm of A as follows:

$$\mathbb{E} \|A\| \le \sqrt{m} + C\sqrt{n},$$

where C is a constant. (See Exercise 4.6.3.) We now use the Sudakov–Fernique inequality to improve upon this bound for *Gaussian* random matrices, showing that it holds with a sharp constant $C = 1$.

Theorem 7.3.1 (Norms of Gaussian random matrices) *Let A be an $m \times n$ matrix with independent $N(0,1)$ entries. Then*

$$\mathbb{E} \|A\| \le \sqrt{m} + \sqrt{n}.$$

Proof We can realize the norm of A as a supremum of a Gaussian process. Indeed,

$$\|A\| = \max_{u \in S^{n-1},\, v \in S^{m-1}} \langle Au, v \rangle = \max_{(u,v) \in T} X_{uv},$$

where T denotes the product set $S^{n-1} \times S^{m-1}$ and

$$X_{uv} := \langle Au, v \rangle \sim N(0, 1).$$

(Check!)

To apply the Sudakov–Fernique comparison inequality (Theorem 7.2.11), let us compute the increments of the process (X_{uv}). For any $(u, v), (w, z) \in T$, we have

$$\begin{aligned}
\mathbb{E}(X_{uv} - X_{wz})^2 &= \mathbb{E}\left(\langle Au, v\rangle - \langle Aw, z\rangle\right)^2 = \mathbb{E}\Big(\sum_{i,j} A_{ij}(u_j v_i - w_j z_i)\Big)^2 \\
&= \sum_{i,j} (u_j v_i - w_j z_i)^2 \quad \text{(by independence, mean 0, variance 1)} \\
&= \|uv^\mathsf{T} - wz^\mathsf{T}\|_F^2 \\
&\le \|u - w\|_2^2 + \|v - z\|_2^2 \quad \text{(see Exercise 7.3.2 below).}
\end{aligned}$$

Let us define a simpler Gaussian process (Y_{uv}) with similar increments, as follows:

$$Y_{uv} := \langle g, u \rangle + \langle h, v \rangle, \quad (u, v) \in T,$$

where

$$g \sim N(0, I_n), \quad h \sim N(0, I_m)$$

are independent Gaussian vectors. The increments of this process are

$$\begin{aligned}
\mathbb{E}(Y_{uv} - Y_{wz})^2 &= \mathbb{E}\left(\langle g, u - w\rangle + \langle h, v - z\rangle\right)^2 \\
&= \mathbb{E}\,\langle g, u - w\rangle^2 + \mathbb{E}\,\langle h, v - z\rangle^2 \quad \text{(by independence, mean 0)} \\
&= \|u - w\|_2^2 + \|v - z\|_2^2 \quad \text{(since } g, h \text{ are standard normal).}
\end{aligned}$$

Comparing the increments of the two processes, we see that

$$\mathbb{E}(X_{uv} - X_{wz})^2 \le \mathbb{E}(Y_{uv} - Y_{wz})^2 \quad \text{for all } (u, v), (w, z) \in T,$$

as required in the Sudakov–Fernique inequality. Applying Theorem 7.2.11, we obtain

$$\begin{aligned}
\mathbb{E}\,\|A\| &= \mathbb{E} \sup_{(u,v)\in T} X_{uv} \le \mathbb{E} \sup_{(u,v)\in T} Y_{uv} \\
&= \mathbb{E} \sup_{u \in S^{n-1}} \langle g, u\rangle + \mathbb{E} \sup_{v \in S^{m-1}} \langle h, v \rangle \\
&= \mathbb{E}\,\|g\|_2 + \mathbb{E}\,\|h\|_2 \\
&\le (\mathbb{E}\,\|g\|_2^2)^{1/2} + (\mathbb{E}\,\|h\|_2^2)^{1/2} \quad \text{(by inequality (1.3) for } L^p \text{ norms)} \\
&= \sqrt{n} + \sqrt{m} \quad \text{(recall Lemma 3.2.4).}
\end{aligned}$$

This completes the proof. ■

Exercise 7.3.2☕☕☕ Prove the following bound, used in the proof of Theorem 7.3.1. For any vectors $u, w \in S^{n-1}$ and $v, z \in S^{m-1}$, we have

$$\|uv^\mathsf{T} - wz^\mathsf{T}\|_F^2 \le \|u - w\|_2^2 + \|v - z\|_2^2.$$

While Theorem 7.3.1 does not give any tail bound for $\|A\|$, we can automatically deduce a tail bound using the concentration inequalities that we studied in Section 5.2.

Corollary 7.3.3 (Norms of Gaussian random matrices: tails) *Let A be an $m \times n$ matrix with independent $N(0,1)$ entries. Then, for every $t \geq 0$, we have*

$$\mathbb{P}\left\{\|A\| \geq \sqrt{m} + \sqrt{n} + t\right\} \leq 2\exp(-ct^2).$$

Proof This result follows by combining Theorem 7.3.1 with the concentration inequality in Gaussian space, Theorem 5.2.2.

To use concentration, let us view A as a long random vector in $\mathbb{R}^{m\times n}$ by concatenating the rows. This makes A a standard normal random vector, i.e. $A \sim N(0, I_{nm})$. Consider the function $f(A) := \|A\|$ that assigns to the vector A the operator norm of the matrix A. We have

$$f(A) \leq \|A\|_2,$$

where $\|A\|_2$ is the Euclidean norm in $\mathbb{R}^{m\times n}$. (This is the same as the Frobenius norm of A, which dominates the operator norm of A.) This shows that $A \mapsto \|A\|$ is a Lipschitz function on $\mathbb{R}^{m\times n}$ and that its Lipschitz norm is bounded by 1. (Why?) Then Theorem 5.2.2 yields

$$\mathbb{P}\left\{\|A\| \geq \mathbb{E}\,\|A\| + t\right\} \leq 2\exp(-ct^2).$$

The bound on $\mathbb{E}\,\|A\|$ from Theorem 7.3.1 completes the proof. ∎

Exercise 7.3.4 (Smallest singular values) ☕☕☕ Use Gordon's inequality, stated in Exercise 7.2.14, to obtain a sharp bound on the smallest singular value of an $m \times n$ random matrix A with independent $N(0,1)$ entries:

$$\mathbb{E}\, s_n(A) \geq \sqrt{m} - \sqrt{n}.$$

Combine this result with concentration to show the tail bound

$$\mathbb{P}\left\{\|A\| \leq \sqrt{m} - \sqrt{n} - t\right\} \leq 2\exp(-ct^2).$$

☞

Exercise 7.3.5 (Symmetric random matrices) ☕☕☕ Modify the arguments above to bound the norm of a *symmetric* $n \times n$ Gaussian random matrix A whose entries above the diagonal are independent $N(0,1)$ random variables and whose diagonal entries are independent $N(0,2)$ random variables. This distribution of random matrices is called a *Gaussian orthogonal ensemble* (GOE). Show that

$$\mathbb{E}\,\|A\| \leq 2\sqrt{n}.$$

Next, deduce the tail bound

$$\mathbb{P}\left\{\|A\| \geq 2\sqrt{n} + t\right\} \leq 2\exp(-ct^2).$$

7.4 Sudakov's Minoration Inequality

Let us return to studying general mean-zero Gaussian processes $(X_t)_{t \in T}$. As we observed in Remark 7.1.7, the increments

$$d(t,s) := \|X_t - X_s\|_{L^2} = \left(\mathbb{E}(X_t - X_s)^2\right)^{1/2} \tag{7.13}$$

define a metric on the (otherwise abstract) index set T, which we called the *canonical metric*.

The canonical metric $d(t,s)$ determines the covariance function $\Sigma(t,s)$, which in turn determines the distribution of the process $(X_t)_{t \in T}$ (recall Exercise 7.1.8 and Remark 7.1.11). So, in principle, we should be able to answer any question about the distribution of a Gaussian process $(X_t)_{t \in T}$ by looking at the geometry of the metric space (T, d). Put plainly, we should be able to study probability via geometry.

Let us then ask an important specific question. How can we evaluate the overall magnitude of the process, namely

$$\mathbb{E} \sup_{t \in T} X_t, \tag{7.14}$$

to the geometry of (T, d)? This turns out to be a difficult problem, which we start to study here and continue in Chapter 8.

In this section we prove a useful lower bound on (7.14) in terms of the *metric entropy* of the metric space (T, d). Recall from Section 4.2 that, for $\varepsilon > 0$, the *covering number*

$$\mathcal{N}(T, d, \varepsilon)$$

is defined to be the smallest cardinality of an ε-net of T in the metric d. Equivalently, $\mathcal{N}(T, d, \varepsilon)$ is the smallest number[6] of closed balls of radius ε whose union covers T. Recall also that the logarithm of the covering number,

$$\log_2 \mathcal{N}(T, d, \varepsilon),$$

is called the *metric entropy* of T.

Theorem 7.4.1 (Sudakov's minoration inequality) *Let $(X_t)_{t \in T}$ be a mean-zero Gaussian process. Then, for any $\varepsilon \geq 0$, we have*

$$\mathbb{E} \sup_{t \in T} X_t \geq c\varepsilon \sqrt{\log \mathcal{N}(T, d, \varepsilon)},$$

where d is the canonical metric defined in (7.13).

Proof Let us deduce this result from the Sudakov–Fernique comparison inequality (Theorem 7.2.11). Assume that

$$\mathcal{N}(T, d, \varepsilon) =: N$$

is finite; the infinite case will be considered in Exercise 7.4.2. Let $\mathcal{N}$ be a maximal ε-separated subset of T. Then $\mathcal{N}$ is an ε-net of T (recall Lemma 4.2.6), and thus

$$|\mathcal{N}| \geq N.$$

[6] If T does not admit a finite ε-net, we set $N(T, d, \varepsilon) = \infty$.

Restricting the process to $\mathcal{N}$, we see that it suffices to show that

$$\mathbb{E} \sup_{t \in \mathcal{N}} X_t \geq c\varepsilon \sqrt{\log N}.$$

We can do this by comparing $(X_t)_{t \in \mathcal{N}}$ with a simpler Gaussian process $(Y_t)_{t \in \mathcal{N}}$, which we define as follows:

$$Y_t := \frac{\varepsilon}{\sqrt{2}} g_t \text{ where } g_t \text{ are independent } N(0, 1) \text{ random variables.}$$

To use the Sudakov–Fernique comparison inequality (Theorem 7.2.11), we need to compare the increments of the two processes. Fix two different points $t, s \in \mathcal{N}$. By definition, we have

$$\mathbb{E}(X_t - X_s)^2 = d(t, s)^2 \geq \varepsilon^2$$

while

$$\mathbb{E}(Y_t - Y_s)^2 = \frac{\varepsilon^2}{2} \mathbb{E}(g_t - g_s)^2 = \varepsilon^2.$$

(In the last line, we used the fact that $g_t - g_s \sim N(0, 2)$.) This implies that

$$\mathbb{E}(X_t - X_s)^2 \geq \mathbb{E}(Y_t - Y_s)^2 \quad \text{for all } t, s \in \mathcal{N}.$$

Applying Theorem 7.2.11, we obtain

$$\mathbb{E} \sup_{t \in \mathcal{N}} X_t \geq \mathbb{E} \sup_{t \in \mathcal{N}} Y_t = \frac{\varepsilon}{\sqrt{2}} \mathbb{E} \max_{t \in \mathcal{N}} g_t \geq c\varepsilon \sqrt{\log N}.$$

In the last inequality we used that the expected maximum of N standard normal random variables is at least $c\sqrt{\log N}$; see Exercise 2.5.11. The proof is complete. ■

Exercise 7.4.2 (Sudakov's minoration for non-compact sets) ☕☕ Show that if (T, d) is not compact, that is, if $N(T, d, \varepsilon) = \infty$ for some $\varepsilon > 0$, then

$$\mathbb{E} \sup_{t \in T} X_t = \infty.$$

7.4.1 Application for Covering Numbers in $\mathbb{R}^n$

Sudakov's minoration inequality can be used to estimate the covering numbers of sets $T \subset \mathbb{R}^n$. To see how to do this, consider a canonical Gaussian process on T, namely

$$X_t := \langle g, t \rangle, \quad t \in T \text{ where } g \sim N(0, I_n).$$

As we observed in Section 7.1.2, the canonical distance for this process is the Euclidean distance in $\mathbb{R}^n$, i.e.,

$$d(t, s) = \|X_t - X_s\|_2 = \|t - s\|_2.$$

Thus Sudakov's inequality can be stated as follows.

Corollary 7.4.3 (Sudakov's minoration inequality in $\mathbb{R}^n$) *Let $T \subset \mathbb{R}^n$. Then, for any $\varepsilon > 0$, we have*

$$\mathbb{E} \sup_{t \in T} \langle g, t \rangle \geq c\varepsilon \sqrt{\log \mathcal{N}(T, \varepsilon)}.$$

Here $\mathcal{N}(T, \varepsilon)$ is the covering number of T for covering by Euclidean balls; it is the smallest number of Euclidean balls with radii ε and centers in T that cover T, just as in Section 4.2.1.

To give an illustration of Sudakov's minoration, note that it yields (up to an absolute constant) the same bound on the covering numbers of polytopes in $\mathbb{R}^n$ as that given in Corollary 0.0.4:

Corollary 7.4.4 (Covering numbers of polytopes) *Let P be a polytope in $\mathbb{R}^n$ with N vertices and whose diameter is bounded by 1. Then, for every $\varepsilon > 0$, we have*

$$\mathcal{N}(P, \varepsilon) \leq N^{C/\varepsilon^2}.$$

Proof As before, by translation, we may assume that the radius of P is bounded by 1. Denote by $x_1, \dots, x_N$ the vertices of P. Then

$$\mathbb{E} \sup_{t \in P} \langle g, t \rangle = \mathbb{E} \sup_{i \leq N} \langle g, x_i \rangle \leq C\sqrt{\log N}.$$

The above equality follows since the maximum of a linear function on the convex set P is attained at an extreme point, i.e., at a vertex of P. The bound is due to Exercise 2.5.10, since $\langle g, x \rangle \sim N(0, \|x\|_2^2)$ and $\|x\|_2 \leq 1$. Substituting this into Sudakov's minoration inequality of Corollary 7.4.3 and simplifying, we complete the proof. ■

Exercise 7.4.5 (Volume of polytopes) ☕☕☕ Let P be a polytope in $\mathbb{R}^n$ which has N vertices and is contained in the unit Euclidean ball B_2^n. Show that

$$\frac{\mathrm{Vol}(P)}{\mathrm{Vol}(B_2^n)} \leq \left(\frac{C \log N}{n}\right)^{Cn}.$$

☞

7.5 Gaussian Width

In the previous section, we encountered an important quantity associated with a general set $T \subset \mathbb{R}^n$. This was the magnitude of the canonical Gaussian process on T, i.e.,

$$\mathbb{E} \sup_{t \in T} \langle g, t \rangle$$

where the expectation is taken with respect to the Gaussian random vector $g \sim N(0, I_n)$. This quantity plays a central role in high-dimensional probability and its applications. Let us give it a name and study its basic properties.

Definition 7.5.1 The *Gaussian width* of a subset $T \subset \mathbb{R}^n$ is defined as

$$w(T) := \mathbb{E} \sup_{x \in T} \langle g, x \rangle \quad \text{where} \quad g \sim N(0, I_n).$$

One can think about the Gaussian width $w(T)$ as one of the basic geometric quantities associated with subsets of $T \subset \mathbb{R}^n$, such as volume and surface area. Several variants of the definition of the Gaussian width such as

$$\mathbb{E} \sup_{x \in T} |\langle g, x \rangle|, \quad \left(\mathbb{E} \sup_{x \in T} \langle g, x \rangle^2\right)^{1/2}, \quad \mathbb{E} \sup_{x, y \in T} \langle g, x - y \rangle, \quad \text{etc.,}$$

can be found in the literature. These versions are equivalent, or almost equivalent, to $w(T)$, as we will see in Section 7.6.

7.5.1 Basic Properties

Proposition 7.5.2 (Gaussian width)

(i) *The Gaussian width $w(T)$ is finite if and only if T is bounded.*
(ii) *The Gaussian width is invariant under affine unitary transformations. Thus, for every orthogonal matrix U and any vector y, we have*

$$w(UT + y) = w(T).$$

(iii) *The Gaussian width is invariant under the taking of convex hulls. Thus,*

$$w(\operatorname{conv}(T)) = w(T).$$

(iv) *The Gaussian width respects the Minkowski addition of sets and scaling. Thus, for $T, S \subset \mathbb{R}^n$ and $a \in \mathbb{R}$, we have*

$$w(T + S) = w(T) + w(S), \quad w(aT) = |a|\, w(T).$$

(v) *We have*

$$w(T) = \tfrac{1}{2} w(T - T) = \tfrac{1}{2} \mathbb{E} \sup_{x, y \in T} \langle g, x - y \rangle.$$

(vi) *(Gaussian width and diameter). We have*[7]

$$\frac{1}{\sqrt{2\pi}} \operatorname{diam}(T) \le w(T) \le \frac{\sqrt{n}}{2} \operatorname{diam}(T).$$

Proof Properties (i)–(iv) are simple and will be checked in Exercise 7.5.3 below.

To prove property (v), we use property (iv) twice and get

$$w(T) = \tfrac{1}{2}\left(w(T) + w(T)\right) = \tfrac{1}{2}\left(w(T) + w(-T)\right) = \tfrac{1}{2} w(T - T),$$

as claimed.

To prove the lower bound in property (vi), fix a pair of points $x, y \in T$. Then both $x - y$ and $y - x$ lie in $T - T$, so by property (v) we have

$$\begin{aligned} w(T) &\ge \tfrac{1}{2} \mathbb{E} \max\left(\langle x - y, g \rangle, \langle y - x, g \rangle\right) \\ &= \tfrac{1}{2} \mathbb{E} |\langle x - y, g \rangle| = \tfrac{1}{2}\sqrt{2/\pi}\, \|x - y\|_2. \end{aligned}$$

[7] Recall that the diameter of a set $T \subset \mathbb{R}^n$ is defined as $\operatorname{diam}(T) := \sup\{\|x - y\|_2 : x, y \in T\}$.

The last identity follows since $\langle x-y, g\rangle \sim N(0, \|x-y\|_2)$ and since $\mathbb{E}|X| = \sqrt{2/\pi}$ for $X \sim N(0,1)$. (Check!) It remains to take the supremum over all $x, y \in T$, and the lower bound in property (vi) follows.

To prove the upper bound in property (vi), we again use property (v) to get

$$\begin{aligned} w(T) &= \tfrac{1}{2}\mathbb{E} \sup_{x,y\in T} \langle g, x-y\rangle \\ &\le \tfrac{1}{2}\mathbb{E} \sup_{x,y\in T} \|g\|_2\|x-y\|_2 \le \tfrac{1}{2}\mathbb{E}\,\|g\|_2 \operatorname{diam}(T). \end{aligned}$$

It remains to recall that $\mathbb{E}\,\|g\|_2 \le (\mathbb{E}\,\|g\|_2^2)^{1/2} = \sqrt{n}$. ■

Exercise 7.5.3 ☕☕ Prove Properties (i)–(iv) in Proposition 7.5.2. ☞

Exercise 7.5.4 (Gaussian width under linear transformations) ☕☕☕ Show that, for any $m \times n$ matrix A, we have

$$w(AT) \le \|A\|\, w(T).$$ ☞

7.5.2 Geometric Meaning of Width

The notion of the Gaussian width of a set $T \subset \mathbb{R}^n$ has a nice geometric meaning. The width of T in the direction of a vector $\theta \in S^{n-1}$ is the smallest width of the slab that is formed by parallel hyperplanes orthogonal to θ and that contains T; see Figure 7.3. Analytically, the width in the direction of θ can be expressed as

$$\sup_{x,y\in T} \langle \theta, x-y\rangle .$$

(Check!) If we average the width over all unit directions θ, we obtain the quantity

$$\mathbb{E} \sup_{x,y\in T} \langle \theta, x-y\rangle . \tag{7.15}$$

Definition 7.5.5 (Spherical width) The *spherical width*[8] of a subset $T \subset R^n$ is defined as

$$w_s(T) := \mathbb{E} \sup_{x\in T} \langle \theta, x\rangle \quad \text{where} \quad \theta \sim \text{Unif}(S^{n-1}).$$

The quantity in (7.15) clearly equals $w_s(T-T)$.

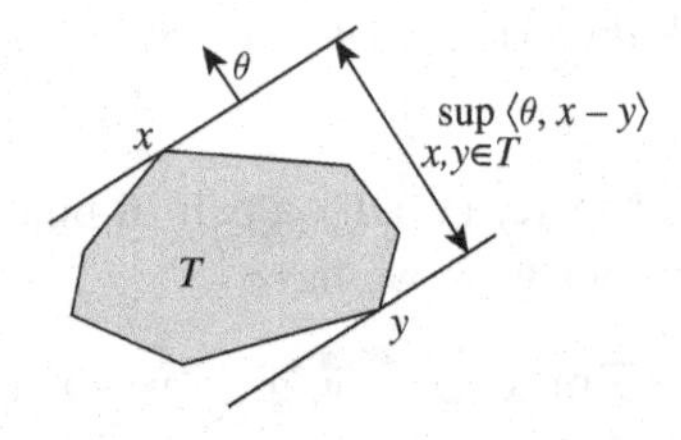

Figure 7.3 The width of a set $T \subset \mathbb{R}^n$ in the direction of a unit vector θ.

[8] The spherical width is also called the *mean width* in the literature.

How different are the Gaussian and spherical widths of T? The difference is in the random vectors we use to do the averaging; they are $g \sim N(0, I_n)$ for the Gaussian width and $\theta \sim \text{Unif}(S^{n-1})$ for the spherical width. Both g and θ are rotation invariant, and, as we know, g is approximately $\sqrt{n}$ longer than θ. This makes the Gaussian width just a scaling of the spherical width by approximately $\sqrt{n}$. Let us make this relation more precise.

Lemma 7.5.6 (Gaussian vs. spherical width) *We have*

$$(\sqrt{n} - C)\, w_s(T) \le w(T) \le (\sqrt{n} + C)\, w_s(T).$$

Proof Let us express the Gaussian vector g through its length and direction:

$$g = \|g\|_2 \frac{g}{\|g\|_2} =: r\theta.$$

As we observed in Section 3.3.3, r and θ are independent and $\theta \sim \text{Unif}(S^{n-1})$. Thus

$$w(T) = \mathbb{E} \sup_{x \in T} \langle r\theta, x \rangle = (\mathbb{E}\, r)\, \mathbb{E} \sup_{x \in T} \langle \theta, x \rangle = \mathbb{E}\, \|g\|_2\, w_s(T).$$

It remains to recall that the concentration of the norm implies that

$$\left| \mathbb{E}\, \|g\|_2 - \sqrt{n} \right| \le C;$$

see Exercise 3.1.4. ■

7.5.3 Examples

Example 7.5.7 (Euclidean ball and sphere) The Gaussian width of the Euclidean unit sphere and ball is

$$w(S^{n-1}) = w(B_2^n) = \mathbb{E}\, \|g\|_2 = \sqrt{n} \pm C, \tag{7.16}$$

where we have used the result of Exercise 3.1.4. The spherical widths of these sets equal 2, of course.

Example 7.5.8 (Cube) The unit ball of the ℓ_∞ norm in $\mathbb{R}^n$ is $B_\infty^n = [-1, 1]^n$. We have

$$\begin{aligned} w(B_\infty^n) &= \mathbb{E}\, \|g\|_1 \quad \text{(check!)} \\ &= \mathbb{E}\, |g_1|\, n = \sqrt{\frac{2}{\pi}}\, n. \end{aligned} \tag{7.17}$$

Comparing with (7.16), we see that the Gaussian widths of the cube B_∞^n and of its circumscribed ball $\sqrt{n} B_2^n$ have the same order n; see Figure 7.4a.

Example 7.5.9 (ℓ_1 ball) The unit ball of the ℓ_1 norm in $\mathbb{R}^n$ is the set

$$B_1^n = \{x \in \mathbb{R}^n : \ \|x\|_1 \le 1\},$$

which is sometimes called a *cross-polytope*; see Figure 7.5 for an illustration. The Gaussian width of the ℓ_1 ball can be bounded as follows:

$$c\sqrt{\log n} \le w(B_1^n) \le C\sqrt{\log n}. \tag{7.18}$$

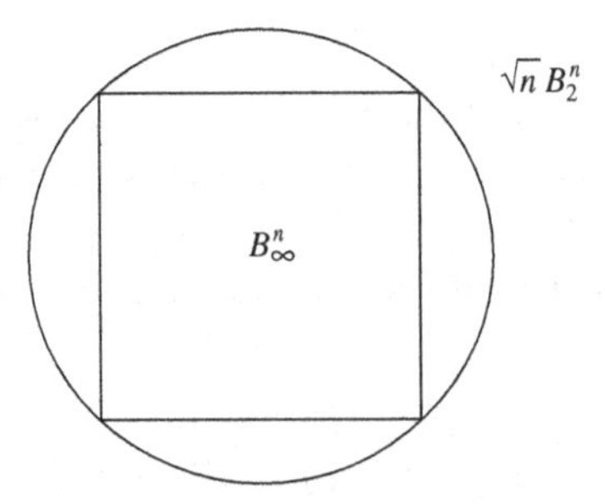

(a) The Gaussian widths of the cube and its circumscribed ball are of the same order n.

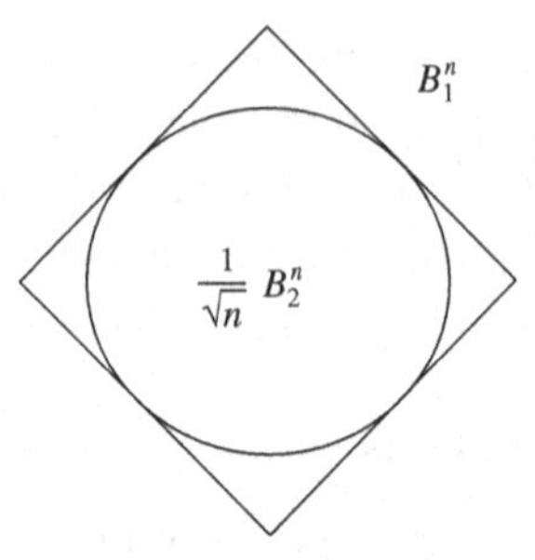

(b) The Gaussian widths of B_1^n and its inscribed ball are almost of the same order.

Figure 7.4 Gaussian widths of some classical sets in $\mathbb{R}^n$.

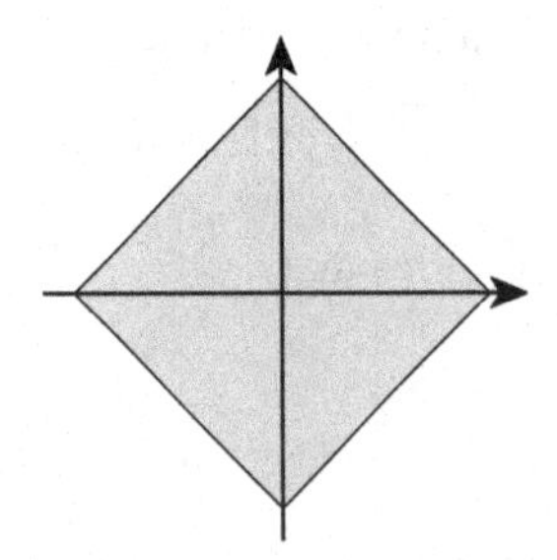

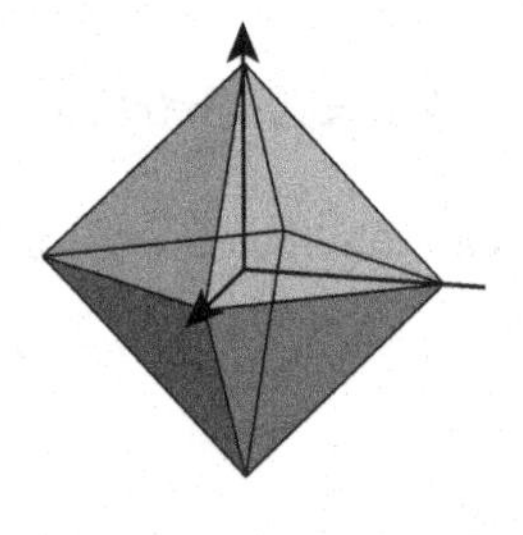

Figure 7.5 The unit ball of the ℓ_1 norm in $\mathbb{R}^n$, denoted B_1^n, is a diamond in dimension $n = 2$ (left) and a regular octahedron in dimension $n = 3$ (right).

To see this, check that

$$w(B_1^n) = \mathbb{E}\,\|g\|_\infty = \mathbb{E}\max_{i\le n}|g_i|.$$

Then the bounds (7.18) follow from Exercises 2.5.10 and 2.5.11. Note that the Gaussian widths of the ℓ_1 ball B_1^n and its inscribed ball $(1/\sqrt{n})B_2^n$ have almost the same order (up to a logarithmic factor); see Figure 7.4b.

Exercise 7.5.10 (Finite point sets) Let T be a finite set of points in $\mathbb{R}^n$. Check that

$$w(T) \le C\sqrt{\log|T|}\;\mathrm{diam}(T).$$

☞

Exercise 7.5.11 (ℓ_p balls) Let $1 \le p < \infty$. Consider the unit ball of the ℓ_p norm in $\mathbb{R}^n$:

$$B_p^n := \left\{x \in \mathbb{R}^n :\ \|x\|_p \le 1\right\}.$$

Check that

$$w(B_p^n) \le C\sqrt{p'}\,n^{1/p'}.$$

Here p' denotes the *conjugate exponent* for p, which is defined by the equation $1/p + 1/p' = 1$.

7.5.4 Surprising Behavior of the Width in High Dimensions

According to our computation in Example 7.5.9, the *spherical* width of B_1^n is

$$w_s(B_1^n) \asymp \sqrt{\frac{\log n}{n}}.$$

Surprisingly, it is much smaller than the diameter of B_1^n, which equals 2! Furthermore, as we have already noted, the Gaussian width of B_1^n is roughly the same (up to a logarithmic factor) as the Gaussian width of its inscribed Euclidean ball $(1/\sqrt{n})\, B_2^n$. This again might look strange. Indeed, the cross-polytope B_1^n looks much larger than its inscribed ball whose diameter is $2/\sqrt{n}$! Why does the Gaussian width behave in this way?

Let us try to give an intuitive explanation. In high dimensions, the cube has so many vertices (2^n) that most of the volume is concentrated near them. In fact, the volumes of the cube and its circumscribed ball are both of order C^n, so these sets are not far from each other from the volumetric point of view. Thus it should not be very surprising to see that the Gaussian widths of the cube and its circumscribed ball are also of the same order.

The octahedron B_1^n has many fewer vertices ($2n$) than the cube. A random direction θ in $\mathbb{R}^n$ is likely to be almost orthogonal to all of them. So, the width of B_1^n in the direction of θ is not significantly influenced by the presence of the vertices. What really determines the width of B_1^n is its "bulk", which is the inscribed Euclidean ball.

A similar picture can be seen from the volumetric viewpoint. There are so few vertices in B_1^n that the regions near them contain very little volume. The bulk of the volume of B_1^n lies much closer to the origin, not far from the inscribed Euclidean ball. Indeed, one can check that the volumes of B_1^n and its inscribed ball are both of order $(C/n)^n$. So, from the volumetric point of view, the octahedron B_1^n is similar to its inscribed ball; the Gaussian width gives the same result.

We illustrate this phenomenon in Figure 7.6b, which shows a "hyperbolic" picture of the B_1^n that is due to V. Milman. Such pictures capture the bulk and outliers very well, but unfortunately they may not accurately show the convexity.

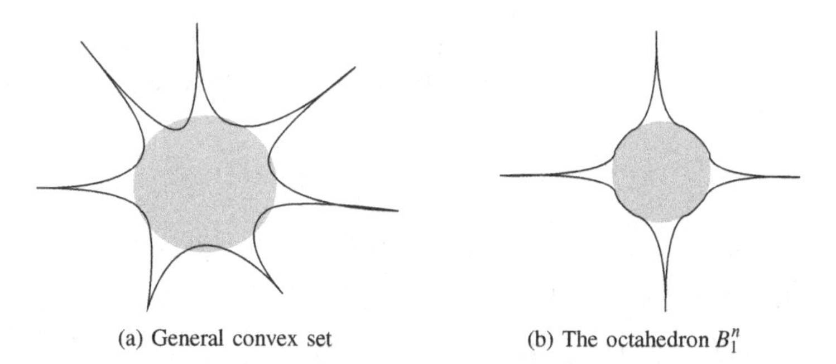

(a) General convex set

(b) The octahedron B_1^n

Figure 7.6 Intuitive hyperbolic pictures of convex sets in $\mathbb{R}^n$. The bulk is a round ball that contains most of the volume.

7.6 Stable Dimension, Stable Rank, and Gaussian Complexity

The notion of the Gaussian width helps us to introduce a more robust version of the classical notion of dimension. The usual linear-algebraic dimension $\dim T$ of a subset $T \subset \mathbb{R}^n$ is the smallest dimension of an affine subspace $E \subset \mathbb{R}^n$ that contains T. The linear-algebraic dimension is unstable: it can significantly change (usually upwards) under a small perturbation of T. A more stable version of the dimension can be defined on the basis of the Gaussian width.

In this section it will be more convenient to work with a closely related *squared* version of the Gaussian width:

$$h(T)^2 := \mathbb{E} \sup_{t \in T} \langle g, t \rangle^2 \quad \text{where} \quad g \sim N(0, I_n). \tag{7.19}$$

It is not difficult to see that the squared and usual versions of the Gaussian width are equivalent up to a constant factor.

Exercise 7.6.1 (Equivalence) Check that

$$w(T-T) \le h(T-T) \le w(T-T) + C_1 \operatorname{diam}(T) \le Cw(T-T).$$

In particular, we have

$$2w(T) \le h(T-T) \le 2Cw(T). \tag{7.20}$$

☞

Definition 7.6.2 (Stable dimension) For a bounded set $T \subset \mathbb{R}^n$, the *stable dimension* of T is defined as

$$d(T) := \frac{h(T-T)^2}{\operatorname{diam}(T)^2} \asymp \frac{w(T)^2}{\operatorname{diam}(T)^2}.$$

The stable dimension is always bounded by the algebraic dimension:

Lemma 7.6.3 *For any set $T \subset \mathbb{R}^n$, we have*

$$d(T) \le \dim(T).$$

Proof Let $\dim T = k$; this means that T lies in some subspace $E \subset \mathbb{R}^n$ of dimension k. By rotation invariance we can assume that E is the coordinate subspace, i.e., $E = \mathbb{R}^k$. (Why?) By definition, we have

$$h(T-T)^2 = \mathbb{E} \sup_{x,y \in T} \langle g, x-y \rangle^2.$$

Since $x - y \in \mathbb{R}^k$ and $\|x-y\|_2 \le \operatorname{diam}(T)$, we have $x - y = \operatorname{diam}(T)\, z$ for some $z \in B_2^k$. Thus the quantity above is bounded by

$$\operatorname{diam}(T)^2 \; \mathbb{E} \sup_{z \in B_2^k} \langle g, z \rangle^2 = \operatorname{diam}(T)^2 \; \mathbb{E} \|g'\|_2^2 = \operatorname{diam}(T)^2 \, k,$$

where $g' \sim N(0, I_k)$ is a standard Gaussian random vector in $\mathbb{R}^k$. The proof is complete. ■

The inequality $d(T) \le \dim(T)$ is in general sharp:

Exercise 7.6.4 Show that if T is a Euclidean ball in any subspace of $\mathbb{R}^n$ then

$$d(T) = \dim(T).$$

However, in many cases the stable dimension can be much smaller than the algebraic dimension:

Example 7.6.5 Let T be a finite set of points in $\mathbb{R}^n$. Then

$$d(T) \le C \log |T|.$$

This follows from the bound on the Gaussian width of T in Exercise 7.5.10.

7.6.1 Stable Rank

The stable dimension is more robust than the algebraic dimension. Indeed, a small perturbation of a set T leads to a small perturbation of the Gaussian width and the diameter of T and thus of the stable dimension $d(T)$.

To give an example, consider the unit Euclidean ball B_2^n, for which both the algebraic and stable dimensions equal n. Let us decrease one axis of B_2^n gradually from 1 to 0. The algebraic dimension will stay at n through this process and then instantly jump to $n-1$. The stable dimension instead decreases gradually from n to $n-1$. To see how exactly it decreases, do the following computation.

Exercise 7.6.6 (Ellipsoids) Let A be an $m \times n$ matrix, and let B_2^n denote the unit Euclidean ball. Check that the squared mean width of the ellipsoid AB_2^n is the Frobenius norm of A, i.e.,

$$h(AB_2^n) = \|A\|_F.$$

Deduce that the stable dimension of the ellipsoid AB_2^n equals

$$d(AB_2^n) = \frac{\|A\|_F^2}{\|A\|^2}. \tag{7.21}$$

This example relates the stable dimension to the notion of the *stable rank* of a matrix, which is a robust version of the classical, linear, algebraic rank.

Definition 7.6.7 (Stable rank) The *stable rank* of an $m \times n$ matrix A is defined as

$$r(A) := \frac{\|A\|_F^2}{\|A\|^2}.$$

The robustness of the stable rank makes it a useful quantity in numerical linear algebra. The usual, algebraic, rank is the algebraic dimension of the image of A; in particular

$$\operatorname{rank}(A) = \dim(AB_2^n).$$

Similarly, (7.21) shows that the stable rank is the *statistical* dimension of the image:

$$r(A) = d(AB_2^n).$$

Finally, note that the stable rank is always bounded by the usual rank:

$$r(A) \le \operatorname{rank}(A).$$

(Check this!)

7.6.2 Gaussian Complexity

Let us mention one more cousin of the Gaussian width where, instead of squaring $\langle g, x\rangle$ as in (7.19), we take the absolute value.

Definition 7.6.8 The *Gaussian complexity* of a subset $T \subset \mathbb{R}^n$ is defined as

$$\gamma(T) := \mathbb{E} \sup_{x\in T} |\langle g, x\rangle| \quad \text{where} \quad g \sim N(0, I_n).$$

Obviously, we have

$$w(T) \le \gamma(T),$$

and equality holds if T is origin-symmetric, i.e., if $T = -T$. Since $T - T$ is origin-symmetric, property (v) of Proposition 7.5.2 implies that

$$w(T) = \tfrac{1}{2} w(T - T) = \tfrac{1}{2}\gamma(T - T). \tag{7.22}$$

In general, the Gaussian width and the Gaussian complexity may be quite different. For example, if T consists of a single point, $w(T) = 0$ but $\gamma(T) > 0$. Nevertheless, these two quantities are very closely related:

Exercise 7.6.9 (Gaussian width vs. Gaussian complexity)☕☕☕ Consider a set $T \subset \mathbb{R}^n$ and a point $y \in T$. Show that

$$\tfrac{1}{3}\left[w(T) + \|y\|_2\right] \le \gamma(T) \le 2\big(w(T) + \|y\|_2\big)$$

This implies in particular that the Gaussian width and the Gaussian complexity are equivalent for any set T that contains the origin:

$$w(T) \le \gamma(T) \le 2w(T).$$

(It is fine if you prove the inequalities above with other absolute constants instead of 2 and $\frac{1}{3}$.)

7.7 Random Projections of Sets

This section will illustrate the importance of the notion of the Gaussian (and spherical) width in dimension reduction problems. Consider a set $T \subset \mathbb{R}^n$ and project it onto a random m-dimensional subspace in $\mathbb{R}^n$ (chosen uniformly from the Grassmannian $G_{n,m}$); see Figure 5.2 for an illustration. In applications, we might think of T as a data set and P as a means of dimension reduction. What can we say about the size (diameter) of the projected set PT?

For a *finite* set T, the Johnson–Lindenstrauss lemma (Theorem 5.3.1) states that, as long as

$$m \gtrsim \log |T|, \tag{7.23}$$

the random projection P acts essentially as a scaling of T. Namely, P shrinks all distances between points in T by a factor $\approx \sqrt{m/n}$. In particular,

$$\operatorname{diam}(PT) \approx \sqrt{\frac{m}{n}} \operatorname{diam}(T). \tag{7.24}$$

If the cardinality of T is too large or infinite then (7.24) may fail. For example, if $T = B_2^n$ is a Euclidean ball then no projection can shrink the size of T, and we have

$$\operatorname{diam}(PT) = \operatorname{diam}(T). \tag{7.25}$$

What happens for a general set T? The following result states that a random projection shrinks T as in (7.24), but it cannot shrink it beyond the spherical width of T.

Theorem 7.7.1 (Sizes of random projections of sets) *Consider a bounded set $T \subset \mathbb{R}^n$. Let P be a projection in $\mathbb{R}^n$ onto a random m-dimensional subspace $E \sim \operatorname{Unif}(G_{n,m})$. Then, with probability at least $1 - 2e^{-m}$, we have*

$$\operatorname{diam}(PT) \le C\Big(w_s(T) + \sqrt{\frac{m}{n}}\, \operatorname{diam}(T)\Big).$$

To prove this result, we pass to an equivalent probabilistic model, just as we did in the proof of the Johnson–Lindenstrauss lemma (see the proof of Proposition 5.3.2). First, a random subspace $E \subset \mathbb{R}^n$ can be realized by a random rotation of some fixed subspace, such as $\mathbb{R}^m$. Next, instead of fixing T and randomly rotating the subspace, we can fix the subspace and randomly rotate T. The following exercise makes this reasoning more formal.

Exercise 7.7.2 (Equivalent models for random projections) Let P be a projection in $\mathbb{R}^n$ onto a random m-dimensional subspace $E \sim \operatorname{Unif}(G_{n,m})$. Let Q be an $m \times n$ matrix obtained by choosing the first m rows of a random $n \times n$ matrix $U \sim \operatorname{Unif}(O(n))$ drawn uniformly from the orthogonal group.

(a) Show that, for any fixed point $x \in \mathbb{R}^n$,

$$\|Px\|_2 \text{ and } \|Qx\|_2 \text{ have the same distribution.}$$

(b) Show that, for any fixed point $z \in S^{m-1}$,

$$Q^{\mathsf{T}} z \sim \operatorname{Unif}(S^{n-1}).$$

In other words, the map Q^{T} acts as a random isometric embedding of $\mathbb{R}^m$ into $\mathbb{R}^n$.

Proof of Theorem 7.7.1 Our argument here is another example of an ε-net argument. Without loss of generality, we may assume that $\operatorname{diam}(T) \le 1$. (Why?)

Step 1: Approximation. By Exercise 7.7.2, it suffices to prove the theorem for Q instead of P. So we are going to bound

$$\operatorname{diam}(QT) = \sup_{x \in T - T} \|Qx\|_2 = \sup_{x \in T - T} \max_{z \in S^{m-1}} \langle Qx, z\rangle .$$

In a similar way to our previous arguments (for example, in the proof of Theorem 4.4.5 on random matrices), we discretize the sphere S^{m-1}. Choose a $1/2$-net $\mathcal{N}$ of S^{m-1} such that

$$|\mathcal{N}| \le 5^m;$$

it is possible to do this by Corollary 4.2.13. We can replace the supremum over the sphere S^{m-1} by a supremum over the net $\mathcal{N}$, paying a factor 2:

$$\operatorname{diam}(QT) \le 2 \sup_{x \in T-T} \max_{z \in \mathcal{N}} \langle Qx, z \rangle = 2 \max_{z \in \mathcal{N}} \sup_{x \in T-T} \langle Q^{\mathsf{T}} z, x \rangle. \tag{7.26}$$

(Recall Exercise 4.4.2.) We first control the quantity

$$\sup_{x \in T-T} \langle Q^{\mathsf{T}} z, x \rangle \tag{7.27}$$

for a fixed $z \in \mathcal{N}$ and with high probability, and then take the union bound over all z.

Step 2: Concentration. So, let us fix $z \in \mathcal{N}$. By Exercise 7.7.2, $Q^{\mathsf{T}} z \sim \operatorname{Unif}(S^{n-1})$. The expectation of (7.27) can be realized as the spherical width:

$$\mathbb{E} \sup_{x \in T-T} \langle Q^{\mathsf{T}} z, x \rangle = w_s(T-T) = 2w_s(T).$$

(The last identity is the spherical version of a similar property of the Gaussian width; see part (v) of Proposition 7.5.2.)

Next, let us check that (7.27) concentrates nicely around its mean $2w_s(T)$. For this, we can use the concentration inequality (5.6) for Lipschitz functions on the sphere. Since we assumed that $\operatorname{diam}(T) \le 1$ at the beginning, we can quickly check that the function

$$\theta \mapsto \sup_{x \in T-T} \langle \theta, x \rangle$$

is a Lipschitz function on the sphere S^{n-1} and that its Lipschitz norm is at most 1. (Do this!) Therefore, applying the concentration inequality (5.6), we obtain

$$\mathbb{P}\left\{ \sup_{x \in T-T} \langle Q^{\mathsf{T}} z, x \rangle \ge 2w_s(T) + t \right\} \le 2\exp(-cnt^2).$$

Step 3: Union bound. Now we unfix $z \in \mathcal{N}$ by taking the union bound over $\mathcal{N}$. We get

$$\mathbb{P}\left\{ \max_{z \in \mathcal{N}} \sup_{x \in T-T} \langle Q^{\mathsf{T}} z, x \rangle \ge 2w_s(T) + t \right\} \le |\mathcal{N}|\, 2\exp(-cnt^2) \tag{7.28}$$

Recall that $|\mathcal{N}| \le 5^m$. Then, if we choose

$$t = C\sqrt{\frac{m}{n}}$$

with C large enough, the probability in (7.28) can be bounded by $2e^{-m}$. Then (7.28) and (7.26) yield

$$\mathbb{P}\left\{ \frac{1}{2} \operatorname{diam}(QT) \ge 2w(T) + C\sqrt{\frac{m}{n}} \right\} \le e^{-m}.$$

This proves Theorem 7.7.1. ■

Exercise 7.7.3 (Gaussian projection) ☕☕☕ Prove a version of Theorem 7.7.1 for an $m \times n$ Gaussian random matrix G with independent $N(0, 1)$ entries. Specifically, show that, for any bounded set $T \subset \mathbb{R}^n$, we have

$$\operatorname{diam}(GT) \le C\left(w(T) + \sqrt{m}\,\operatorname{diam}(T)\right)$$

with probability at least $1 - 2e^{-m}$. Here $w(T)$ is the Gaussian width of T.

Exercise 7.7.4 (The reverse bound) Show that the bound in Theorem 7.7.1 is optimal: prove the reverse bound

$$\mathbb{E}\operatorname{diam}(PT) \ge c\left(w_s(T) + \sqrt{\frac{m}{n}}\operatorname{diam}(T)\right)$$

for all bounded sets $T \subset \mathbb{R}^n$.

Exercise 7.7.5 (Random projections of matrices) Let A be an $n \times k$ matrix.

(a) Let P be a projection in $\mathbb{R}^n$ onto a random m-dimensional subspace chosen uniformly in $G_{n,m}$. Show that, with probability at least $1 - 2e^{-m}$, we have

$$\|PA\| \le C\left(\frac{1}{\sqrt{n}}\|A\|_F + \sqrt{\frac{m}{n}}\|A\|\right).$$

(b) Let G be an $m \times n$ Gaussian random matrix with independent $N(0,1)$ entries. Show that, with probability at least $1 - 2e^{-m}$, we have

$$\|GA\| \le C\left(\|A\|_F + \sqrt{m}\,\|A\|\right).$$

7.7.1 The Phase Transition

Let us pause to take a closer look at the bound given by Theorem 7.7.1. We can write it equivalently as

$$\operatorname{diam}(PT) \le C \max\left(w_s(T),\ \sqrt{\frac{m}{n}}\operatorname{diam}(T)\right).$$

Let us compute the dimension m for which a "phase transition" occurs between the two terms $w_s(T)$ and $\sqrt{m/n}\operatorname{diam}(T)$. Setting them equal to each other and solving for m, we find that the phase transition happens when

$$\begin{aligned} m &= \frac{(\sqrt{n}\,w_s(T))^2}{\operatorname{diam}(T)^2} \\ &\asymp \frac{w(T)^2}{\operatorname{diam}(T)^2} \quad \text{(pass to the Gaussian width using Lemma 7.5.6)} \\ &\asymp d(T) \quad \text{(by Definition 7.6.2 of the stable dimension).} \end{aligned}$$

So, we can express the conclusion of Theorem 7.7.1 as follows:

$$\operatorname{diam}(PT) \le \begin{cases} C\sqrt{\dfrac{m}{n}}\operatorname{diam}(T), & \text{if } m \ge d(T), \\ C w_s(T), & \text{if } m \le d(T). \end{cases}$$

Figure 7.7 shows a graph of $\operatorname{diam}(PT)$ as a function of the dimension m.

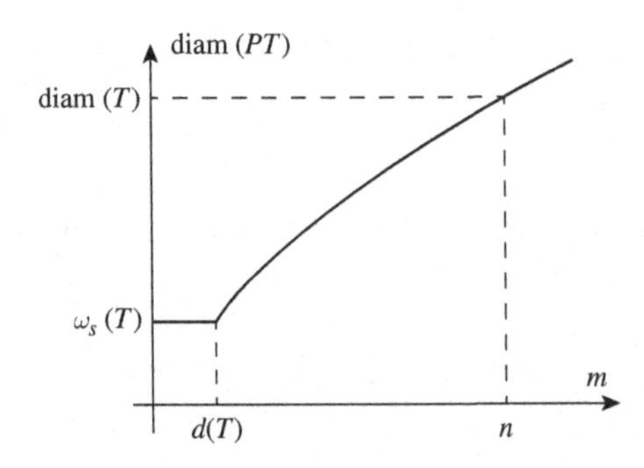

Figure 7.7 The diameter of a random m-dimensional projection of a set T as a function of m.

For large m, the random m-dimensional projection shrinks T by a factor $\sim \sqrt{m/n}$, just as we saw in (7.24) in the context of the Johnson–Lindenstrauss lemma. However, when the dimension m drops below the stable dimension $d(T)$, the shrinking stops – it levels off at the spherical width $w_s(T)$. We saw an example of this in (7.25), where a Euclidean ball cannot be shrunk by a projection.

7.8 Notes

There are several introductory books on random processes (also called stochastic processes) and in particular on Brownian motion, for example [37, 123, 178, 152].

Slepian's inequality (Theorem 7.2.1) was originally due to D. Slepian [181, 182]; modern proofs can be found in e.g. [126, Corollary 3.12], [3, Section 2.2], [206, Section 6.1], [103], and [108]. The Sudakov–Fernique inequality (Theorem 7.2.11) is attributed to V. N. Sudakov [187, 188] and X. Fernique [73]. Our presentation of the proofs of Slepian's and the Sudakov–Fernique inequalities in Section 7.2 is based on the approach of J.-P. Kahane [108] and a smoothing argument of S. Chatterjee (see [3, Section 2.2]), and it follows [206, Section 6.1]. A more general version of the Gaussian contraction inequality in Exercise 7.2.13 can be found in [126, Corollary 3.17].

Gordon's inequality, stated in Exercise 7.2.14, and its extensions can be found in [82, 83, 86, 108]. Applications of Gordon's inequality for convex optimization can be found in e.g. [197, 194, 196].

The relevance of comparison inequalities to random matrix theory was noticed by S. Szarek. The applications we presented in Section 7.3 can be derived from the work of Gordon [82]. Our presentation there follows the argument in [58, Section II.c], which is also reproduced in [216, Section 5.3.1].

Sudakov's minoration inequality (Theorem 7.4.1) was originally proved by V. N. Sudakov. Our presentation follows [126, Theorem 3.18]; see [11, Section 4.2] for an alternative proof via duality. The volume bound in Exercise 7.4.5 is almost best possible, but not quite. A slightly stronger bound

$$\frac{\mathrm{Vol}(P)}{\mathrm{Vol}(B_2^n)} \le \Big(\frac{C\log(1+N/n)}{n}\Big)^{n/2}$$

can be deduced in exactly the same way if we use the stronger bound on the covering numbers given in Exercise 0.0.6. This result is known and is best possible up to a constant C [49, Section 3].

The Gaussian width and its cousins, which we discussed in Section 7.5, was originally introduced in geometric functional analysis and asymptotic convex geometry [11, 146]. More recently, starting from [174], the role of the Gaussian width has been recognized in applications to signal processing and high-dimensional statistics [183, 157, 184, 51, 165, 9, 195, 159]; see also [217, Section 3.5] and [128]. In Section 7.5.4 we noted some surprising geometric phenomena in high dimensions; to learn more about them see the preface of [11] and see also [13].

The notion of the stable dimension $d(T)$ of a set $T \subset \mathbb{R}^n$, introduced in Section 7.6, seems to be new. In the special case where T is a closed convex cone, the squared version of the Gaussian width $h(T)$ defined in (7.19) is often called the *statistical dimension* of T in the literature on signal recovery [136, 9, 159].

The notion of the stable rank $r(A) = \|A\|_F^2/\|A\|^2$ of a matrix A (also called the effective, or numerical rank) seems to have appeared for the first time in [173]. In some literature (e.g. [216, 115]) the quantity

$$k(\Sigma) = \frac{\operatorname{tr} \Sigma}{\|\Sigma\|}$$

is also called the stable rank of a positive-semidefinite matrix Σ; we call $k(\Sigma)$ the *intrinsic dimension* following [201, Definition 7.1.1]. Note that we used the quantity $k(\Sigma)$ in covariance estimation (see Remark 5.6.3). Clearly, if $\Sigma = A^{\mathsf{T}}A$ or $\Sigma = AA^{\mathsf{T}}$ then

$$k(\Sigma) = r(A).$$

Theorem 7.7.1 and its improvement, which we will give in Section 9.2.2, is due to V. Milman [145]; see also [11, Proposition 5.7.1].

8

Chaining

This chapter presents some central concepts and methods for bounding random processes. Chaining is a powerful and general technique that can be used to prove uniform bounds on a random process $(X_t)_{t\in T}$. We present a basic version of the chaining method in Section 8.1. There we prove Dudley's bound on random processes in terms of the covering numbers of the set T. In Section 8.2, we give applications of Dudley's inequality to Monte-Carlo integration and a uniform law of large numbers.

In Section 8.3 we show how to find bounds for random processes in terms of the VC dimension of T. Unlike covering numbers, the VC dimension is a combinatorial rather than geometric quantity. It plays an important role in problems of statistical learning theory, which we discuss in Section 8.4.

As we will see in Section 8.1.2, the bounds on empirical processes in terms of covering numbers – Sudakov's inequality from Section 7.4 and Dudley's inequality – are sharp up to a logarithmic factor. The logarithmic gap is insignificant in many applications but it cannot be removed in general. A sharper bound on random processes, without any logarithmic gap, can be given in terms of the so-called Talagrand functional $\gamma_2(T)$, which captures the geometry of T better than the covering numbers. We prove a sharp upper bound in Section 8.5 by a refined chaining argument, often called "generic chaining".

The matching lower bound, due to M. Talagrand, is more difficult to obtain; we will state it without proof in Section 8.6. The resulting sharp two-sided bound on random processes is known as the *majorizing measure theorem* (Theorem 8.6.1). A very useful consequence of this result is *Talagrand's comparison inequality* (Corollary 8.6.2), which generalizes the Sudakov–Fernique inequality for all sub-gaussian random processes.

Talagrand's comparison inequality has many applications. One of them, *Chevet's inequality*, will be discussed in Section 8.7; others will appear later.

8.1 Dudley's Inequality

Sudakov's minoration inequality, which we studied in Section 7.4, gives a *lower bound* on the magnitude

$$\mathbb{E} \sup_{t\in T} X_t$$

of a Gaussian random process $(X_t)_{t\in T}$ in terms of the metric entropy of T. In this section, we obtain a similar *upper bound.*

This time, we are able to work not just with Gaussian processes but with more general processes having sub-gaussian increments.

Definition 8.1.1 (Sub-gaussian increments) Consider a random process $(X_t)_{t\in T}$ on a metric space (T, d). We say that the process has *sub-gaussian increments* if there exists $K \geq 0$ such that

$$\|X_t - X_s\|_{\psi_2} \leq K d(t,s) \quad \text{for all } t, s \in T. \tag{8.1}$$

Example 8.1.2 Let $(X_t)_{t\in T}$ be a mean-zero Gaussian process on an abstract set T. Define a metric on T by

$$d(t,s) := \|X_t - X_s\|_{L^2}, \quad t, s \in T.$$

Then $(X_t)_{t\in T}$ is obviously a process with sub-gaussian increments, and K is an absolute constant.

We now state Dudley's inequality, which gives a bound on a general sub-gaussian random process $(X_t)_{t\in T}$ in terms of the metric entropy $\log \mathcal{N}(T, d, \varepsilon)$ of T.

Theorem 8.1.3 (Dudley's integral inequality) *Let $(X_t)_{t\in T}$ be a mean-zero random process on a metric space (T, d) with sub-gaussian increments as in* (8.1). *Then*

$$\mathbb{E} \sup_{t\in T} X_t \leq CK \int_0^\infty \sqrt{\log \mathcal{N}(T, d, \varepsilon)}\, d\varepsilon.$$

Before we prove Dudley's inequality, it is helpful to compare it with Sudakov's inequality (Theorem 7.4.1), which, for Gaussian processes, states that

$$\mathbb{E} \sup_{t\in T} X_t \geq c \sup_{\varepsilon>0} \varepsilon \sqrt{\log \mathcal{N}(T, d, \varepsilon)}.$$

Figure 8.1 illustrates Dudley's and Sudakov's bounds. There is an obvious gap between these two bounds. It cannot be closed in terms of the entropy numbers alone; we will explore this point later.

The right-hand side of Dudley's inequality might suggest to us that $\mathbb{E} \sup_{t\in T} X_t$ is a *multi-scale* quantity, in that we have to examine T at all possible scales ε in order to bound the process. This is indeed so, and our proof will indeed be multi-scale. We now state and prove a discrete version of Dudley's inequality in which the integral over all positive ε is replaced

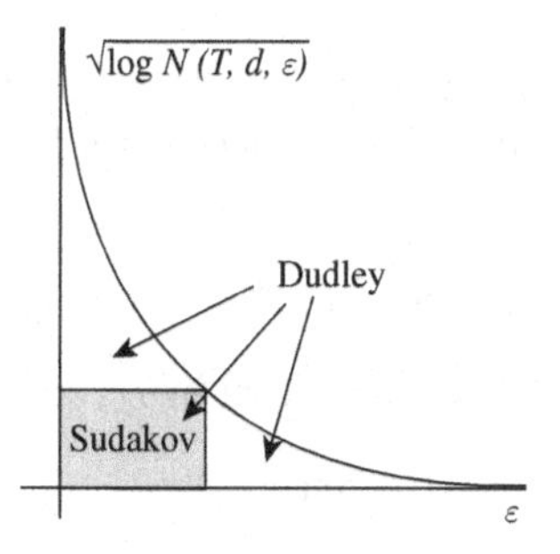

Figure 8.1 Dudley's inequality bounds $\mathbb{E} \sup_{t\in T} X_t$ by the area under the curve. Sudakov's inequality bounds it below by the largest area of a rectangle under the curve, up to constants.

by a sum over dyadic values $\varepsilon = 2^{-k}$, which somewhat resembles a Riemann sum. Later we will quickly pass to the original form of Dudley's inequality.

Theorem 8.1.4 (Discrete version of Dudley's inequality) *Let $(X_t)_{t\in T}$ be a mean-zero random process on a metric space (T, d) with sub-gaussian increments as in* (8.1). *Then*

$$\mathbb{E}\sup_{t\in T} X_t \le CK \sum_{k\in\mathbb{Z}} 2^{-k}\sqrt{\log \mathcal{N}(T, d, 2^{-k})}. \tag{8.2}$$

Our proof of this theorem will be based on the important technique of *chaining*, which can be useful in many other problems. Chaining is a *multi-scale* version of the ε-net argument that we have previously used successfully, for example, in the proofs of Theorems 4.4.5 and 7.7.1.

In the familiar, single-scale, ε-net argument we discretize T by choosing an ε-net $\mathcal{N}$ of T. Then every point $t \in T$ can be approximated by a closest point from the net $\pi(t) \in \mathcal{N}$ with accuracy ε, so that $d(t, \pi(t)) \le \varepsilon$. The increment condition (8.1) yields

$$\|X_t - X_{\pi(t)}\|_{\psi_2} \le K\varepsilon. \tag{8.3}$$

This gives

$$\mathbb{E}\sup_{t\in T} X_t \le \mathbb{E}\sup_{t\in T} X_{\pi(t)} + \mathbb{E}\sup_{t\in T}(X_t - X_{\pi(t)}).$$

The first term can be controlled by a union bound over $|\mathcal{N}| = \mathcal{N}(T, d, \varepsilon)$ points $\pi(t)$.

To bound the second term, we would like to use (8.3). But it only holds for fixed $t \in T$, and it is not clear how to control the supremum over $t \in T$. To overcome this difficulty, we do not stop here but continue to run the ε-net argument further, building progressively finer approximations $\pi_1(t), \pi_2(t), \ldots$ to t with finer nets. Let us now develop formally this technique of chaining.

Proof of Theorem 8.1.4 Step 1: Chaining set-up. Without loss of generality, we may assume that $K = 1$ and that T is finite. (Why?) Let us set the dyadic scale

$$\varepsilon_k = 2^{-k}, \quad k \in \mathbb{Z}, \tag{8.4}$$

and choose ε_k-nets T_k of T such that

$$|T_k| = \mathcal{N}(T, d, \varepsilon_k). \tag{8.5}$$

Only a part of the dyadic scale will be needed. Indeed, since T is finite, there exists a small enough number $\kappa \in \mathbb{Z}$ (defining the coarsest net) and a large enough number K (defining the finest net) such that

$$T_\kappa = \{t_0\} \text{ for some } t_0 \in T, \quad T_K = T. \tag{8.6}$$

For a point $t \in T$, let $\pi_k(t)$ denote a closest point in T_k, so we have

$$d(t, \pi_k(t)) \le \varepsilon_k. \tag{8.7}$$

Since $\mathbb{E} X_{t_0} = 0$, we have

$$\mathbb{E}\sup_{t\in T} X_t = \mathbb{E}\sup_{t\in T}(X_t - X_{t_0}).$$

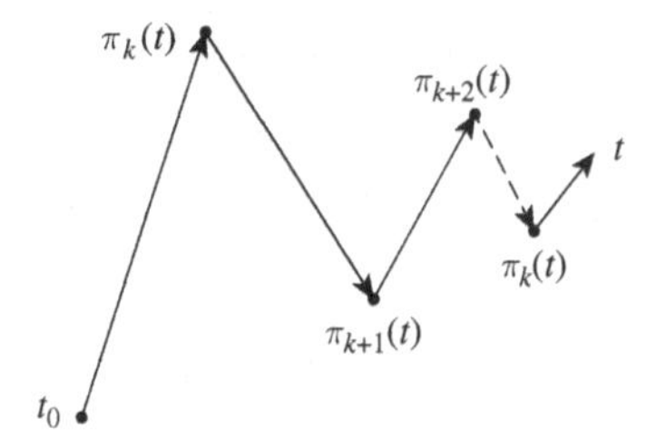

Figure 8.2 Chaining: a walk from a fixed point t_0 to an arbitrary point t in T along elements $\pi_k(T)$ of progressively finer nets of T

We can express $X_t - X_{t_0}$ as a telescoping sum; think about walking from t_0 to t along a chain of points $\pi_k(t)$ that mark progressively finer approximations to t:

$$X_t - X_{t_0} = (X_{\pi_\kappa(t)} - X_{t_0}) + (X_{\pi_{\kappa+1}(t)} - X_{\pi_\kappa(t)}) + \cdots + (X_t - X_{\pi_K(t)}), \tag{8.8}$$

see Figure 8.2 for an illustration. The first and last terms of this sum are zero by (8.6), so we have

$$X_t - X_{t_0} = \sum_{k=\kappa+1}^{K} (X_{\pi_k(t)} - X_{\pi_{k-1}(t)}). \tag{8.9}$$

Since the supremum of the sum is bounded by the sum of the suprema, this yields

$$\mathbb{E} \sup_{t \in T} (X_t - X_{t_0}) \le \sum_{k=\kappa+1}^{K} \mathbb{E} \sup_{t \in T} (X_{\pi_k(t)} - X_{\pi_{k-1}(t)}). \tag{8.10}$$

Step 2: Controlling the increments. Although each term in the bound (8.10) still has a supremum over the entire set T, a closer look reveals that it is actually a maximum over a much smaller set, namely the set all possible pairs $(\pi_k(t), \pi_{k-1}(t))$. The number of such pairs is

$$|T_k|\,|T_{k-1}| \le |T_k|^2,$$

a number that we can control through (8.5).

Next, for a fixed t, the increments in (8.10) can be bounded as follows:

$$\begin{aligned}
\|X_{\pi_k(t)} - X_{\pi_{k-1}(t)}\|_{\psi_2} &\le d(\pi_k(t), \pi_{k-1}(t)) \quad \text{(by (8.1) and since } K = 1\text{)} \\
&\le d(\pi_k(t), t) + d(t, \pi_{k-1}(t)) \quad \text{(by triangle inequality)} \\
&\le \varepsilon_k + \varepsilon_{k-1} \quad \text{(by (8.7))} \\
&\le 2\varepsilon_{k-1}.
\end{aligned}$$

Recall from Exercise 2.5.10 that the expected maximum of N sub-gaussian random variables is at most $CL\sqrt{\log N}$, where L is the maximal ψ_2 norm. Thus we can bound each term in (8.10) as follows:

$$\mathbb{E} \sup_{t \in T} (X_{\pi_k(t)} - X_{\pi_{k-1}(t)}) \le C\varepsilon_{k-1}\sqrt{\log |T_k|}. \tag{8.11}$$

Step 3: Summing up the increments. We have shown that

$$\mathbb{E} \sup_{t \in T} (X_t - X_{t_0}) \le C \sum_{k=\kappa+1}^{K} \varepsilon_{k-1} \sqrt{\log |T_k|}. \tag{8.12}$$

It remains to substitute the values $\varepsilon_k = 2^{-k}$ from (8.4) and the bounds (8.5) on $|T_k|$, and conclude that

$$\mathbb{E} \sup_{t \in T} (X_t - X_{t_0}) \le C_1 \sum_{k=\kappa+1}^{K} 2^{-k} \sqrt{\log \mathcal{N}(T, d, 2^{-k})}.$$

Theorem 8.1.4 is proved. ■

Let us now deduce the integral form of Dudley's inequality.

Proof of Dudley's integral inequality, Theorem 8.1.3 To convert the sum (8.2) into an integral, we express 2^{-k} as $2 \int_{2^{-k-1}}^{2^{-k}} d\varepsilon$. Then

$$\sum_{k \in \mathbb{Z}} 2^{-k} \sqrt{\log \mathcal{N}(T, d, 2^{-k})} = 2 \sum_{k \in \mathbb{Z}} \int_{2^{-k-1}}^{2^{-k}} \sqrt{\log \mathcal{N}(T, d, 2^{-k})} \, d\varepsilon.$$

Within the limits of the integral $2^{-k} \ge \varepsilon$, so $\log \mathcal{N}(T, d, 2^{-k}) \le \log \mathcal{N}(T, d, \varepsilon)$ and the sum is bounded by

$$2 \sum_{k \in \mathbb{Z}} \int_{2^{-k-1}}^{2^{-k}} \sqrt{\log \mathcal{N}(T, d, \varepsilon)} \, d\varepsilon = 2 \int_0^\infty \sqrt{\log \mathcal{N}(T, d, \varepsilon)} \, d\varepsilon.$$

The proof is complete. ■

Remark 8.1.5 (Supremum of increments) A quick glance at the proof reveals that the chaining method actually yields the bound

$$\mathbb{E} \sup_{t \in T} |X_t - X_{t_0}| \le CK \int_0^\infty \sqrt{\log \mathcal{N}(T, d, \varepsilon)} \, d\varepsilon$$

for any fixed $t_0 \in T$. Combining it with a similar bound for $X_s - X_{t_0}$ and using the triangle inequality, we deduce that

$$\mathbb{E} \sup_{t,s \in T} |X_t - X_s| \le CK \int_0^\infty \sqrt{\log \mathcal{N}(T, d, \varepsilon)} \, d\varepsilon.$$

Note that in either of these two bounds, we need not require the mean-zero assumption $\mathbb{E} X_t = 0$. It is required, however, in Dudley's theorem 8.1.3; otherwise the theorem may fail. (Why?)

Dudley's inequality gives a bound on the expectation only, but adapting the argument yields a nice tail bound as well.

Theorem 8.1.6 (Dudley's integral inequality: tail bound) *Let $(X_t)_{t\in T}$ be a random process on a metric space (T,d) with sub-gaussian increments as in* (8.1). *Then, for every $u \geq 0$, the event*

$$\sup_{t,s\in T} |X_t - X_s| \leq CK\Big(\int_0^\infty \sqrt{\log \mathcal{N}(T,d,\varepsilon)}\,d\varepsilon + u \operatorname{diam}(T)\Big)$$

holds with probability at least $1 - 2\exp(-u^2)$.

Exercise 8.1.7 Prove Theorem 8.1.6. To this end, first obtain a high-probability version of (8.11):

$$\sup_{t\in T}(X_{\pi_k(t)} - X_{\pi_{k-1}(t)}) \leq C\varepsilon_{k-1}\big(\sqrt{\log |T_k|} + z\big)$$

with probability at least $1 - 2\exp(-z^2)$.

Use this inequality with $z = z_k$ to control all such terms simultaneously. Summing them up, deduce a bound on $\sup_{t\in T} |X_t - X_{t_0}|$ with probability at least $1 - 2\sum_k \exp(-z_k^2)$. Finally, choose values for z_k that give you a good bound; one can set $z_k = u + \sqrt{k-\kappa}$ for example.

Exercise 8.1.8 (Equivalence of Dudley's integral and sum) In the proof of Theorem 8.1.3 we bounded Dudley's sum by an integral. Show the reverse bound:

$$\int_0^\infty \sqrt{\log \mathcal{N}(T,d,\varepsilon)}\,d\varepsilon \leq C \sum_{k\in\mathbb{Z}} 2^{-k}\sqrt{\log \mathcal{N}(T,d,2^{-k})}.$$

8.1.1 Remarks and Examples

Remark 8.1.9 (Limits of Dudley's integral) Although Dudley's integral is formally over $[0,\infty]$, we can clearly make the upper bound equal the diameter of T in the metric d; thus

$$\mathbb{E}\sup_{t\in T} X_t \leq CK \int_0^{\operatorname{diam}(T)} \sqrt{\log \mathcal{N}(T,d,\varepsilon)}\,d\varepsilon. \tag{8.13}$$

Indeed, if $\varepsilon > \operatorname{diam}(T)$ then a single point (any point in T) is an ε-net of T, which shows that $\log \mathcal{N}(T,d,\varepsilon) = 0$ for such ε.

Let us apply Dudley's inequality for the canonical Gaussian process, just as we did with Sudakov's inequality in Section 7.4.1. We immediately obtain the following bound.

Theorem 8.1.10 (Dudley's inequality for sets in $\mathbb{R}^n$) *For any set $T \subset R^n$, we have*

$$w(T) \leq C \int_0^\infty \sqrt{\log \mathcal{N}(T,\varepsilon)}\,d\varepsilon.$$

Example 8.1.11 Let us test Dudley's inequality for the unit Euclidean ball $T = B_2^n$. Recall from (4.10) that

$$\mathcal{N}(B_2^n,\varepsilon) \leq \Big(\frac{3}{\varepsilon}\Big)^n \quad \text{for } \varepsilon \in (0,1]$$

and $\mathcal{N}(B_2^n,\varepsilon) = 1$ for $\varepsilon > 1$. Then Dudley's inequality yields a converging integral

$$w(B_2^n) \le C \int_0^1 \sqrt{n \log \frac{3}{\varepsilon}}\, d\varepsilon \le C_1\sqrt{n}.$$

This is optimal: indeed, as we know from (7.16), the Gaussian width of B_2^n is equivalent to $\sqrt{n}$ up to a constant factor.

Exercise 8.1.12 (Dudley's inequality can be loose)☕☕☕ Let $e_1, \dots, e_n$ denote the canonical basis vectors in $\mathbb{R}^n$. Consider the set

$$T := \left\{ \frac{e_k}{\sqrt{1+\log k}},\ k = 1, \dots, n \right\}.$$

(a) Show that

$$w(T) \le C,$$

where as usual C denotes an absolute constant. ☞

(b) Show that

$$\int_0^\infty \sqrt{\log \mathcal{N}(T, d, \varepsilon)}\, d\varepsilon \to \infty$$

as $n \to \infty$. ☞

8.1.2 Two-Sided Sudakov Inequality*

This subsection is optional; further material is not based on it.

As we have just seen in Exercise 8.1.12, in general there is a gap between Sudakov's and Dudley's inequalities. Fortunately, this gap is only logarithmically large. Let us make this statement more precise and show that Sudakov's inequality in $\mathbb{R}^n$ (Corollary 7.4.3) is optimal up to a factor $\log n$.

Theorem 8.1.13 (Two-sided Sudakov inequality) *Let $T \subset \mathbb{R}^n$ and set*

$$s(T) := \sup_{\varepsilon \ge 0} \varepsilon \sqrt{\log \mathcal{N}(T, \varepsilon)}.$$

Then

$$c\, s(T) \le w(T) \le C \log(n)\, s(T).$$

Proof The lower bound is a form of Sudakov's inequality (Corollary 7.4.3). To prove the upper bound, the main idea is that the chaining process converges exponentially fast, and thus $O(\log n)$ steps should suffice to walk from t_0 to somewhere very near t.

As we already noted in (8.13), the coarsest scale in the chaining sum (8.9) can be chosen as the diameter of T. In other words, we can start the chaining at κ, the smallest integer such that

$$2^{-\kappa} < \operatorname{diam}(T).$$

This is not different from what we did before. What will be different is the finest scale. Instead of going all the way down, let us stop chaining at K, which is the largest integer for which

$$2^{-K} \geq \frac{w(T)}{4\sqrt{n}}.$$

(It will soon be clear why we have made this choice.)

Then the last term in (8.8) may not be zero as before, and instead of (8.9) we need the bound

$$w(T) \leq \sum_{k=\kappa+1}^{K} \mathbb{E} \sup_{t \in T} (X_{\pi_k(t)} - X_{\pi_{k-1}(t)}) + \mathbb{E} \sup_{t \in T} (X_t - X_{\pi_K(t)}). \tag{8.14}$$

To control the last term, recall that $X_t = \langle g, t \rangle$ is the canonical process, so

$$\begin{aligned}
\mathbb{E} \sup_{t \in T} (X_t - X_{\pi_K(t)}) &= \mathbb{E} \sup_{t \in T} \langle g, t - \pi_K(t) \rangle \\
&\leq 2^{-K} \, \mathbb{E} \, \|g\|_2 \quad (\text{since } \|t - \pi_K(t)\|_2 \leq 2^{-K}) \\
&\leq 2^{-K} \sqrt{n} \\
&\leq \frac{w(T)}{2\sqrt{n}} \sqrt{n} \quad (\text{by definition of } K) \\
&\leq \frac{1}{2} w(T).
\end{aligned}$$

Putting this into (8.14) and subtracting $\frac{1}{2} w(T)$ from both sides, we conclude that

$$w(T) \leq 2 \sum_{k=\kappa+1}^{K} \mathbb{E} \sup_{t \in T} (X_{\pi_k(t)} - X_{\pi_{k-1}(t)}). \tag{8.15}$$

Thus, we have removed the last term from (8.14). Each remaining term can be bounded as before. The number of terms in this sum is

$$\begin{aligned}
K - \kappa &\leq \log_2 \frac{\operatorname{diam}(T)}{w(T)/4\sqrt{n}} \quad (\text{by the definitions of } K \text{ and } \kappa) \\
&\leq \log_2 \left(4\sqrt{n} \sqrt{2\pi} \right) \quad (\text{by property (vi) of Proposition 7.5.2}) \\
&\leq C \log n.
\end{aligned}$$

Thus we can replace the sum by the maximum in (8.15) by paying a factor $C \log n$. This completes the argument as in the proof of Theorem 8.1.4. ■

Exercise 8.1.14 (Limits in Dudley's integral) ☕☕☕ Prove the following improvement of Dudley's inequality (Theorem 8.1.10). For any set $T \subset R^n$, we have

$$w(T) \leq C \int_a^b \sqrt{\log \mathcal{N}(T, \varepsilon)} \, d\varepsilon \quad \text{where} \quad a = \frac{c w(T)}{\sqrt{n}}, \quad b = \operatorname{diam}(T).$$

8.2 Application: Empirical Processes

We now give an application of Dudley's inequality to *empirical processes*, which are certain random processes indexed by functions. The theory of empirical processes comprises a large branch of probability theory, and we will only scratch its surface here. Let us consider a motivating example.

8.2.1 Monte-Carlo Method

Suppose that we want to evaluate the integral of a function $f : \Omega \to \mathbb{R}$ with respect to some probability measure μ on some domain $\Omega \subset \mathbb{R}^d$, i.e.,

$$\int_\Omega f \, d\mu;$$

see Figure 8.3a. For example, we might be interested in computing $\int_0^1 f(x)\,dx$ for a function f: $[0, 1] \to \mathbb{R}$.

We will use probability to evaluate this integral. Consider a random point X that takes values in Ω according to the law μ, i.e.,

$$\mathbb{P}\left\{X \in A\right\} = \mu(A) \quad \text{for any measurable set } A \subset \Omega.$$

(For example, to evaluate $\int_0^1 f(x)\,dx$, we take $X \sim \text{Unif}[0, 1]$.) Then we may interpret the integral as an expectation:

$$\int_\Omega f \, d\mu = \mathbb{E}\, f(X).$$

Let $X_1, X_2, \ldots$ be i.i.d. copies of X. The law of large numbers (Theorem 1.3.1) yields that

$$\frac{1}{n}\sum_{i=1}^{n} f(X_i) \to \mathbb{E}\, f(X) \quad \text{almost surely} \tag{8.16}$$

as $n \to \infty$. This means that we can approximate the integral by a sum:

$$\int_\Omega f \, d\mu \approx \frac{1}{n}\sum_{i=1}^{n} f(X_i), \tag{8.17}$$

where the points X_i are drawn at random from the domain Ω; see Figure 8.3b for an illustration. This way of numerically computing integrals is called the *Monte-Carlo method.*

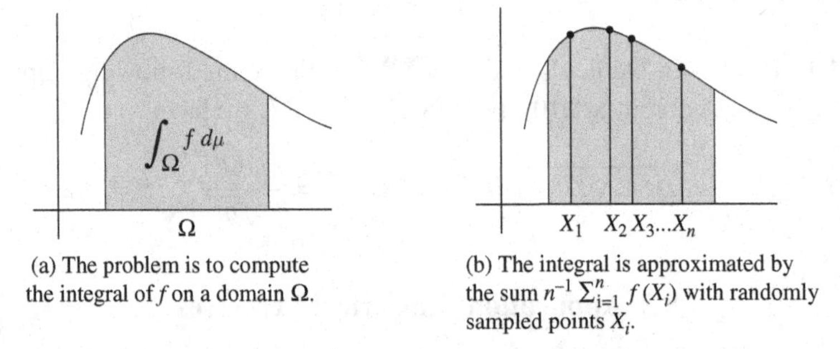

(a) The problem is to compute the integral of f on a domain Ω.

(b) The integral is approximated by the sum $n^{-1} \sum_{i=1}^{n} f(X_i)$ with randomly sampled points X_i.

Figure 8.3 Monte-Carlo method for randomized numerical integration.

Remark 8.2.1 (Error rate) Note that the average error in (8.17) is $O(1/\sqrt{n})$. Indeed, as we noted in (1.5), the rate of convergence in the law of large numbers is

$$\mathbb{E}\left|\frac{1}{n}\sum_{i=1}^{n} f(X_i) - \mathbb{E} f(X)\right| \le \left(\mathrm{Var}\left(\frac{1}{n}\sum_{i=1}^{n} f(X_i)\right)\right)^{1/2} = O\left(\frac{1}{\sqrt{n}}\right). \tag{8.18}$$

Remark 8.2.2 Note that we do not even need to know the measure μ in order to evaluate the integral $\int_\Omega f\, d\mu$; it suffices to be able to draw random samples X_i according to μ. Similarly, we do not even need to know f at all points in the domain; a few random points suffice.

8.2.2 A Uniform Law of Large Numbers

Can we use the same sample $X_1, \dots, X_n$ to evaluate the integral of *any* function $f\colon \Omega \to \mathbb{R}$? Of course not. For a given sample, we could choose a function that oscillates in the wrong way between the sample points, and then the approximation (8.17) would fail.

Will it help if we consider only those functions f that do not oscillate wildly – for example, Lipschitz functions? It will. Our next theorem states that the Monte-Carlo method (8.17) does work well over the whole class of Lipschitz functions

$$\mathcal{F} := \left\{f\colon [0,1] \to \mathbb{R},\ \|f\|_{\mathrm{Lip}} \le L\right\}, \tag{8.19}$$

where L is any fixed number.

Theorem 8.2.3 (Uniform law of large numbers) *Let $X, X_1, X_2, \dots, X_n$ be i.i.d. random variables taking values in* $[0,1]$. *Then*

$$\mathbb{E} \sup_{f \in \mathcal{F}} \left|\frac{1}{n}\sum_{i=1}^{n} f(X_i) - \mathbb{E} f(X)\right| \le \frac{CL}{\sqrt{n}}. \tag{8.20}$$

Remark 8.2.4 Before we prove this result, let us pause to emphasize its key point: the supremum over $f \in \mathcal{F}$ appears *inside* the expectation. By Markov's inequality, this means that, with high probability, a random sample $X_1, \dots, X_n$ is good; here "good" means that using this sample, we can approximate the integral of *any* function $f \in \mathcal{F}$ with error bounded by the same quantity, $CL/\sqrt{n}$. This is the same rate of convergence that the classical law of large numbers (8.18) guarantees for a *single* function f. So we have

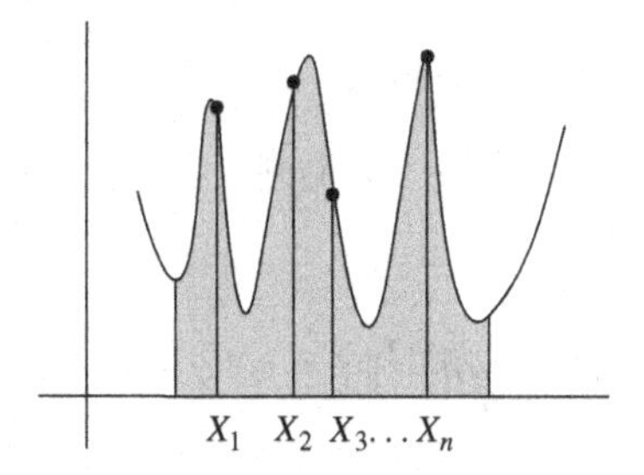

Figure 8.4 One cannot use the same sample $X_1, \dots, X_n$ to approximate the integral of *any* function f.

paid essentially nothing for making the law of large numbers uniform over the class of functions $\mathcal{F}$.

To prepare for the proof of Theorem 8.2.3, it will be useful to view the left-hand side of (8.20) as the magnitude of a random process indexed by functions $f \in \mathcal{F}$. Such random processes are called *empirical processes*.

Definition 8.2.5 Let $\mathcal{F}$ be a class of real-valued functions $f\colon \Omega \to \mathbb{R}$, where (Ω, Σ, μ) is a probability space. Let X be a random point in Ω distributed according to the law μ, and let $X_1, X_2, \dots, X_n$ be independent copies of X. The random process $(X_f)_{f\in\mathcal{F}}$ defined by

$$X_f := \frac{1}{n}\sum_{i=1}^{n} f(X_i) - \mathbb{E}\, f(X) \tag{8.21}$$

is called an *empirical process* indexed by $\mathcal{F}$.

Proof of Theorem 8.2.3 Without loss of generality, it is enough to prove the theorem for the class

$$\mathcal{F} := \left\{f\colon [0,1] \to [0,1],\ \|f\|_{\mathrm{Lip}} \le 1\right\}. \tag{8.22}$$

(Why?) We would like to bound the magnitude

$$\mathbb{E} \sup_{f\in\mathcal{F}} |X_f|$$

of the empirical process $(X_f)_{f\in\mathcal{F}}$ defined in (8.21).

Step 1: Checking for sub-gaussian increments. We can do this using Dudley's inequality, Theorem 8.1.3. To apply this result, we just need to check that the empirical process has sub-gaussian increments. So, fix a pair of functions $f, g \in \mathcal{F}$ and consider

$$\|X_f - X_g\|_{\psi_2} = \frac{1}{n}\Big\|\sum_{i=1}^{n} Z_i\Big\|_{\psi_2} \quad \text{where} \quad Z_i := (f-g)(X_i) - \mathbb{E}(f-g)(X).$$

The random variables Z_i are independent and have zero mean. So, by Proposition 2.6.1, we have

$$\|X_f - X_g\|_{\psi_2} \lesssim \frac{1}{n}\Big(\sum_{i=1}^{n} \|Z_i\|_{\psi_2}^2\Big)^{1/2}.$$

Now, using centering (Lemma 2.6.8) we have

$$\|Z_i\|_{\psi_2} \lesssim \|(f-g)(X_i)\|_{\psi_2} \lesssim \|f-g\|_\infty.$$

It follows that

$$\|X_f - X_g\|_{\psi_2} \lesssim \frac{1}{n} n^{1/2}\|f-g\|_\infty = \frac{1}{\sqrt{n}}\|f-g\|_\infty.$$

Step 2: Applying Dudley's inequality. We found that the empirical process $(X_f)_{f\in\mathcal{F}}$ has sub-gaussian increments with respect to the L^∞ norm. This allows us to apply Dudley's inequality. Note that (8.22) implies that the diameter of $\mathcal{F}$ in L^∞ metric is bounded by 1. Thus

$$\mathbb{E} \sup_{f \in \mathcal{F}} |X_f| = \mathbb{E} \sup_{f \in \mathcal{F}} |X_f - X_0| \lesssim \frac{1}{\sqrt{n}} \int_0^1 \sqrt{\log \mathcal{N}(\mathcal{F}, \|\cdot\|_\infty, \varepsilon)}\, d\varepsilon.$$

(Here we have used the fact that the zero function belongs to $\mathcal{F}$ and have also used the version of Dudley's inequality from Remark 8.1.5; see also (8.13).)

Since all functions in $f \in \mathcal{F}$ are Lipschitz with $\|f\|_{\mathrm{Lip}} \le 1$, it is not difficult to bound the covering numbers of $\mathcal{F}$ as follows:

$$\mathcal{N}(\mathcal{F}, \|\cdot\|_\infty, \varepsilon) \le \left(\frac{C}{\varepsilon}\right)^{C/\varepsilon};$$

we will show this in Exercise 8.2.6 below. This bound makes Dudley's integral converge, and we conclude that

$$\mathbb{E} \sup_{f \in \mathcal{F}} |X_f| \lesssim \frac{1}{\sqrt{n}} \int_0^1 \sqrt{\frac{C}{\varepsilon} \log \frac{C}{\varepsilon}}\, d\varepsilon \lesssim \frac{1}{\sqrt{n}}.$$

Theorem 8.2.3 is proved. ■

Exercise 8.2.6 (Metric entropy of the class of Lipschitz functions) Consider the class of functions

$$\mathcal{F} := \left\{ f\colon [0,1] \to [0,1],\ \|f\|_{\mathrm{Lip}} \le 1 \right\}.$$

Show that

$$\mathcal{N}(\mathcal{F}, \|\cdot\|_\infty, \varepsilon) \le \left(\frac{2}{\varepsilon}\right)^{2/\varepsilon} \quad \text{for any } \varepsilon > 0.$$

Exercise 8.2.7 (An improved bound on the metric entropy) Improve the bound in Exercise 8.2.6 to

$$\mathcal{N}(\mathcal{F}, \|\cdot\|_\infty, \varepsilon) \le e^{C/\varepsilon} \quad \text{for any } \varepsilon > 0.$$

Exercise 8.2.8 (Higher dimensions) Consider the class of functions

$$\mathcal{F} := \left\{ f\colon [0,1]^d \to \mathbb{R},\ f(0) = 0,\ \|f\|_{\mathrm{Lip}} \le 1 \right\}$$

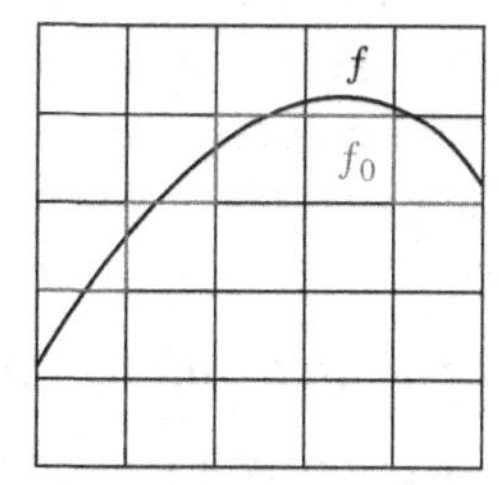

Figure 8.5 Bounding the metric entropy of the class of Lipschitz functions in Exercise 8.2.6. A Lipschitz function f is approximated by a function f_0 on a mesh.

for some dimension $d \geq 1$. Show that

$$\mathcal{N}(\mathcal{F}, \|\cdot\|_\infty, \varepsilon) \leq e^{C/\varepsilon^d} \quad \text{for any } \varepsilon > 0.$$

8.2.3 Empirical Measure

Let us take one more look at Definition 8.2.5 concerning empirical processes. Consider a probability measure μ_n that is uniformly distributed on the sample $X_1, \ldots, X_N$, that is,

$$\mu_n(\{X_i\}) = \frac{1}{n} \quad \text{for every } i = 1, \ldots, n. \tag{8.23}$$

Note that μ_n is a *random* measure. It is called the *empirical measure*.

While the integral of f with respect to the original measure μ is $\mathbb{E} f(X)$ (the "population" average of f), the integral of f with respect to the empirical measure is $\frac{1}{n}\sum_{i=1}^n f(X_i)$ (the "sample", or empirical, average of f). In the literature on empirical processes, the population expectation of f is denoted by μf and the empirical expectation by $\mu_n f$:

$$\mu f = \int f \, d\mu = \mathbb{E} f(X), \qquad \mu_n f = \int f \, d\mu_n = \frac{1}{n}\sum_{i=1}^n f(X_i).$$

The empirical process X_f in (8.21) thus measures the deviation of the population expectation from the empirical expectation:

$$X_f = \mu f - \mu_n f.$$

The uniform law of large numbers (8.20) gives a uniform bound on the deviation

$$\mathbb{E} \sup_{f \in \mathcal{F}} |\mu_n f - \mu f| \tag{8.24}$$

over the class of Lipschitz functions $\mathcal{F}$ defined in (8.19).

The quantity (8.24) can be thought of as the distance between the measures μ_n and μ. It is called the *Wasserstein distance* $W_1(\mu, \mu_n)$. The Wasserstein distance has an equivalent formulation as the *transportation cost* of measure μ into measure μ_n, where the cost of moving a mass (probability) $p > 0$ is proportional to p and to the distance moved. The equivalence between the transportation cost and (8.24) is provided by the Kantorovich–Rubinstein duality theorem.

8.3 VC Dimension

In this section, we introduce the notion of the VC dimension, which plays a major role in statistical learning theory. We relate the VC dimension to covering numbers and then, through Dudley's inequality, to random processes and the uniform law of large numbers. Applications to statistical learning theory will be given in the next section.

8.3.1 Definition and Examples

The Vapnik–Chervonenkis (VC) dimension is a measure of the complexity of classes of Boolean functions. By a class of Boolean functions we mean any collection $\mathcal{F}$ of functions $f\colon \Omega \to \{0, 1\}$ defined on a common domain Ω.

Definition 8.3.1 (VC dimension) Consider a class $\mathcal{F}$ of Boolean functions on some domain Ω. We say that a subset $\Lambda \subseteq \Omega$ is *shattered* by $\mathcal{F}$ if any function $g\colon \Lambda \to \{0, 1\}$ can be obtained by restricting some function $f \in \mathcal{F}$ to Λ. The *VC dimension* of $\mathcal{F}$, denoted $\mathrm{vc}(\mathcal{F})$, is the largest cardinality[1] of a subset $\Lambda \subseteq \Omega$ shattered by $\mathcal{F}$.

The definition of the VC dimension may take some time to fully comprehend. We work out a few examples to illustrate this notion.

Example 8.3.2 (Intervals) Let $\mathcal{F}$ be the class of indicators of all closed intervals in $\mathbb{R}$, that is,

$$\mathcal{F} := \left\{\mathbf{1}_{[a,b]} \colon a, b \in \mathbb{R},\ a \le b\right\}.$$

We claim that there exists a two-point set $\Lambda \subset \mathbb{R}$ that is shattered by $\mathcal{F}$, and thus

$$\mathrm{vc}(\mathcal{F}) \ge 2.$$

Take, for example, $\Lambda := \{3, 5\}$. It is not too difficult to see that each of the four possible functions $g\colon \Lambda \to \{0, 1\}$ is a restriction of some indicator function $f = \mathbf{1}_{[a,b]}$ onto Λ. For example, the function g defined by $g(3) = 1$, $g(5) = 0$ is a restriction of $f = \mathbf{1}_{[2,4]}$ onto Λ, since $f(3) = g(3) = 1$ and $f(5) = g(5) = 0$. The three other possible functions g can be treated similarly; see Figure 8.6. Thus $\Lambda = \{3, 5\}$ is indeed shattered by $\mathcal{F}$, as claimed.

Next, we claim that no three-point set $\Lambda = \{p, q, r\}$ can be shattered by $\mathcal{F}$, and thus

$$\mathrm{vc}(\mathcal{F}) = 2.$$

To see this, assume that $p < q < r$ and define the function $g\colon \Lambda \to \{0, 1\}$ by $g(p) = 1$, $g(q) = 0$, $g(r) = 1$. Then g cannot be a restriction of any indicator $\mathbf{1}_{[a,b]}$ onto Λ, for otherwise $[a, b]$ would contain two points p and r but not the point q that lies between them, which is impossible.

Example 8.3.3 (Half-planes) Let $\mathcal{F}$ be the class of indicators of all closed half-planes in $\mathbb{R}^2$. We claim that there is a three-point set $\Lambda \subset \mathbb{R}^2$ that is shattered by $\mathcal{F}$, and thus

$$\mathrm{vc}(\mathcal{F}) \ge 3.$$

To see this, let Λ be a set of three points in general position, such as in Figure 8.7. Then each of the $2^3 = 8$ functions $g\colon \Lambda \to \{0, 1\}$ is a restriction of the indicator function of some half-plane. To see this, arrange the half-plane to contain exactly those points of Λ where g takes the value 1, which can always be done – see Figure 8.7.

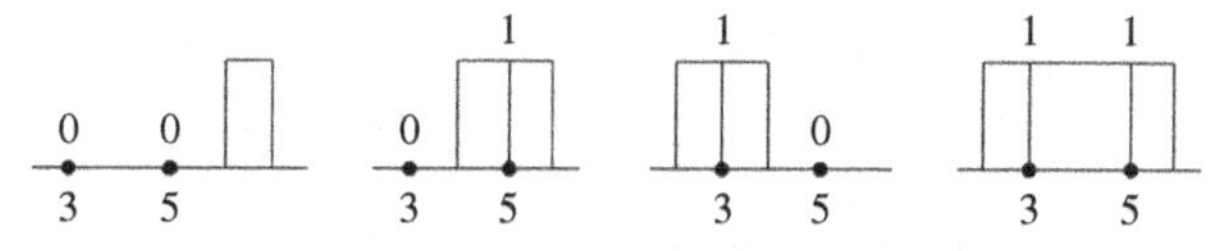

Figure 8.6 The function $g(3) = g(5) = 0$ is a restriction of $\mathbf{1}_{[6,7]}$ onto $\Lambda = \{3, 5\}$ (far left). The function $g(3) = 0$, $g(5) = 1$ is a restriction of $\mathbf{1}_{[4,6]}$ onto Λ (middle left). The function $g(3) = 1$, $g(5) = 0$ is a restriction of $\mathbf{1}_{[2,4]}$ onto Λ (middle right). The function $g(3) = g(5) = 1$ is a restriction of $\mathbf{1}_{[2,6]}$ onto Λ (right).

[1] If the largest cardinality does not exist, we set $\mathrm{vc}(\mathcal{F}) = \infty$.

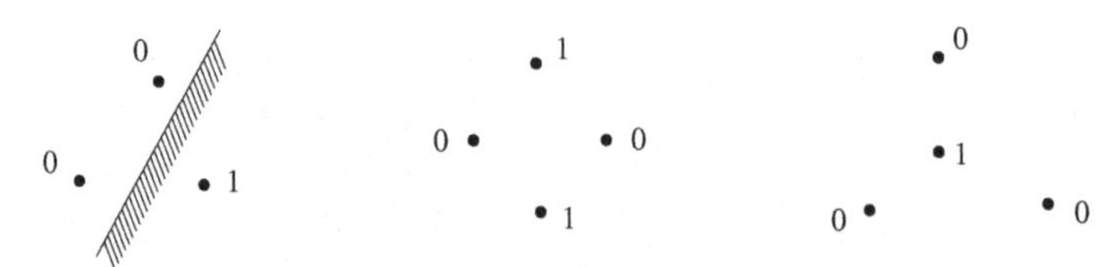

Figure 8.7 Left: a three-point set Λ and function $g\colon \Lambda \to \{0, 1\}$. Such a g is a restriction of the indicator function of the shaded half-plane. Middle and right: two kinds of four-point sets Λ in general position, and functions $g\colon \Lambda \to \{0, 1\}$. In each case, no half-plane can contain exactly the points with value 1. Thus, g is not a restriction of the indicator function of any half-plane.

Next, we claim that no four-point set can be shattered by $\mathcal{F}$, and thus

$$\operatorname{vc}(\mathcal{F}) = 3.$$

There are two possible arrangements of the four-point sets Λ in general position, shown in Figure 8.7. (What if Λ is not in general position? Analyze this case.) In each of the two cases, there exists a $0/1$ labeling of the points such that no half-plane can contain exactly the points labeled 1; see Figure 8.7. This means that in each case there exists a function $g\colon \Lambda \to \{0, 1\}$ that is not a restriction of any function $f \in \mathcal{F}$ onto Λ, and thus the four-point sets Λ are not shattered by $\mathcal{F}$ as claimed.

Example 8.3.4 Let $\Omega = \{1, 2, 3\}$. We can conveniently represent Boolean functions on Ω as binary strings of length three. Consider the class

$$\mathcal{F} := \{001, 010, 100, 111\}.$$

The set $\Lambda = \{1, 3\}$ is shattered by $\mathcal{F}$. Indeed, restricting the functions in $\mathcal{F}$ onto Λ amounts to dropping the second digit, thus producing strings $00, 01, 10, 11$. Thus, the restriction produces all possible binary strings of length two or, equivalently, all possible functions $g\colon \Lambda \to \{0, 1\}$. Hence Λ is shattered by $\mathcal{F}$, and thus $\operatorname{vc}(\mathcal{F}) \geq |\Lambda| = 2$. However, the (only) three-point set $\{1, 2, 3\}$ is not shattered by $\mathcal{F}$, as this would require all eight binary digits of length three to appear in $\mathcal{F}$, which is not the case.

Exercise 8.3.5 (Pairs of intervals) Let $\mathcal{F}$ be the class of indicators of sets of the form $[a, b] \cup [c, d]$ in $\mathbb{R}$. Show that

$$\operatorname{vc}(\mathcal{F}) = 4.$$

Exercise 8.3.6 (Circles) Let $\mathcal{F}$ be the class of indicators of all circles in $\mathbb{R}^2$. Show that

$$\operatorname{vc}(\mathcal{F}) = 3.$$

Exercise 8.3.7 (Rectangles) Let $\mathcal{F}$ be the class of indicators of all closed axis-aligned rectangles, i.e., product sets $[a, b] \times [c, d]$, in $\mathbb{R}^2$. Show that

$$\operatorname{vc}(\mathcal{F}) = 4.$$

Exercise 8.3.8 (Squares) Let $\mathcal{F}$ be the class of indicators of all closed axis-aligned squares, i.e. product sets $[a, a+d] \times [b, b+d]$, in $\mathbb{R}^2$. Show that

$$\mathrm{vc}(\mathcal{F}) = 3.$$

Exercise 8.3.9 (Polygons) Let $\mathcal{F}$ be the class of indicators of all convex polygons in $\mathbb{R}^2$, without any restriction on the number of vertices. Show that

$$\mathrm{vc}(\mathcal{F}) = \infty.$$

Remark 8.3.10 (VC dimension of classes of sets) We can talk about the VC dimension of classes of *sets* instead of functions. This is due to the natural correspondence between the two: a Boolean function f on Ω determines the subset $\{x \in \Omega \colon f(x) = 1\}$ and, vice versa, a subset $\Omega_0 \subset \Omega$ determines the Boolean function $f = \mathbf{1}_{\Omega_0}$. In this language, the VC dimension of the set of intervals in $\mathbb{R}$ equals 2, the VC dimension of the set of half-planes in $\mathbb{R}^2$ equals 3, and so on.

Exercise 8.3.11 Give the definition of the VC dimension of a class of subsets of Ω without mentioning any functions.

Remark 8.3.12 (More examples) It can be shown that the VC dimension of the class of all rectangles on the plane (not necessarily axis-aligned) equals 7. For the class of all polygons with k vertices on the plane, the VC dimension is $2k+1$. For the class of half-spaces in $\mathbb{R}^n$, the VC dimension is $n+1$.

8.3.2 Pajor's Lemma

Consider a class of Boolean functions $\mathcal{F}$ on a *finite* set Ω. We will study a remarkable connection between the cardinality $|\mathcal{F}|$ and the VC dimension of $\mathcal{F}$. Somewhat oversimplifying, we can say that $|\mathcal{F}|$ is exponential in $\mathrm{vc}(\mathcal{F})$. A lower bound is trivial:

$$|\mathcal{F}| \geq 2^{\mathrm{vc}(\mathcal{F})}.$$

(Check!) We now pass to upper bounds; they are less trivial. The following lemma states that there are at least as many shattered subsets of Ω as the functions in $\mathcal{F}$.

Lemma 8.3.13 (Pajor's lemma) *Let $\mathcal{F}$ be a class of Boolean functions on a finite set Ω. Then*

$$|\mathcal{F}| \leq \left|\left\{\Lambda \subseteq \Omega \colon \Lambda \text{ is shattered by } \mathcal{F}\right\}\right|.$$

We include the empty set $\Lambda = \emptyset$ in the counting on the right-hand side.

Before we prove Pajor's lemma, let us pause to give a quick illustration using Example 8.3.4. In that example, $|\mathcal{F}| = 4$ and there are six subsets Λ that are shattered by $\mathcal{F}$, namely $\{1\}$, $\{2\}$, $\{3\}$, $\{1,2\}$, $\{1,3\}$, and $\{2,3\}$. (Check!) Thus the inequality in Pajor's lemma reads $4 \leq 6$ in this case.

Proof of Lemma 8.3.13 We proceed by induction on the cardinality of Ω. The case $|\Omega| = 1$ is trivial, since we include the empty set in the counting. Now assume that the lemma holds for any n-point set Ω, and let us prove it for Ω with $|\Omega| = n + 1$.

Chopping out one (arbitrary) point from the set Ω, we can express this as

$$\Omega = \Omega_0 \cup \{x_0\}, \quad \text{where} \quad |\Omega_0| = n.$$

The class $\mathcal{F}$ then naturally breaks into two sub-classes,

$$\mathcal{F}_0 := \{f \in \mathcal{F} : f(x_0) = 0\} \quad \text{and} \quad \mathcal{F}_1 := \{f \in \mathcal{F} : f(x_0) = 1\}.$$

By the induction hypothesis, the counting function

$$S(\mathcal{F}) = \left|\left\{\Lambda \subseteq \Omega : \Lambda \text{ is shattered by } \mathcal{F}\right\}\right|$$

satisfies[2]

$$S(\mathcal{F}_0) \geq |\mathcal{F}_0| \quad \text{and} \quad S(\mathcal{F}_1) \geq |\mathcal{F}_1|. \tag{8.25}$$

To complete the proof, all we need to check is

$$S(\mathcal{F}) \geq S(\mathcal{F}_0) + S(\mathcal{F}_1), \tag{8.26}$$

for then (8.25) would give $S(\mathcal{F}) \geq |\mathcal{F}_0| + |\mathcal{F}_1| = |\mathcal{F}|$, as needed.

Inequality (8.26) may seem trivial. Any set Λ that is shattered by $\mathcal{F}_0$ or $\mathcal{F}_1$ is automatically shattered by the larger class $\mathcal{F}$, and thus each set Λ counted by $S(\mathcal{F}_0)$ or $S(\mathcal{F}_1)$ is automatically counted by $S(\mathcal{F})$. The problem, however, lies in the double counting. Assume that the same set Λ is shattered by *both* $\mathcal{F}_0$ and $\mathcal{F}_1$. The counting function $S(\mathcal{F})$ will not count Λ twice. However, a different set will be counted by $S(\mathcal{F})$, which was not counted by either $S(\mathcal{F}_0)$ or $S(\mathcal{F}_1)$, namely, $\Lambda \cup \{x_0\}$. A moment's thought reveals that this set is indeed shattered by $\mathcal{F}$. (Check!) This establishes inequality (8.26) and completes the proof of Pajor's lemma. ∎

It may be helpful to illustrate the key point in the proof of Pajor's lemma with a specific example.

Example 8.3.14 Let us again go back to Example 8.3.4. Following the proof of Pajor's lemma, we chop out $x_0 = 3$ from $\Omega = \{1, 2, 3\}$, making $\Omega_0 = \{1, 2\}$. The class $\mathcal{F} = \{001, 010, 100, 111\}$ then breaks into two sub-classes

$$\mathcal{F}_0 = \{010, 100\} \quad \text{and} \quad \mathcal{F}_1 = \{001, 111\}.$$

There are exactly two subsets Λ shattered by $\mathcal{F}_0$, namely $\{1\}$ and $\{2\}$, and the *same subsets* are shattered by $\mathcal{F}_1$, making $S(\mathcal{F}_0) = S(\mathcal{F}_1) = 2$. Of course, the same two subsets are also shattered by $\mathcal{F}$, but we need two more shattered subsets to make $S(\mathcal{F}) \geq 4$ for the key inequality (8.26). Here is how we construct them: append $x_0 = 3$ to the already counted subsets Λ. The resulting sets $\{1, 3\}$ and $\{2, 3\}$ are also shattered by $\mathcal{F}$, and we have not counted them yet. Now have at least *four* subsets shattered by $\mathcal{F}$, making the key inequality (8.26) in the proof of Pajor's lemma true.

[2] Here, to use the induction hypothesis properly, restrict the functions in $\mathcal{F}_0$ and $\mathcal{F}_1$ onto the n-point set Ω_0.

Exercise 8.3.15 (Sharpness of Pajor's lemma) Show that Pajor's lemma 8.3.13 is sharp.

8.3.3 Sauer–Shelah Lemma

We now deduce a remarkable upper bound on the cardinality of a function class in terms of the VC dimension.

Theorem 8.3.16 (Sauer–Shelah lemma) *Let $\mathcal{F}$ be a class of Boolean functions on an n-point set Ω. Then*

$$|\mathcal{F}| \le \sum_{k=0}^{d} \binom{n}{k} \le \left(\frac{en}{d}\right)^d,$$

where $d = \mathrm{vc}(\mathcal{F})$.

Proof Pajor's lemma states that $|\mathcal{F}|$ is bounded by the number of subsets $\Lambda \subseteq \Omega$ that are shattered by $\mathcal{F}$. The cardinality of each such set Λ is bounded by $d = \mathrm{vc}(\mathcal{F})$, according to the definition of the VC dimension. Thus

$$|\mathcal{F}| \le \left|\left\{\Lambda \subseteq \Omega \colon |\Lambda| \le d\right\}\right| = \sum_{k=0}^{d} \binom{n}{k},$$

since the sum on the right-hand side gives the total number of subsets of an n-element set with cardinalities at most k. This proves the first inequality of the Sauer–Shelah lemma. The second inequality follows from the bound on the binomial sum proved in Exercise 0.0.5. ■

Exercise 8.3.17 (Sharpness of Sauer–Shelah lemma) Show that the Sauer–Shelah lemma is sharp for all n and d.

8.3.4 Covering Numbers via the VC Dimension

The Sauer–Shelah lemma is sharp, but it can be used only for finite function classes $\mathcal{F}$. What about *infinite* function classes $\mathcal{F}$, for example, the indicator functions of half-planes in Example 8.3.3? It turns out that we can always bound the *covering numbers* of $\mathcal{F}$ in terms of the VC dimension.

Let $\mathcal{F}$ be a class of Boolean functions on a set Ω as before, and let μ be any probability measure on Ω. Then $\mathcal{F}$ can be considered as a metric space under the $L^2(\mu)$ norm, with the metric on $\mathcal{F}$ given by

$$d(f,g) = \|f-g\|_{L^2(\mu)} = \left(\int_\Omega |f-g|^2 \, d\mu\right)^{1/2}, \quad f,g \in \mathcal{F}.$$

Then we can talk about the covering numbers of the class $\mathcal{F}$ in the $L^2(\mu)$ norm, which we denote[3] $\mathcal{N}(\mathcal{F}, L^2(\mu), \varepsilon)$.

[3] If you are not completely comfortable with measure theory, it may be helpful to consider a discrete case, which is all we need for applications in the next section. Let Ω be an N-point set, say $\Omega = \{1, \ldots, N\}$, and μ be the

Theorem 8.3.18 (Covering numbers via VC dimension) *Let $\mathcal{F}$ be a class of Boolean functions on a probability space (Ω, Σ, μ). Then, for every $\varepsilon \in (0, 1)$, we have*

$$\mathcal{N}(\mathcal{F}, L^2(\mu), \varepsilon) \le \left(\frac{2}{\varepsilon}\right)^{Cd}$$

where $d = \mathrm{vc}(\mathcal{F})$.

This result should be compared with the volumetric bound (4.10), which also states that the covering numbers scale exponentially with the dimension. The important difference is that the VC dimension captures a combinatorial rather than a linear algebraic complexity of sets.

For a first attempt at proving Theorem 8.3.18, let us assume for a moment that Ω is finite, say $|\Omega| = n$. Then the Sauer–Shelah lemma (Theorem 8.3.16) yields

$$\mathcal{N}(\mathcal{F}, L^2(\mu), \varepsilon) \le |\mathcal{F}| \le \left(\frac{en}{d}\right)^d.$$

This is not quite what Theorem 8.3.18 claims, but it comes close. To improve the bound, we need to remove the dependence on the size n of Ω. Can we reduce the domain Ω to a much smaller subset without harming the covering numbers? It turns out that we can; this will be based on the following lemma.

Lemma 8.3.19 (Dimension reduction) *Let $\mathcal{F}$ be a class of N Boolean functions on a probability space (Ω, Σ, μ). Assume that all functions in $\mathcal{F}$ are ε-separated, that is,*

$$\|f - g\|_{L^2(\mu)} > \varepsilon \quad \text{for all distinct } f, g \in \mathcal{F}.$$

Then there exist a number $n \le C\varepsilon^{-4} \log N$ and an n-point subset $\Omega_n \subset \Omega$ such that the uniform probability measure μ_n on Ω_n satisfies[4]

$$\|f - g\|_{L^2(\mu_n)} > \frac{\varepsilon}{2} \quad \text{for all distinct } f, g \in \mathcal{F}.$$

Proof Our argument will be based on the probabilistic method. We choose the subset Ω_n at random and show that it satisfies the conclusion of the theorem with positive probability. This will automatically imply the existence of at least one suitable choice of Ω_n.

Let $X, X_1, \dots, X_n$ be independent random points in Ω distributed[5] according to the law μ. Fix a pair of distinct functions $f, g \in \mathcal{F}$ and denote $h := (f - g)^2$ for convenience. We would like to bound the deviation

$$\|f - g\|^2_{L^2(\mu_n)} - \|f - g\|^2_{L^2(\mu)} = \frac{1}{n}\sum_{i=1}^{n} h(X_i) - \mathbb{E}\, h(X).$$

uniform measure on Ω; thus $\mu(i) = 1/N$ for every $i = 1, \dots, N$. In this case, the $L^2(\mu)$ norm of a function $f\colon \Omega \to \mathbb{R}$ is simply $\|f\|_{L^2(\mu)} = ((1/N)\sum_{i=1}^{N} f(i)^2)^{1/2}$. Equivalently, one can think of f as a vector in $\mathbb{R}^N$. The $L^2(\mu)$ norm is just the scaled Euclidean norm $\|\cdot\|_2$ on $\mathbb{R}^n$, i.e., $\|f\|_{L^2(\mu)} = (1/\sqrt{N})\|f\|_2$.

[4] To express the conclusion more conveniently, let $\Omega_n = \{x_1, \dots, x_n\}$. Then $\|f - g\|^2_{L^2(\mu_n)} = (1/n)\sum_{i=1}^{n}(f - g)(x_i)^2$.

[5] For example, if $\Omega = \{1, \dots, N\}$ then X is a random variable which takes the values $1, \dots, N$ with probability $1/N$ each.

We have a sum of independent random variables on the right, and we use the general Hoeffding's inequality to bound it. To do this, we first check that these random variables are sub-gaussian. Indeed,[6]

$$\begin{aligned} \|h(X_i) - \mathbb{E}\, h(X)\|_{\psi_2} &\lesssim \|h(X)\|_{\psi_2} \quad \text{(by the centering lemma 2.6.8)} \\ &\lesssim \|h(X)\|_\infty \quad \text{(by (2.17))} \\ &\le 1 \quad \text{(since } h = f - g \text{ with } f, g \text{ Boolean).} \end{aligned}$$

Then the general Hoeffding inequality (Theorem 2.6.2) gives

$$\mathbb{P}\left\{ \left| \|f - g\|^2_{L^2(\mu_n)} - \|f - g\|^2_{L^2(\mu)} \right| > \frac{\varepsilon^2}{4} \right\} \le 2\exp(-cn\varepsilon^4).$$

(Check!) Therefore, with probability at least $1 - 2\exp(-cn\varepsilon^4)$, we have

$$\|f - g\|^2_{L^2(\mu_n)} \ge \|f - g\|^2_{L^2(\mu)} - \frac{\varepsilon^2}{4} \ge \varepsilon^2 - \frac{\varepsilon^2}{4} = \frac{3\varepsilon^2}{4}, \tag{8.27}$$

where we have used the triangle inequality and the assumption of the lemma.

This is a good bound, and even stronger than we need, but we have proved it only for a *fixed* pair $f, g \in \mathcal{F}$ so far. Let us take a union bound over all such pairs; there are at most N^2 of them. Then, with probability at least

$$1 - N^2\, 2\exp(-cn\varepsilon^4), \tag{8.28}$$

the lower bound (8.27) holds simultaneously for all pairs of distinct functions $f, g \in \mathcal{F}$. We can make (8.28) positive by choosing $n := \lceil C\varepsilon^{-4} \log N \rceil$ with a sufficiently large absolute constant C. Thus the random set Ω_n satisfies the conclusion of the lemma with positive probability. ■

Proof of Theorem 8.3.18 Let us choose

$$N \ge \mathcal{N}(\mathcal{F}, L^2(\mu), \varepsilon)$$

ε-separated functions in $\mathcal{F}$. (To see why they exist, recall the covering–packing relationship in Lemma 4.2.8.) Apply Lemma 8.3.19 to those functions. We obtain a subset $\Omega_n \subset \Omega$, with

$$|\Omega_n| = n \le C\varepsilon^{-4} \log N,$$

such that the restrictions of those functions onto Ω_n are still $\varepsilon/2$-separated in $L^2(\mu_n)$. We use a much weaker fact – that these restrictions are just distinct. Summarizing, we have a class $\mathcal{F}_n$ of distinct Boolean functions on Ω_n, obtained as restrictions of certain functions from $\mathcal{F}$.

Apply the Sauer–Shelah lemma (Theorem 8.3.16) for $\mathcal{F}_n$. It gives

$$N \le \left(\frac{en}{d_n}\right)^{d_n} \le \left(\frac{C\varepsilon^{-4} \log N}{d_n}\right)^{d_n},$$

[6] The inequalities "$\lesssim$" given below hide absolute constant factors.

where $d_n = \mathrm{vc}(\mathcal{F}_n)$. Simplifying this bound,[7] we conclude that

$$N \le (C\varepsilon^{-4})^{2d_n}.$$

To complete the proof, replace $d_n = \mathrm{vc}(\mathcal{F}_n)$ in the above bound by the larger quantity $d = \mathrm{vc}(\mathcal{F})$. ■

Remark 8.3.20 (Johnson–Lindenstrauss lemma for coordinate projections) You may spot a similarity between the dimension reduction lemma 8.3.19 and another dimension reduction result, the Johnson–Lindenstrauss lemma (Theorem 5.3.1). Both results state that a random projection of a set of N points onto a subspace of dimension $\log N$ preserves the geometry of the set. The difference is in the distribution of the random subspace. In the Johnson–Lindenstrauss lemma it is uniformly distributed in the Grassmannian, and in Lemma 8.3.19 it is a coordinate subspace.

Exercise 8.3.21 (Dimension reduction for covering numbers) Let $\mathcal{F}$ be a class of functions, on a probability space (Ω, Σ, μ), which are all bounded by 1 in absolute value. Let $\varepsilon \in (0,1)$. Show that there exists a number $n \le C\varepsilon^{-4} \log \mathcal{N}(\mathcal{F}, L^2(\mu), \varepsilon)$ and an n-point subset $\Omega_n \subset \Omega$ such that

$$\mathcal{N}(\mathcal{F}, L^2(\mu), \varepsilon) \le \mathcal{N}(\mathcal{F}, L^2(\mu_n), \varepsilon/4)$$

where μ_n denotes the uniform probability measure on Ω_n. ☞

Exercise 8.3.22 Theorem 8.3.18 is stated for $\varepsilon \in (0,1)$. What bound holds for larger ε?

8.3.5 Empirical Processes via the VC Dimension

Let us turn again to the concept of empirical processes that we first introduced in Section 8.2.2. There we showed how to control one specific example of an empirical process, namely a process on the class of Lipschitz functions. In this section we develop a general bound for an arbitrary class of Boolean functions.

Theorem 8.3.23 (Empirical processes via VC dimension) *Let $\mathcal{F}$ be a class of Boolean functions on a probability space (Ω, Σ, μ) with finite VC dimension $\mathrm{vc}(\mathcal{F}) \ge 1$. Let $X, X_1, X_2, \ldots, X_n$ be independent random points in Ω distributed according to the law μ. Then*

$$\mathbb{E} \sup_{f \in \mathcal{F}} \left| \frac{1}{n} \sum_{i=1}^{n} f(X_i) - \mathbb{E} f(X) \right| \le C \sqrt{\frac{\mathrm{vc}(\mathcal{F})}{n}}. \tag{8.29}$$

We can quickly derive this result from Dudley's inequality combined with the bound on the covering numbers that we have just proved in Section 8.3.4. To carry out this argument, it is helpful to preprocess the empirical process using symmetrization.

[7] To do this, note that $\log N/(2d_n) = \log(N^{1/2d_n}) \le N^{1/2d_n}$.

Exercise 8.3.24 (Symmetrization for empirical processes) Let $\mathcal{F}$ be a class of functions on a probability space (Ω, Σ, μ). Let $X, X_1, X_2, \ldots, X_n$ be random points in Ω distributed according to the law μ. Prove that

$$\mathbb{E} \sup_{f \in \mathcal{F}} \left| \frac{1}{n} \sum_{i=1}^{n} f(X_i) - \mathbb{E} f(X) \right| \leq 2 \mathbb{E} \sup_{f \in \mathcal{F}} \left| \frac{1}{n} \sum_{i=1}^{n} \varepsilon_i f(X_i) \right|$$

where $\varepsilon_1, \varepsilon_2, \ldots$ are independent symmetric Bernoulli random variables (which are also independent of $X_1, X_2, \ldots$). ☞

Proof of Theorem 8.3.23 First we use symmetrization and bound the left-hand side of (8.29) by

$$\frac{2}{\sqrt{n}} \mathbb{E} \sup_{f \in \mathcal{F}} |Z_f| \quad \text{where} \quad Z_f := \frac{1}{\sqrt{n}} \sum_{i=1}^{n} \varepsilon_i f(X_i).$$

Next we condition on (X_i), leaving all randomness in the random signs (ε_i). We are going to use Dudley's inequality to bound the process $(Z_f)_{f \in \mathcal{F}}$. For simplicity, let us drop the absolute values for Z_f for a moment; we will deal with this minor issue in Exercise 8.3.25.

To apply Dudley's inequality, we need to check that the increments of the process $(Z_f)_{f \in \mathcal{F}}$ are sub-gaussian. These are

$$\|Z_f - Z_g\|_{\psi_2} = \frac{1}{\sqrt{n}} \left\| \sum_{i=1}^{n} \varepsilon_i (f - g)(X_i) \right\|_{\psi_2} \lesssim \left(\frac{1}{n} \sum_{i=1}^{n} (f - g)(X_i)^2 \right)^{1/2}.$$

Here we have used Proposition 2.6.1 and the obvious fact that $\|\varepsilon_i\|_{\psi_2} \lesssim 1$.[8] We can interpret the last expression as the $L^2(\mu_n)$ norm of the function $f - g$, where μ_n is the uniform probability measure supported on the subset $\{X_1, \ldots, X_n\} \subset \Omega$.[9] In other words, the increments satisfy

$$\|Z_f - Z_g\|_{\psi_2} \lesssim \|f - g\|_{L^2(\mu_n)}.$$

Now we can use Dudley's inequality (Theorem 8.1.3) conditionally on (X_i) and get[10]

$$\frac{2}{\sqrt{n}} \mathbb{E} \sup_{f \in \mathcal{F}} Z_f \lesssim \frac{1}{\sqrt{n}} \mathbb{E} \int_0^1 \sqrt{\log \mathcal{N}(\mathcal{F}, L^2(\mu_n), \varepsilon)}\, d\varepsilon. \tag{8.30}$$

The expectation on the right-hand side is obviously with respect to (X_i).

Finally, we use Theorem 8.3.18 to bound the covering numbers:

$$\log \mathcal{N}(\mathcal{F}, L^2(\mu_n), \varepsilon) \lesssim \mathrm{vc}(\mathcal{F}) \log \frac{2}{\varepsilon}.$$

[8] Keep in mind that here X_i and thus $(f - g)(X_i)$ are fixed numbers owing to conditioning.

[9] Recall that we have already encountered the *empirical measure* μ_n and the $L^2(\mu_n)$ norm a few times before, in particular in Lemma 8.3.19 and its proof as well as in (8.23).

[10] The diameter of $\mathcal{F}$ gives the upper limit according to (8.13); check that the diameter is indeed bounded by 1.

When we substitute this into (8.30), we get the integral of $\sqrt{\log(2/\varepsilon)}$, which is bounded by an absolute constant. This gives

$$\frac{2}{\sqrt{n}} \mathbb{E} \sup_{f \in \mathcal{F}} Z_f \lesssim \sqrt{\frac{\mathrm{vc}(\mathcal{F})}{n}},$$

as required. ■

Exercise 8.3.25 (Reinstating absolute value)☕☕☕ In the proof above, we bounded $\mathbb{E} \sup_{f \in \mathcal{F}} Z_f$ instead of $\mathbb{E} \sup_{f \in \mathcal{F}} |Z_f|$. Give a bound for the latter quantity. ☞

Let us examine an important application of Theorem 8.3.23, which is called the Glivenko–Cantelli theorem. It addresses one of the most basic problems in statistics: how can we estimate the distribution of a random variable by sampling? Let X be a random variable with unknown cumulative distribution function (CDF)

$$F(x) = \mathbb{P}\left\{X \le x\right\}, \quad x \in \mathbb{R}.$$

Suppose we have a sample $X_1, \ldots, X_n$ of i.i.d. random variables drawn from the same distribution as X. Then we can hope that $F(x)$ could be estimated by computing the fraction of the sample points satisfying $X_i \le x$, i.e. by the *empirical distribution function*

$$F_n(x) := \frac{\left|\{i \in [n] \colon X_i \le x\}\right|}{n}, \quad x \in \mathbb{R}.$$

Note that $F_n(x)$ is a *random* function.

The quantitative law of large numbers gives

$$\mathbb{E} \left|F_n(x) - F(x)\right| \le \frac{C}{\sqrt{n}} \quad \text{for every } x \in \mathbb{R}.$$

(Check this! Recall the variance computation in Section 1.3, but do it for the indicator random variables $\mathbf{1}_{\{X_i \le x\}}$ instead of X_i.)

The Glivenko–Cantelli theorem is a stronger statement, which says that F_n approximates F *uniformly* over $x \in \mathbb{R}$.

Theorem 8.3.26 (Glivenko–Cantelli theorem[11]) *Let $X_1, \ldots, X_n$ be independent random variables with common cumulative distribution function F. Then*

$$\mathbb{E} \left\|F_n - F\right\|_\infty = \mathbb{E} \sup_{x \in \mathbb{R}} \left|F_n(x) - F(x)\right| \le \frac{C}{\sqrt{n}}.$$

Proof This result is a particular case of Theorem 8.3.23. Indeed, let $\Omega = \mathbb{R}$, let $\mathcal{F}$ consist of the indicators of all half-bounded intervals, i.e.

$$\mathcal{F} := \left\{\mathbf{1}_{(-\infty,x]} \colon x \in \mathbb{R}\right\},$$

[11] The classical statement of the Glivenko–Cantelli theorem is about almost sure convergence, and we do not give it here. However, it can be obtained from a high-probability version of the same argument using the Borel–Cantelli lemma.

and let the measure μ be the distribution[12] of X_i. As we know from Example 8.3.2, $\mathrm{vc}(\mathcal{F}) \le 2$. Thus Theorem 8.3.23 immediately implies the conclusion. ■

Example 8.3.27 (Discrepancy) The Glivenko–Cantelli theorem can be easily generalized to random vectors. (Do this!) Let us give an illustration for $\mathbb{R}^2$. Draw a sample of i.i.d. points $X_1, \dots, X_n$ from the uniform distribution on the unit square $[0,1]^2$ on the plane; see Figure 8.8. Consider the class $\mathcal{F}$ of indicators of all circles in that square. From Exercise 8.3.6 we know that $\mathrm{vc}(\mathcal{F}) = 3$. (Why does intersection with the square not affect the VC dimension?)

Apply Theorem 8.3.23. The sum $\sum_{i=1}^n f(X_i)$ is just the number of points in the circle with indicator function f, and the expectation $\mathbb{E} f(X)$ is the area of that circle. Then we can interpret the conclusion of Theorem 8.3.23 as follows. With high probability, a random sample of points $X_1, \dots, X_n$ satisfies the following: for every circle $\mathcal{C}$ in the square $[0,1]^2$,

$$\text{number of points in } \mathcal{C} = \mathrm{Area}(\mathcal{C})\, n + O(\sqrt{n}).$$

This is an example of a result in *geometric discrepancy* theory. The same result holds not only for circles but for half-planes, rectangles, squares, triangles, polygons with $O(1)$ vertices, and any other class with bounded VC dimension.

Remark 8.3.28 (Uniform Glivenko–Cantelli classes) A class of real-valued functions $\mathcal{F}$ on a set Ω is called a *uniform Glivenko–Cantelli class* if, for any $\varepsilon > 0$,

$$\lim_{n\to\infty} \sup_{\mu} \mathbb{P}\left\{ \sup_{f\in\mathcal{F}} \left| \frac{1}{n}\sum_{i=1}^n f(X_i) - \mathbb{E} f(X) \right| > \varepsilon \right\} = 0,$$

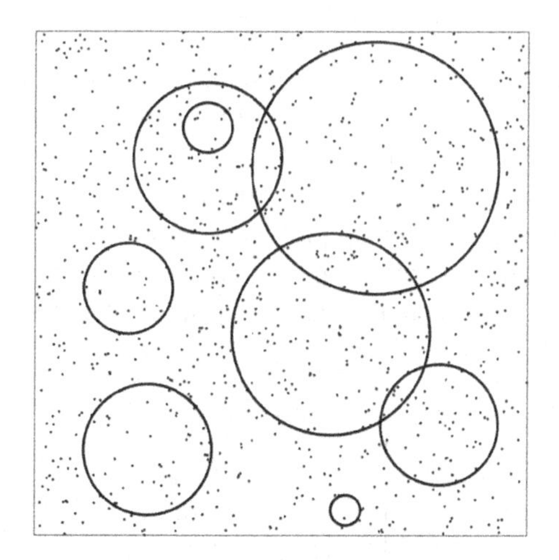

Figure 8.8 According to the uniform deviation inequality from Theorem 8.3.23, all circles have a fair share of the random sample of points. The number of points in each circle is proportional to its area, with $O(\sqrt{n})$ error.

[12] Precisely, we define $\mu(A) := \mathbb{P}\{X \in A\}$ for every (Borel) subset $A \subset \mathbb{R}$.

where the supremum is taken over all probability measures μ on Ω and the points $X, X_1, \ldots, X_n$ are sampled from Ω according to the law μ. Theorem 8.3.23 followed by Markov's inequality yields the conclusion that any class of Boolean functions with finite VC dimension is uniform Glivenko–Cantelli.

Exercise 8.3.29 (Sharpness) Prove that any class of Boolean functions with infinite VC dimension is not uniform Glivenko–Cantelli. ☞

Exercise 8.3.30 (A simpler, weaker, bound) Use the Sauer–Shelah lemma directly, instead of Pajor's lemma, to prove a weaker version of the uniform deviation inequality (8.29) with

$$C\sqrt{\frac{d}{n}\log\frac{en}{d}}$$

on the right-hand side, where $d = \mathrm{vc}(\mathcal{F})$. ☞

8.4 Application: Statistical Learning Theory

Statistical learning theory, or machine learning, allows one to make predictions based on data. A typical problem of statistical learning can be stated mathematically as follows. Consider a function $T : \Omega \to \mathbb{R}$ on some set Ω, which we call a *target function.* Suppose that T is unknown. We would like to learn T from its values on a finite sample of points $X_1, \ldots, X_n \in \Omega$. We assume that these points are independently sampled according to some common probability distribution $\mathbb{P}$ on Ω. Thus, our *training data* is

$$(X_i, T(X_i)), \quad i = 1, \ldots, n. \tag{8.31}$$

Our ultimate goal is to use the training data to make a good *prediction* of $T(X)$ for a new random point $X \in \Omega$, which was not in the training sample but is sampled from the same distribution; see Figure 8.9 for illustration.

You may notice some similarity between learning problems and Monte-Carlo integration, which we studied in Section 8.2.1. In both problems, we are trying to infer something about a function from its values on a random sample of points. But now our task is more difficult, as we are trying to learn the function itself and not just its integral, or average value, on Ω.

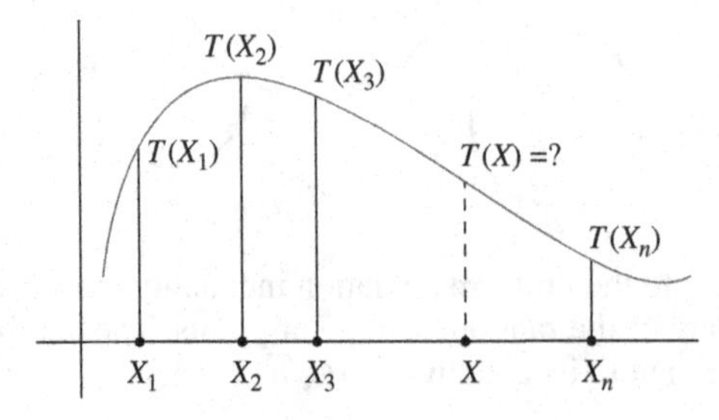

Figure 8.9 In a general learning problem, we are trying to learn an unknown function $T : \Omega \to \mathbb{R}$ (a "target function") from its values on a training sample $X_1, \ldots, X_n$ of i.i.d. points. The goal is to predict $T(X)$ for a new random point X.

8.4.1 Classification Problems

An important class of learning problems comprises classification problems, where the function T is Boolean (takes values 0 or 1); thus T classifies points in Ω into two classes.

Example 8.4.1 Consider a health study on a sample of n patients. We record d various health parameters of each patient, such as blood pressure, body temperature, etc., arranging them as a vector $X_i \in \mathbb{R}^d$. Suppose we also know whether each patient has diabetes, and we encode this information as a binary number $T(X_i) \in \{0, 1\}$ ($0 =$ healthy, $1 =$ sick). Our goal is to learn from this training sample how to diagnose diabetes. We want to learn the target function $T : \mathbb{R}^d \to \{0, 1\}$ which would output a diagnosis for *any* person based on his or her d health parameters.

To extend the example, the vector X_i could contain the d gene expressions of the ith patient. Our goal is to learn to diagnose a certain disease based on the patient's genetic information.

Figure 8.10c illustrates a classification problem where X is a random vector on the plane and the label Y can take the value 0 or 1 as in Example 8.4.1. A solution of this classification problem can be described as a partition of the plane into two regions, one where $f(X) = 0$ (healthy) and another where $f(X) = 1$ (sick). On the basis of this solution, one can diagnose new patients by determining the region into which their parameter vectors X fall.

8.4.2 Risk, Fit, and Complexity

A solution to the learning problem can be expressed as a function $f \colon \Omega \to \mathbb{R}$. We would naturally want f to be as close to the target T as possible, so we would like to choose f that minimizes the *risk*

$$R(f) := \mathbb{E}\,(f(X) - T(X))^2 . \tag{8.32}$$

Here X denotes a random variable with distribution $\mathbb{P}$, i.e. with the same distribution as the sample points $X_1, \ldots, X_n \in \Omega$.

Example 8.4.2 In classification problems, T and f are Boolean functions and thus

$$R(f) = \mathbb{P}\{f(X) \neq T(X)\}. \tag{8.33}$$

(Check!) So the risk is just the probability of misclassification, e.g., the misdiagnosis for a patient.

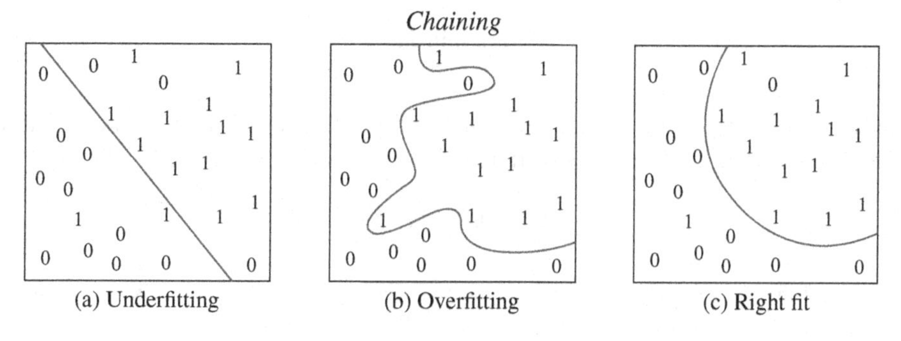

(a) Underfitting (b) Overfitting (c) Right fit

Figure 8.10 Tradeoff between fit and complexity

How much data do we need to learn, i.e how large does the sample size n need to be? This depends on the *complexity* of the problem. We need more data if we believe that the target function $T(X)$ may depend on X in an intricate way; otherwise we need less. Usually we do not know the complexity a priori. So we may restrict the complexity of the candidate functions f, insisting that our solution f must belong to some given class of functions $\mathcal{F}$ called the *hypothesis space*.

But how do we choose the hypothesis space $\mathcal{F}$ for a learning problem at hand? Although there is no general rule, the choice of $\mathcal{F}$ should be based on a *tradeoff between fit and complexity*. Suppose that we choose $\mathcal{F}$ to be too small; for example, we insist that the interface between healthy ($f(x) = 0$) and sick diagnoses ($f(x) = 1$) be a line, as in Figure 8.10a. Although we can learn such a simple function f with less data, we have probably oversimplified the problem. The linear functions do not capture the essential trends in this data, and this will lead to a big risk $R(f)$.

If, however, we choose $\mathcal{F}$ too large then this may result in *overfitting*, where we are essentially fitting f to the noise, as in Figure 8.10b. Furthermore, in this case we need a lot of data to learn such complicated functions.

A good choice of $\mathcal{F}$ is one that avoids either underfitting or overfitting, and captures the essential trends in the data, just as in Figure 8.10c.

8.4.3 Empirical Risk

What would be an optimal solution to the learning problem on the basis of the training data? Ideally, we would like to find a function f^* from the hypothesis space $\mathcal{F}$ which would minimize the risk[13] $R(f) = \mathbb{E}\,(f(X) - T(X))^2$, that is,

$$f^* := \arg\min_{f \in \mathcal{F}} R(f).$$

If we are lucky and choose the hypothesis space $\mathcal{F}$ so that it contains the target function T, then the risk is zero. Unfortunately, we cannot compute the risk $R(f)$ and thus f^* from the training data. But we can try to *estimate* $R(f)$ and f^*.

Definition 8.4.3 The *empirical risk* for a function $f\colon \Omega \to \mathbb{R}$ is defined as

$$R_n(f) := \frac{1}{n} \sum_{i=1}^{n} (f(X_i) - T(X_i))^2. \tag{8.34}$$

Denote by f_n^* a function in the hypothesis space $\mathcal{F}$ which minimizes the empirical risk:

$$f_n^* := \arg\min_{f \in \mathcal{F}} R_n(f).$$

Note that both $R_n(f)$ and f_n^* can be computed from the data. The outcome of learning from the data is thus f_n^*. The main question is: how large is the *excess risk*

$$R(f_n^*) - R(f^*)$$

[13] We assume for simplicity that the minimum is attained; an approximate minimizer could be used as well.

produced by our having to learn from a finite sample of size n? We give an answer in the next section.

8.4.4 Bounding the Excess Risk by the VC Dimension

Let us specialize to classification problems where the target T is a Boolean function.

Theorem 8.4.4 (Excess risk via VC dimension) *Assume that the target T is a Boolean function and that the hypothesis space $\mathcal{F}$ is a class of Boolean functions with finite VC dimension* $\mathrm{vc}(\mathcal{F}) \geq 1$. *Then*

$$\mathbb{E}\, R(f_n^*) \leq R(f^*) + C\sqrt{\frac{\mathrm{vc}(\mathcal{F})}{n}}.$$

We can deduce this theorem from a uniform deviation inequality that we proved in Theorem 8.3.23. The following elementary observation will help us connect these two results.

Lemma 8.4.5 (Excess risk via uniform deviations) *We have*

$$R(f_n^*) - R(f^*) \leq 2 \sup_{f \in \mathcal{F}} |R_n(f) - R(f)|$$

pointwise.

Proof Denote $\varepsilon := \sup_{f \in \mathcal{F}} |R_n(f) - R(f)|$. Then

$$\begin{aligned} R(f_n^*) &\leq R_n(f_n^*) + \varepsilon \quad (\text{since } f_n^* \in \mathcal{F} \text{ by construction}) \\ &\leq R_n(f^*) + \varepsilon \quad (\text{since } f_n^* \text{ minimizes } R_n \text{ in the class } \mathcal{F}) \\ &\leq R(f^*) + 2\varepsilon \quad (\text{since } f^* \in \mathcal{F} \text{ by construction}). \end{aligned}$$

Subtracting $R(f^*)$ from both sides, we get the desired inequality. ■

Proof of Theorem 8.4.4 By Lemma 8.4.5, it will be enough to show that

$$\mathbb{E} \sup_{f \in \mathcal{F}} |R_n(f) - R(f)| \lesssim \sqrt{\frac{\mathrm{vc}(\mathcal{F})}{n}}.$$

Recalling the definitions (8.34) and (8.32) of the empirical and true (population) risk, we express the left-hand side as

$$\mathbb{E} \sup_{\ell \in \mathcal{L}} \left| \frac{1}{n} \sum_{i=1}^{n} \ell(X_i) - \mathbb{E}\, \ell(X) \right| \tag{8.35}$$

where $\mathcal{L}$ is the class of Boolean functions defined as

$$\mathcal{L} = \{(f - T)^2 \colon f \in \mathcal{F}\}.$$

The uniform deviation bound from Theorem 8.3.23 could be used at this point, but it would only give a bound in terms of the VC dimension of $\mathcal{L}$, and it is not clear how to relate that back to the VC dimension of $\mathcal{F}$. Instead, let us recall that in the proof of Theorem 8.3.23, we first bounded (8.35) by

$$\frac{1}{\sqrt{n}} \mathbb{E} \int_0^1 \sqrt{\log \mathcal{N}(\mathcal{L}, L^2(\mu_n), \varepsilon)}\, d\varepsilon, \tag{8.36}$$

up to an absolute constant factor. It is not hard to see that the covering numbers of $\mathcal{L}$ and $\mathcal{F}$ are related by the inequality

$$\mathcal{N}(\mathcal{L}, L^2(\mu_n), \varepsilon) \le \mathcal{N}(\mathcal{F}, L^2(\mu_n), \varepsilon/4) \quad \text{for any } \varepsilon \in (0, 1). \tag{8.37}$$

(We will check this inequality accurately in Exercise 8.4.6.) So we may replace $\mathcal{L}$ by $\mathcal{F}$ in (8.36), paying the price of an absolute constant factor (check!). We then follow the rest of the proof of Theorem 8.3.23 and conclude that (8.36) is bounded by

$$\sqrt{\frac{\mathrm{vc}(\mathcal{F})}{n}},$$

as required. ■

Exercise 8.4.6 Check the inequality (8.37).

8.4.5 Interpretation and Examples

What does Theorem 8.4.4 really say about learning? It quantifies the risk of having to learn from limited data, which we called the excess risk. Theorem 8.4.4 states that on average, the excess risk of learning from a finite sample of size n is proportional to $\sqrt{\mathrm{vc}(\mathcal{F})/n}$. Equivalently, if we want to bound the expected excess risk by ε, all we need to do is to take a sample of size

$$n \asymp \varepsilon^{-2}\, \mathrm{vc}(\mathcal{F}).$$

This result answers the question of how much training data we need for learning. And the answer is: *it is enough if the sample size n exceeds the VC dimension of the hypothesis class* $\mathcal{F}$ (up to some constant factor).

Let us illustrate this principle by thoroughly working out the specific learning problem from Figure 8.10. We are trying to learn an unknown function $T\colon \mathbb{R}^2 \to \{0, 1\}$. This is a classification problem where the function T assigns labels 0 or 1 to the points on the plane, and we are trying to learn those labels.

First, we collect the training data, consisting of n points $X_1, \ldots, X_n$ on the plane whose labels $T(X_i)$ we know. We assume that the points X_i are sampled at random according to some probability distribution $\mathbb{P}$ on the plane.

Next, we need to choose a hypothesis space $\mathcal{F}$. This is a class of Boolean functions on the plane where we will be looking for a solution to our learning problem. We need to make sure that $\mathcal{F}$ is neither too large (to prevent overfitting) nor too small (to prevent underfitting). We may expect that the interface between the two classes of Boolean functions is a nontrivial but not too intricate curve, such as the arc in Figure 8.10c. For example, it may be reasonable to include in $\mathcal{F}$ the indicator functions of all circles on the plane.[14] So let us choose

$$\mathcal{F} := \{\mathbf{1}_C\colon \text{circles } C \subset \mathbb{R}^2\}. \tag{8.38}$$

Recall from Exercise 8.3.6 that $\mathrm{vc}(\mathcal{F}) = 3$.

[14] We can also include all half-spaces, which we can think of as circles with infinite radii centered at infinity.

Next, we set up the empirical risk

$$R_n(f) := \frac{1}{n} \sum_{i=1}^{n} (f(X_i) - T(X_i))^2.$$

We can compute the empirical risk from data for any given function f on the plane. Finally, we minimize the empirical risk over our hypothesis class $\mathcal{F}$, and thus compute

$$f_n^* := \arg\min_{f \in \mathcal{F}} R_n(f).$$

Exercise 8.4.7 Check that f_n^* is a function in $\mathcal{F}$ that minimizes the number of data points X_i where the function disagrees with the labels $T(X_i)$.

We output the function f_n^* as the solution of the learning problem. By computing $f_n^*(x)$, we can make a prediction for the labels of the points x that were not in the training set.

How reliable is this prediction? We quantified the predicting power of a Boolean function f with the concept of the *risk* $R(f)$. It gives the probability that f assigns the wrong label to a random point X sampled from the same distribution on the plane as the data points:

$$R(f) = \mathbb{P}\{f(X) \neq T(X)\}.$$

Using Theorem 8.4.4 and recalling that $\mathrm{vc}(\mathcal{F}) = 3$, we get a bound on the risk for our solution f_n^*:

$$\mathbb{E}\, R(f_n^*) \leq R(f^*) + \frac{C}{\sqrt{n}}.$$

Thus, on average, our solution f_n^* gives correct predictions with almost the same probability – within $1/\sqrt{n}$ error – as the best available function f^* in the hypothesis class $\mathcal{F}$, i.e., the best chosen circle.

Exercise 8.4.8 (Random outputs) Our learning problem model (8.31) postulates that the output $T(X)$ must be completely determined by the input X. This is rarely the case in practice. For example, it is not realistic to assume that the diagnosis $T(X) \in \{0, 1\}$ of a disease is completely determined by the available genetic information X. What is more often true is that the output Y is a random variable, which is correlated with the input X; the goal of learning is still to predict Y from X as well as possible.

Extend the theory of learning leading up to Theorem 8.4.4 for training data of the form

$$(X_i, Y_i), \quad i = 1, \ldots, n,$$

where (X_i, Y_i) are independent copies of a pair (X, Y) consisting of an input random point $X \in \Omega$ and an output random variable Y.

Exercise 8.4.9 (Learning in the class of Lipschitz functions) Consider the hypothesis class of Lipschitz functions

$$\mathcal{F} := \left\{ f : [0, 1] \to \mathbb{R},\ \|f\|_{\mathrm{Lip}} \leq L \right\}$$

and a target function $T : [0, 1] \to [0, 1]$.

(a) Show that the random process $X_f := R_n(f) - R(f)$ has sub-gaussian increments:

$$\|X_f - X_g\|_{\psi_2} \leq \frac{CL}{\sqrt{n}}\|f - g\|_\infty \quad \text{for all } f, g \in \mathcal{F}.$$

(b) Use Dudley's inequality to deduce that

$$\mathbb{E} \sup_{f \in \mathcal{F}} |R_n(f) - R(f)| \leq \frac{C(L+1)}{\sqrt{n}}.$$

☞

(c) Conclude that the excess risk satisfies

$$R(f_n^*) - R(f^*) \leq \frac{C(L+1)}{\sqrt{n}}.$$

8.5 Generic Chaining

Dudley's inequality is a simple and useful tool for bounding a general random process. Unfortunately, as we saw in Exercise 8.1.12, Dudley's inequality can be loose. The reason behind this is that the covering numbers $\mathcal{N}(T, d, \varepsilon)$ do not contain enough information to control the magnitude of $\mathbb{E}\sup_{t\in T} X_t$.

8.5.1 A Makeover of Dudley's Inequality

Fortunately, there *is* a way to obtain accurate two-sided bounds on $\mathbb{E}\sup_{t\in T} X_t$ for sub-gaussian processes $(X_t)_{t\in T}$ in terms of the geometry of T. This method is called *generic chaining*, and it is essentially a sharpening of the chaining method developed in the proof of Dudley's inequality (Theorem 8.1.4). Recall that the outcome of chaining was the bound (8.12):

$$\mathbb{E}\sup_{t\in T} X_t \lesssim \sum_{k=\kappa+1}^{\infty} \varepsilon_{k-1}\sqrt{\log |T_k|}. \tag{8.39}$$

Here the ε_k are decreasing positive numbers and the T_k are ε_k-nets of T such that $|T_\kappa| = 1$. To be specific, in the proof of Theorem 8.1.4 we chose

$$\varepsilon_k = 2^{-k} \quad \text{and} \quad |T_k| = \mathcal{N}(T, d, \varepsilon_k),$$

so the $T_k \subset T$ were the smallest ε_k-nets of T.

In preparation for generic chaining, let us now turn around our choice of ε_k and T_k. Instead of fixing ε_k and operating with the smallest possible cardinality of T_k, let us fix the cardinality of T_k and operate with the smallest possible ε_k. Namely, let us fix some subsets $T_k \subset T$ such that

$$|T_0| = 1, \quad |T_k| \leq 2^{2^k}, \quad k = 1, 2, \ldots \tag{8.40}$$

Such a sequence of sets $(T_k)_{k=0}^{\infty}$ is called an *admissible sequence*. Put

$$\varepsilon_k = \sup_{t\in T} d(t, T_k),$$

where $d(t, T_k)$ denotes the distance[15] from the point t to the set T_k. Then each T_k is an ε_k-net of T. With this choice of ε_k and T_k, the chaining bound (8.39) becomes

$$\mathbb{E} \sup_{t \in T} X_t \lesssim \sum_{k=1}^{\infty} 2^{k/2} \sup_{t \in T} d(t, T_{k-1}).$$

After re-indexing we conclude that

$$\mathbb{E} \sup_{t \in T} X_t \lesssim \sum_{k=0}^{\infty} 2^{k/2} \sup_{t \in T} d(t, T_k). \tag{8.41}$$

8.5.2 Talagrand's γ_2-Functional and Generic Chaining

So far, nothing has really happened. The bound (8.41) is just an equivalent way to state Dudley's inequality. The important step will come now. Generic chaining will allow us to pull the supremum *outside* the sum in (8.41). The resulting important quantity has a name:

Definition 8.5.1 (Talagrand's γ_2-functional) Let (T, d) be a metric space. A sequence of subsets $(T_k)_{k=0}^{\infty}$ of T is called an *admissible sequence* if the cardinalities of T_k satisfy (8.40). The γ_2*-functional* of T is defined as

$$\gamma_2(T, d) = \inf_{(T_k)} \sup_{t \in T} \sum_{k=0}^{\infty} 2^{k/2} d(t, T_k),$$

where the infimum is taken with respect to all admissible sequences.

Since the supremum in the γ_2-functional is outside the sum, it is smaller than the Dudley's inequality sum in (8.41). The difference between the γ_2-functional and Dudley's sum can look minor, but sometimes it is real:

Exercise 8.5.2 (γ_2-functional and Dudley's sum) ☕☕☕ Consider the same set $T \subset \mathbb{R}^n$ as in Exercise 8.1.12, i.e.,

$$T := \left\{ \frac{e_k}{\sqrt{1 + \log k}}, \quad k = 1, \ldots, n \right\}.$$

(a) Show that the γ_2-functional of T (with respect to the Euclidean metric) is bounded, i.e.,

$$\gamma_2(T, d) = \inf_{(T_k)} \sup_{t \in T} \sum_{k=0}^{\infty} 2^{k/2} d(t, T_k) \leq C.$$

☞

(b) Check that Dudley's sum is unbounded, i.e.,

$$\inf_{(T_k)} \sum_{k=0}^{\infty} 2^{k/2} \sup_{t \in T} d(t, T_k) \to \infty$$

as $n \to \infty$.

[15] Formally, the distance from a point $t \in T$ to a subset $A \subset T$ in a metric space T is defined as $d(t, A) := \inf\{d(t, a) \colon a \in A\}$.

We now state an improvement of Dudley's inequality, in which Dudley's sum (or integral) is replaced by a tighter quantity, the γ_2-functional.

Theorem 8.5.3 (Generic chaining bound) *Let $(X_t)_{t\in T}$ be a mean-zero random process on a metric space (T, d) with sub-gaussian increments as in* (8.1). *Then*

$$\mathbb{E} \sup_{t\in T} X_t \le CK\gamma_2(T, d).$$

Proof We proceed with the same chaining method as was introduced in the proof of Dudley's inequality (Theorem 8.1.4), but we will do the chaining more accurately.

Step 1: Chaining set-up. As before, we may assume that $K = 1$ and that T is finite. Let (T_k) be an admissible sequence of subsets of T, and denote $T_0 = \{t_0\}$. We now walk from t_0 to a general point $t \in T$ along the chain

$$t_0 = \pi_0(t) \to \pi_1(t) \to \cdots \to \pi_K(t) = t$$

of points $\pi_k(t) \in T_k$ that are chosen as best approximations to t in T_k, i.e.

$$d(t, \pi_k(t)) = d(t, T_k).$$

The displacement $X_t - X_{t_0}$ can be expressed as a telescoping sum similar to (8.9):

$$X_t - X_{t_0} = \sum_{k=1}^{K} (X_{\pi_k(t)} - X_{\pi_{k-1}(t)}). \tag{8.42}$$

Step 2: Controlling the increments. This is where we need to be more accurate than in Dudley's inequality. We would like to have a uniform bound on the increments, a bound that would state with high probability that

$$\left|X_{\pi_k(t)} - X_{\pi_{k-1}(t)}\right| \le 2^{k/2} d(t, T_k) \quad \forall k \in \mathcal{N}, \quad \forall t \in T. \tag{8.43}$$

Summing these inequalities over all k would lead to a desired bound in terms of $\gamma_2(T, d)$.

To prove (8.43), let us fix k and t first. The sub-gaussian assumption tells us that

$$\left\|X_{\pi_k(t)} - X_{\pi_{k-1}(t)}\right\|_{\psi_2} \le d(\pi_k(t), \pi_{k-1}(t)).$$

This means that, for every $u \ge 0$, the event

$$\left|X_{\pi_k(t)} - X_{\pi_{k-1}(t)}\right| \le Cu2^{k/2} d(\pi_k(t), \pi_{k-1}(t)) \tag{8.44}$$

holds with probability at least

$$1 - 2\exp(-8u^2 2^k).$$

(To obtain the constant 8, choose the absolute constant C large enough.)

We can now unfix $t \in T$ by taking a union bound over

$$|T_k|\,|T_{k-1}| \le |T_k|^2 = 2^{2^{k+1}}$$

possible pairs $(\pi_k(t), \pi_{k-1}(t))$. Similarly, we can unfix k by a union bound over all $k \in \mathbb{N}$. Then the probability that the bound (8.44) holds simultaneously for all $t \in T$ and $k \in \mathbb{N}$ is at least

$$1 - \sum_{k=1}^{\infty} 2^{2^{k+1}} \times 2\exp(-8u^2 2^k) \geq 1 - 2\exp(-u^2).$$

if $u > c$. (Check the last inequality!)

Step 3: Summing up the increments. In the event that the bound (8.44) does hold for all $t \in T$ and $k \in \mathbb{N}$, we can sum the inequalities over $k \in \mathcal{N}$ and substitute the result into the chaining sum (8.42). This yields

$$|X_t - X_{t_0}| \leq Cu \sum_{k=1}^{\infty} 2^{k/2} d(\pi_k(t), \pi_{k-1}(t)). \tag{8.45}$$

By the triangle inequality we have

$$d(\pi_k(t), \pi_{k-1}(t)) \leq d(t, \pi_k(t)) + d(t, \pi_{k-1}(t)).$$

Using this bound and re-indexing, we find that the right-hand side of (8.45) can be bounded by $\gamma_2(T, d)$, that is,

$$|X_t - X_{t_0}| \leq C_1 u \gamma_2(T, d).$$

(Check!) Taking the supremum over T yields

$$\sup_{t \in T} |X_t - X_{t_0}| \leq C_2 u \gamma_2(T, d).$$

Recall that this inequality holds with probability at least $1 - 2\exp(-u^2)$ for any $u > c$. This means that the magnitude in question is a sub-gaussian random variable:

$$\left\| \sup_{t \in T} |X_t - X_{t_0}| \right\|_{\psi_2} \leq C_3 \gamma_2(T, d).$$

This quickly implies the conclusion of Theorem 8.5.3. (Check!) ∎

Remark 8.5.4 (Supremum of increments) Like Dudley's inequality (Remark 8.1.5), generic chaining gives the uniform bound

$$\mathbb{E} \sup_{t,s \in T} |X_t - X_s| \leq CK\gamma_2(T, d).$$

which is valid even without the mean-zero assumption $\mathbb{E} X_t = 0$.

The argument above gives not only a bound on the expectation but also a tail bound for $\sup_{t \in T} X_t$. Let us now give a better tail bound, similar to that in Theorem 8.1.6 for Dudley's inequality.

Theorem 8.5.5 (Generic chaining: tail bound) *Let $(X_t)_{t \in T}$ be a random process on a metric space (T, d) with sub-gaussian increments as in* (8.1)*. Then, for every $u \geq 0$, the event*

$$\sup_{t,s \in T} |X_t - X_s| \le CK\Big(\gamma_2(T,d) + u \operatorname{diam}(T)\Big)$$

holds with probability at least $1 - 2\exp(-u^2)$.

Exercise 8.5.6 ☕☕☕☕ Prove Theorem 8.5.5. To this end, state and use a variant of the increment bound (8.44) with $u + 2^{k/2}$ instead of $u2^{k/2}$. At the end of the argument you will need a bound on the sum of steps $\sum_{k=1}^{\infty} d(\pi_k(t), \pi_{k-1}(t))$. For this, modify the chain $\{\pi_k(t)\}$ by doing a "lazy walk" on it. Stay at the current point $\pi_k(t)$ for a few steps (say, $q-1$) until the distance to t improves by a factor 2, that is, until

$$d(t, \pi_{k+q}(t)) \le \tfrac{1}{2} d(t, \pi_k(t)),$$

then jump to $\pi_{k+q}(t)$. This will make the sum of the steps geometrically convergent.

Exercise 8.5.7 (Dudley's integral vs. γ_2-functional) ☕☕☕ Show that the γ_2-functional is bounded by Dudley's integral. Namely, show that, for any metric space (T,d), one has

$$\gamma_2(T,d) \le C \int_0^{\infty} \sqrt{\log \mathcal{N}(T,d,\varepsilon)}\, d\varepsilon.$$

8.6 Talagrand's Majorizing Measure and Comparison Theorems

Talagrand's γ_2-functional introduced in Definition 8.5.1 has some advantages and disadvantages over Dudley's integral. A disadvantage is that $\gamma_2(T,d)$ is usually harder to compute than the metric entropy that defines Dudley's integral. Indeed, it could take a real effort to construct a good admissible sequence of sets. However, unlike Dudley's integral, the γ_2-functional gives a bound on Gaussian processes that is *optimal* up to an absolute constant. This is the content of the following theorem.

Theorem 8.6.1 (Talagrand's majorizing measure theorem) *Let $(X_t)_{t\in T}$ be a mean-zero Gaussian process on a set T. Consider the canonical metric defined on T by* (7.13), *i.e.,* $d(t,s) = \|X_t - X_s\|_{L^2}$. *Then*

$$c\,\gamma_2(T,d) \le \mathbb{E} \sup_{t\in T} X_t \le C\,\gamma_2(T,d).$$

The upper bound in Theorem 8.6.1 follows directly from generic chaining (Theorem 8.5.3). The lower bound is harder to obtain. Its proof, which we do not present in this book, can be thought of as a far-reaching multi-scale strengthening of Sudakov's inequality (Theorem 7.4.1).

Note that the upper bound, as we know from Theorem 8.5.3, holds for any *sub-gaussian* process. Therefore, by combining the upper and lower bounds, we can deduce that any sub-gaussian process is bounded (via a γ_2-functional) by a Gaussian process. Let us state this important comparison result.

Corollary 8.6.2 (Talagrand's comparison inequality) *Let $(X_t)_{t\in T}$ be a mean-zero random process on a set T and let $(Y_t)_{t\in T}$ be a mean-zero Gaussian process. Assume that, for all $t,s \in T$, we have*

$$\|X_t - X_s\|_{\psi_2} \le K \|Y_t - Y_s\|_{L^2}.$$

Then

$$\mathbb{E} \sup_{t \in T} X_t \le CK\, \mathbb{E} \sup_{t \in T} Y_t.$$

Proof Consider the canonical metric on T given by $d(t,s) = \|Y_t - Y_s\|_{L^2}$. Apply the generic chaining bound (Theorem 8.5.3) followed by the lower bound in the majorizing measure Theorem 8.6.1. Thus we get

$$\mathbb{E} \sup_{t \in T} X_t \le CK \gamma_2(T,d) \le CK\, \mathbb{E} \sup_{t \in T} Y_t.$$

The proof is complete. ■

Corollary 8.6.2 extends the Sudakov–Fernique inequality (Theorem 7.2.11) to sub-gaussian processes. All we pay for such an extension is an absolute constant factor.

Let us apply Corollary 8.6.2 for a canonical Gaussian process

$$Y_x = \langle g, x \rangle, \quad x \in T,$$

defined on a subset $T \subset \mathbb{R}^n$. Recall from Section 7.5 that the magnitude of this process,

$$w(T) = \mathbb{E} \sup_{x \in T} \langle g, x \rangle,$$

is the *Gaussian width* of T. We immediately obtain the following corollary.

Corollary 8.6.3 (Talagrand's comparison inequality: geometric form) *Let $(X_x)_{x \in T}$ be a mean-zero random process on a subset $T \subset \mathbb{R}^n$. Assume that, for all $x, y \in T$, we have*

$$\|X_x - X_y\|_{\psi_2} \le K \|x - y\|_2.$$

Then

$$\mathbb{E} \sup_{x \in T} X_x \le CKw(T).$$

Exercise 8.6.4 (Bound on absolute values) Let $(X_x)_{x \in T}$ be a random process (not necessarily mean-zero) on a subset $T \subset \mathbb{R}^n$. Assume that[16] $X_0 = 0$, and for all $x, y \in T \cup \{0\}$ we have

$$\|X_x - X_y\|_{\psi_2} \le K \|x - y\|_2.$$

Prove that[17]

$$\mathbb{E} \sup_{x \in T} |X_x| \le CK\gamma(T).$$

Exercise 8.6.5 (Tail bound) Show that, in the setting of Exercise 8.6.4, for every $u \ge 0$ we have[18]

$$\sup_{x \in T} |X_x| \le CK\big(w(T) + u\, \mathrm{rad}(T)\big)$$

with probability at least $1 - 2\exp(-u^2)$.

[16] If $0 \notin T$, then simply define $X_0 := 0$.

[17] Recall from Section 7.6.2 that $\gamma(T)$ is the Gaussian complexity of T.

[18] Here as usual $\mathrm{rad}(T)$ denotes the radius of T.

Exercise 8.6.6 (Higher moments of the deviation) Check that, in the setting of Exercise 8.6.4,

$$\left(\mathbb{E}\sup_{x\in T}|X_x|^p\right)^{1/p} \le C\sqrt{p}\,K\gamma(T).$$

8.7 Chevet's Inequality

Talagrand's comparison inequality (Corollary 8.6.2) has several important consequences. We cover one application now; others will appear later in this book.

In this section we look for a uniform bound for a random quadratic form, i.e., a bound on the quantity

$$\sup_{x\in T,\, y\in S} \langle Ax, y\rangle, \tag{8.46}$$

where A is a random matrix and T and S are general sets.

We have already encountered problems of this type when we analyzed the norms of random matrices, namely in the proofs of Theorems 4.4.5 and 7.3.1. In those situations the sets T and S were Euclidean balls. This time, we let T and S be arbitrary geometric sets. Our bound on (8.46) will depend on just two geometric parameters of T and S: the *Gaussian width* and the *radius*, which we define as

$$\mathrm{rad}(T) := \sup_{x\in T}\|x\|_2. \tag{8.47}$$

Theorem 8.7.1 (Sub-gaussian Chevet inequality) *Let A be an $m\times n$ random matrix whose entries A_{ij} are independent mean-zero sub-gaussian random variables. Let $T\subset\mathbb{R}^n$ and $S\subset\mathbb{R}^m$ be arbitrary bounded sets. Then*

$$\mathbb{E}\sup_{x\in T,\, y\in S}\langle Ax, y\rangle \le CK\left(w(T)\,\mathrm{rad}(S) + w(S)\,\mathrm{rad}(T)\right)$$

where $K=\max_{ij}\|A_{ij}\|_{\psi_2}$.

Before we prove this theorem, let us make one simple illustration of its use. Setting $T=S^{n-1}$ and $S=S^{m-1}$, we recover a bound on the operator norm of A,

$$\mathbb{E}\|A\| \le CK(\sqrt{n}+\sqrt{m}),$$

which we obtained in Section 4.4.2 using a different method.

Proof of Theorem 8.7.1 We use the same method as in our proof of the sharp bound on Gaussian random matrices (Theorem 7.3.1). That argument was based on the Sudakov–Fernique comparison inequality; this time, we use the more general Talagrand's comparison inequality.

Without loss of generality, set $K=1$. We would like to bound the random process

$$X_{uv} := \langle Au, v\rangle, \quad u\in T,\ v\in S.$$

First we show that this process has sub-gaussian increments. For any $(u,v), (w,z)\in T\times S$, we have

$$
\begin{aligned}
\|X_{uv} - X_{wz}\|_{\psi_2} &= \Big\| \sum_{i,j} A_{ij}(u_i v_j - w_i z_j) \Big\|_{\psi_2} \\
&\le \Big(\sum_{i,j} \|A_{ij}(u_i v_j - w_i z_j)\|_{\psi_2}^2 \Big)^{1/2} \quad \text{(by Proposition 2.6.1)} \\
&\le \Big(\sum_{i,j} \|u_i v_j - w_i z_j\|_2^2 \Big)^{1/2} \quad \text{(since } \|A_{ij}\|_{\psi_2} \le K = 1) \\
&= \|uv^{\mathsf T} - wz^{\mathsf T}\|_F \\
&= \|(uv^{\mathsf T} - wv^{\mathsf T}) + (wv^{\mathsf T} - wz^{\mathsf T})\|_F \quad \text{(adding, subtracting)} \\
&\le \|(u-w)v^{\mathsf T}\|_F + \|w(v-z)^{\mathsf T}\|_F \quad \text{(by the triangle inequality)} \\
&= \|u-w\|_2 \|v\|_2 + \|v-z\|_2 \|w\|_2 \\
&\le \|u-w\|_2 \,\mathrm{rad}(S) + \|v-z\|_2 \,\mathrm{rad}(T).
\end{aligned}
$$

To apply Talagrand's comparison inequality we need to choose a Gaussian process (Y_{uv}) to which to compare the process (X_{uv}). The outcome of our calculation of the increments of (X_{uv}) suggests the following definition for (Y_{uv}):

$$\|Y_{uv} - Y_{wz}\|_2^2 = \|u-w\|_2^2 \,\mathrm{rad}(S)^2 + \|v-z\|_2^2 \,\mathrm{rad}(T)^2.$$

where

$$g \sim N(0, I_n), \quad h \sim N(0, I_m)$$

are independent Gaussian vectors. The increments of this process are

$$\|Y_{uv} - Y_{wz}\|_2^2 = \|u-w\|_2^2 \,\mathrm{rad}(T)^2 + \|v-z\|_2^2 \,\mathrm{rad}(S)^2.$$

(Check this as in the proof of Theorem 7.3.1.)

Comparing the increments of the two processes, we find that

$$\|X_{uv} - X_{wz}\|_{\psi_2} \lesssim \|Y_{uv} - Y_{wz}\|_2.$$

(Check!) Applying Talagrand's comparison inequality (Corollary 8.6.3), we conclude that

$$
\begin{aligned}
\mathbb{E} \sup_{u \in T,\, v \in S} X_{uv} &\lesssim \mathbb{E} \sup_{u \in T,\, v \in S} Y_{uv} \\
&= \mathbb{E} \sup_{u \in T} \langle g, u \rangle \,\mathrm{rad}(S) + \mathbb{E} \sup_{v \in S} \langle h, v \rangle \,\mathrm{rad}(T) \\
&= w(T)\,\mathrm{rad}(S) + w(S)\,\mathrm{rad}(T),
\end{aligned}
$$

as claimed. ■

Chevet's inequality is optimal, up to an absolute constant factor.

Exercise 8.7.2 (Sharpness of Chevet's inequality) Let A be an $m \times n$ random matrix whose entries A_{ij} are independent $N(0,1)$ random variables. Let $T \subset \mathbb{R}^n$ and $S \subset \mathbb{R}^m$ be arbitrary bounded sets. Show that the reverse of Chevet's inequality holds:

$$\mathbb{E} \sup_{x \in T,\, y \in S} \langle Ax, y \rangle \ge c\,(w(T)\,\mathrm{rad}(S) + w(S)\,\mathrm{rad}(T)).$$

Exercise 8.7.3 (High-probability version of Chevet) Under the assumptions of Theorem 8.7.1, prove a tail bound for $\sup_{x\in T,\, y\in S} \langle Ax, y\rangle$. ☞

Exercise 8.7.4 (Gaussian Chevet inequality) Suppose that the entries of A are $N(0,1)$. Show that Theorem 8.7.1 holds with the sharp constant 1, that is,

$$\mathbb{E} \sup_{x\in T,\, y\in S} \langle Ax, y\rangle \le w(T)\,\mathrm{rad}(S) + w(S)\,\mathrm{rad}(T).$$ ☞

8.8 Notes

The idea of chaining appeared in Kolmogorov's proof of his continuity theorem for Brownian motion; see e.g. [152, Chapter 1]. Dudley's integral inequality (Theorem 8.1.3) can indeed be traced to the work of R. Dudley. Our exposition in Section 8.1 mostly follows [126, Chapter 11], [193, Section 1.2], and [206, Section 5.3]. The upper bound in Theorem 8.1.13 (a reverse Sudakov's inequality) seems to be a folklore result.

The Monte-Carlo methods mentioned in Section 8.2 are extremely popular in scientific computing, especially when combined with the power of Markov chains; see e.g. [36]. In the same section we introduced the concept of empirical processes. A rich theory of empirical processes has applications to statistics and machine learning; see [205, 204, 167, 139]. In the terminology of empirical processes, Theorem 8.2.3 yields that the class of Lipschitz functions $\mathcal{F}$ is *uniform Glivenko–Cantelli*. Our presentation of this result (as well as the relation to Wasserstein's distance and transportation) is loosely based on [206, Example 5.15]. For a deep introduction to the transportation of measures, see [218].

The concept of the VC dimension studied in Section 8.3 goes back to the foundational work of V. Vapnik and A. Chervonenkis [211]; modern treatments can be found in e.g. [205, Section 2.6.1], [126, Section 14.3], [206, Section 7.2], [134, Sections 10.2 and 10.3], [139, Section 2.2], and [205, Section 2.6]. Pajor's lemma 8.3.13 was originally due to A. Pajor [160]; see [77], [126, Proposition 14.11], [206, Theorem 7.19], and [205, Lemma 2.6.2].

What we now call the Sauer–Shelah lemma (Theorem 8.3.16) was proved independently by V. Vapnik and A. Chervonenkis [211], N. Sauer [176], and M. Perles and S. Shelah (see Shelah [179]). Various proofs of the Sauer–Shelah lemma can be found in the literature, e.g. in [24, Chapter 17], [134, Sections 10.2 and 10.3], and [126, Section 14.3]. A number of variants of the Sauer–Shelah lemma are known; see e.g. [99, 190, 191, 6, 213].

Theorem 8.3.18 is due to R. Dudley [68]; see [126, Section 14.3] and [205, Theorem 2.6.4]. The dimension reduction lemma 8.3.19 is implicit in Dudley's proof; it was stated explicitly in [142] and reproduced in [206, Lemma 7.17]. For the generalization of VC theory from $\{0, 1\}$ to general real-valued function classes, see [142, 172] and [206, Sections 7.3 and 7.4].

Since the foundational work of V. Vapnik and A. Chervonenkis [211], bounds on empirical processes via the VC dimension such as in Theorem 8.3.23 have been in the spotlight of statistical learning theory; see e.g. [139, 17, 205, 172] and [206, Chapter 7]. Our presentation of Theorem 8.3.23 was based on [206, Corollary 7.18]. Although an explicit statement of this result is difficult to find in the earlier literature, it can be derived from [17, Theorem 6] and [32, Section 5].

The Glivenko–Cantelli theorem (Theorem 8.3.26) is a result from 1933 [80, 47] which predated and partly motivated the later development of VC theory; see [126, Section 14.2] and [205, 69] for more on Glivenko–Cantelli theorems and other uniform results in probability theory. Example 8.3.27 discusses a basic problem in discrepancy theory; see [133] for a comprehensive treatment of discrepancy theory.

In Section 8.4 we scratched the surface of statistical learning theory, which is a large area at the intersection of probability, statistics, and theoretical computer science. For a deeper introduction to this subject, see e.g. the tutorials [29, 139] and the books [104, 97, 119].

Generic chaining, which we presented in Section 8.5, has been put forward by M. Talagrand since 1985 (after an earlier work of X. Fernique [73]) as a sharp method for obtaining bounds on Gaussian processes. Our presentation was based on the book [193], which discusses ramifications, applications, and history of generic chaining in great detail. The upper bound on sub-gaussian processes (Theorem 8.5.3) can be found in [193, Theorem 2.2.22]; the lower bound (the majorizing measure Theorem 8.6.1) can be found in [193, Theorem 2.4.1]. Talagrand's comparison inequality (Corollary 8.6.2) was borrowed from [193, Theorem 2.4.12]. Another presentation of generic chaining can be found in [206, Chapter 6]. A different proof of the majorizing measure theorem was recently given by R. van Handel in [208, 209]. A high-probability version of the generic chaining bound (Theorem 8.5.5) is from [193, Theorem 2.2.27]; it was also proved by a different method by S. Dirksen [61].

In Section 8.7 we presented Chevet's inequality for sub-gaussian processes. In the existing literature, this inequality is stated only for Gaussian processes. It goes back to S. Chevet [53]; the constants were then improved by Y. Gordon [82], leading to the result we stated in Exercise 8.7.4. An exposition of this result can be found in [11, Section 9.4]. For variants and applications of Chevet's inequality, see [199, 2].

9

Deviations of Random Matrices and Geometric Consequences

This chapter is devoted to a remarkably useful uniform deviation inequality for random matrices. Given an $m \times n$ random matrix A, our goal is to show that, with high probability, the approximate equality

$$\|Ax\|_2 \approx \mathbb{E}\,\|Ax\|_2 \tag{9.1}$$

holds *simultaneously for many vectors* $x \in \mathbb{R}^n$. To quantify how many, we may choose an arbitrary subset $T \subset \mathbb{R}^n$ and ask whether (9.1) holds simultaneously for all $x \in T$. The answer turns out to be remarkably simple: with high probability, we have

$$\|Ax\|_2 = \mathbb{E}\,\|Ax\|_2 + O\big(\gamma(T)\big) \quad \text{for all } x \in T. \tag{9.2}$$

Recall that $\gamma(T)$ is the Gaussian complexity of T, which is a cousin of the Gaussian width we introduced in Section 7.6.2. In Section 9.1 we will deduce the uniform matrix deviation inequality (9.2) from Talagrand's comparison inequality.

The uniform matrix deviation inequality has many consequences. Some are results that we proved earlier by different methods: in Sections 9.2 and 9.3 we will quickly deduce two-sided bounds on random matrices, bounds on random projections of geometric sets, guarantees for covariance estimation for lower-dimensional distributions, and the Johnson–Lindenstrauss lemma and its generalization for infinite sets. New consequences will be proved starting from Section 9.4, where we deduce two classical results about geometric sets in high dimensions: the M^* bound and the escape theorem. Applications to sparse signal recovery will follow in Chapter 10.

9.1 Matrix Deviation Inequality

The following theorem is the main result of this chapter.

Theorem 9.1.1 (Matrix deviation inequality) *Let A be an $m \times n$ matrix whose rows A_i are independent, isotropic, and sub-gaussian random vectors in $\mathbb{R}^n$. Then, for any subset $T \subset \mathbb{R}^n$, we have*

$$\mathbb{E} \sup_{x \in T} \Big| \|Ax\|_2 - \sqrt{m}\|x\|_2 \Big| \le CK^2 \gamma(T).$$

Here $\gamma(T)$ is the Gaussian complexity introduced in Section 7.6.2, and $K = \max_i \|A_i\|_{\psi_2}$.

Before we proceed to the proof of this theorem, let us pause to check that $\mathbb{E}\,\|Ax\|_2 \approx \sqrt{m}\|x\|_2$, so that Theorem 9.1.1 does indeed yield (9.2).

Exercise 9.1.2 (Deviation around expectation) Deduce from Theorem 9.1.1 that

$$\mathbb{E} \sup_{x \in T} \Big| \|Ax\|_2 - \mathbb{E} \|Ax\|_2 \Big| \le CK^2 \gamma(T).$$

We will deduce Theorem 9.1.1 from Talagrand's comparison inequality (Corollary 8.6.3). To apply the comparison inequality, all we have to check is that the random process

$$X_x := \|Ax\|_2 - \sqrt{m}\|x\|_2$$

indexed by $x \in \mathbb{R}^n$ has sub-gaussian increments. Let us state this.

Theorem 9.1.3 (Sub-gaussian increments) *Let A be an $m \times n$ matrix whose rows A_i are independent, isotropic, and sub-gaussian random vectors in $\mathbb{R}^n$. Then the random process*

$$X_x := \|Ax\|_2 - \sqrt{m}\|x\|_2$$

has sub-gaussian increments, namely,

$$\|X_x - X_y\|_{\psi_2} \le CK^2 \|x - y\|_2 \quad \text{for all } x, y \in \mathbb{R}^n. \tag{9.3}$$

Here $K = \max_i \|A_i\|_{\psi_2}$.

Proof of matrix deviation inequality (Theorem 9.1.1) By Theorem 9.1.3 and Talagrand's comparison inequality in the form of Exercise 8.6.4, we get

$$\mathbb{E} \sup_{x \in T} |X_x| \le CK^2 \gamma(T),$$

as announced. ■

It remains to prove Theorem 9.1.3. Although the proof is a bit longer than most of the arguments in this book, we will make it easier by working out simpler, partial, cases first and gradually moving toward full generality. We develop this argument in the next few subsections.

9.1.1 Theorem 9.1.3 for Unit Vector x and Zero Vector y

Assume that

$$\|x\|_2 = 1 \quad \text{and} \quad y = 0.$$

In this case, the inequality in (9.3) that we want to prove becomes

$$\Big\| \|Ax\|_2 - \sqrt{m} \Big\|_{\psi_2} \le CK^2. \tag{9.4}$$

Note that Ax is a random vector in $\mathbb{R}^m$ with independent sub-gaussian coordinates $\langle A_i, x \rangle$, which satisfy $\mathbb{E} \langle A_i, x \rangle^2 = 1$ by isotropy. Then the concentration of the norm theorem 3.1.1 yields (9.4).

9.1.2 Theorem 9.1.3 for Unit Vectors x, y and for a Squared Process

Assume now that

$$\|x\|_2 = \|y\|_2 = 1.$$

In this case, the inequality in (9.3) that we want to prove becomes

$$\Big\| \|Ax\|_2 - \|Ay\|_2 \Big\|_{\psi_2} \le CK^2 \|x - y\|_2. \tag{9.5}$$

We first prove a version of this inequality for *squared* Euclidean norms, which are more convenient to handle. Let us guess what form such an inequality should take. We have

$$\begin{aligned} \|Ax\|_2^2 - \|Ay\|_2^2 &= \Big(\|Ax\|_2 + \|Ay\|_2 \Big) \Big(\|Ax\|_2 - \|Ay\|_2 \Big) \\ &\lesssim \sqrt{m}\, \|x - y\|_2. \end{aligned} \tag{9.6}$$

The last bound should hold with high probability because the typical magnitude of $\|Ax\|_2$ and $\|Ay\|_2$ is $\sqrt{m}$, by (9.4), and because we expect (9.5) to hold.

Now that we have guessed the inequality (9.6) for the squared process, let us prove it. We are looking to bound the random variable

$$Z := \frac{\|Ax\|_2^2 - \|Ay\|_2^2}{\|x - y\|_2} = \frac{\langle A(x-y), A(x+y) \rangle}{\|x - y\|_2} = \langle Au, Av \rangle, \tag{9.7}$$

where

$$u := \frac{x - y}{\|x - y\|_2} \quad \text{and} \quad v := x + y.$$

The desired bound is

$$|Z| \lesssim \sqrt{m} \quad \text{with high probability.}$$

The coordinates of the vectors Au and Av are $\langle A_i, u \rangle$ and $\langle A_i, v \rangle$. So we can represent Z as a sum of independent random variables:

$$Z = \sum_{i=1}^{m} \langle A_i, u \rangle \langle A_i, v \rangle.$$

Lemma 9.1.4 *The random variables $\langle A_i, u \rangle \langle A_i, v \rangle$ are independent, mean zero, and sub-exponential; more precisely,*

$$\big\| \langle A_i, u \rangle \langle A_i, v \rangle \big\|_{\psi_1} \le 2K^2.$$

Proof Independence follows from the construction, but the mean-zero property is less obvious. Although both $\langle A_i, u \rangle$ and $\langle A_i, v \rangle$ do have zero means, these variables are not necessarily independent of each other. Still, we can check that they are uncorrelated. Indeed,

$$\mathbb{E} \langle A_i, x - y \rangle \langle A_i, x + y \rangle = \mathbb{E} \big(\langle A_i, x \rangle^2 - \langle A_i, y \rangle^2 \big) = 1 - 1 = 0,$$

by isotropy. By the definitions of u and v, this implies that $\mathbb{E} \langle A_i, u \rangle \langle A_i, v \rangle = 0$.

To finish the proof, recall from Lemma 2.7.7 that the product of two sub-gaussian random variables is sub-exponential. So, we get

$$\begin{aligned}\| \langle A_i, u\rangle\langle A_i, v\rangle \|_{\psi_1} &\le \| \langle A_i, u\rangle \|_{\psi_2} \| \langle A_i, v\rangle \|_{\psi_2} \\ &\le K\|u\|_2 \, K\|v\|_2 \quad \text{(by the sub-gaussian assumption)} \\ &\le 2K^2,\end{aligned}$$

where in the last step we used that $\|u\|_2 = 1$ and $\|v\|_2 \le \|x\|_2 + \|y\|_2 \le 2$. ■

To bound Z, we can use Bernstein's inequality (Corollary 2.8.3); recall that it applies for a sum of independent mean-zero sub-exponential random variables.

Exercise 9.1.5 Apply Bernstein's inequality (Corollary 2.8.3) and simplify the bound. You should get

$$\mathbb{P}\left\{|Z| \ge s\sqrt{m}\right\} \le 2\exp\left(-\frac{cs^2}{K^4}\right)$$

for any $0 \le s \le \sqrt{m}$. ☞

Recalling the definition of Z, we can see that we have obtained the desired bound (9.6).

9.1.3 Theorem 9.1.3 for Unit Vectors x, y and for the Original Process

Next, we would like to remove the squares from $\|Ax\|_2^2$ and $\|Ay\|_2^2$ and deduce inequality (9.5) for unit vectors x and y. Let us state this goal again.

Lemma 9.1.6 (Unit y, original process) *Let $x, y \in S^{n-1}$. Then*

$$\Big\| \|Ax\|_2 - \|Ay\|_2 \Big\|_{\psi_2} \le CK^2\|x - y\|_2.$$

Proof Fix $s \ge 0$. The conclusion we want to prove is that

$$p(s) := \mathbb{P}\left\{\frac{\big|\|Ax\|_2 - \|Ay\|_2\big|}{\|x - y\|_2} \ge s\right\} \le 4\exp\left(-\frac{cs^2}{K^4}\right). \tag{9.8}$$

We proceed differently for small and large s.

Case 1: $s \le 2\sqrt{m}$. In this range we can use our results from the previous subsection. They are stated for the squared process, though. So, to be able to apply those results, we multiply both sides of the inequality defining $p(s)$ by $\|Ax\|_2+\|Ay\|_2$. With the same Z as we defined in (9.7), this gives

$$p(s) = \mathbb{P}\left\{|Z| \ge s\big(\|Ax\|_2 + \|Ay\|_2\big)\right\} \le \mathbb{P}\left\{|Z| \ge s\|Ax\|_2\right\}.$$

From (9.4) we know that $\|Ax\|_2 \approx \sqrt{m}$ with high probability. So it makes sense to break the probability that $|Z| \ge s\|Ax\|_2$ into two cases: one where $\|Ax\|_2 \ge \sqrt{m}/2$ and thus $|Z| \ge s\sqrt{m}/2$, and the other where $\|Ax\|_2 < \sqrt{m}/2$ (and then we do not care about Z). This leads to

$$p(s) \le \mathbb{P}\left\{|Z| \ge \frac{s\sqrt{m}}{2}\right\} + \mathbb{P}\left\{\|Ax\|_2 < \frac{\sqrt{m}}{2}\right\} =: p_1(s) + p_2(s).$$

The result of Exercise 9.1.5 gives

$$p_1(s) \le 2\exp\Big(-\frac{cs^2}{K^4}\Big).$$

Further, the bound (9.4) and the triangle inequality give

$$p_2(s) \le \mathbb{P}\left\{\big|\|Ax\|_2 - \sqrt{m}\big| > \frac{\sqrt{m}}{2}\right\} \le 2\exp\Big(-\frac{cs^2}{K^4}\Big).$$

Summing the two probabilities, we conclude that the desired bound is

$$p(s) \le 4\exp\Big(-\frac{cs^2}{K^4}\Big).$$

Case 2: $s > 2\sqrt{m}$. Let us look again at the inequality (9.8) that defines $p(s)$ and slightly simplify it. By the triangle inequality, we have

$$\big|\|Ax\|_2 - \|Ay\|_2\big| \le \|A(x-y)\|_2.$$

Thus

$$\begin{aligned}
p(s) &\le \mathbb{P}\left\{\|Au\|_2 \ge s\right\} \quad \Big(\text{where } u := \frac{x-y}{\|x-y\|_2} \text{ as before}\Big)\\
&\le \mathbb{P}\left\{\|Au\|_2 - \sqrt{m} \ge s/2\right\} \quad (\text{since } s > 2\sqrt{m})\\
&\le 2\exp\Big(-\frac{cs^2}{K^4}\Big) \quad (\text{using (9.4) again}).
\end{aligned}$$

Therefore, in both cases we have obtained the desired estimate (9.8). This completes the proof of the lemma. ■

9.1.4 Theorem 9.1.3 in Full Generality

Finally, we are ready to prove (9.3) for arbitrary $x, y \in \mathbb{R}^n$. By scaling, we can assume without loss of generality that

$$\|x\|_2 = 1 \quad \text{and} \quad \|y\|_2 \ge 1. \tag{9.9}$$

(Why?) Consider the contraction of y onto the unit sphere, see Figure 9.1,

$$\bar{y} := \frac{y}{\|y\|_2}. \tag{9.10}$$

Use the triangle inequality to break the increment into two parts:

$$\|X_x - X_y\|_{\psi_2} \le \|X_x - X_{\bar{y}}\|_{\psi_2} + \|X_{\bar{y}} - X_y\|_{\psi_2}.$$

Since x and $\bar{y}$ are unit vectors, Lemma 9.1.6 may be used to bound the first part. It gives

$$\|X_x - X_{\bar{y}}\|_{\psi_2} \le CK^2\|x - \bar{y}\|_2.$$

To bound the second part, note that $\bar{y}$ and y are collinear vectors, so that

$$\|X_{\bar{y}} - X_y\|_{\psi_2} = \|\bar{y} - y\|_2\,\|X_{\bar{y}}\|_{\psi_2}.$$

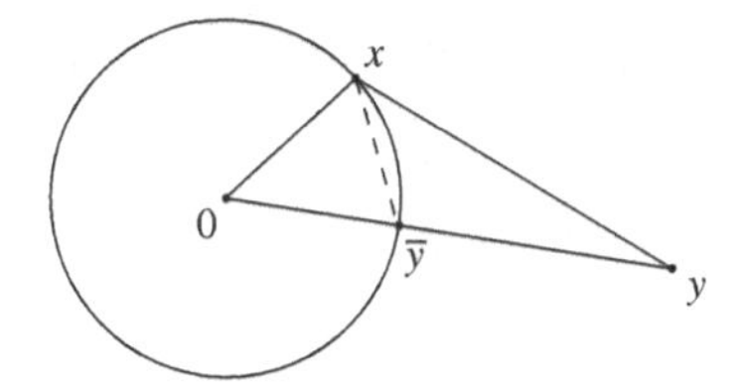

Figure 9.1 Exercise 9.1.7 shows that the triangle inequality can be approximately reversed from these three vectors, and so we have
$\|x - \bar{y}\|_2 + \|\bar{y} - y\|_2 \le \sqrt{2}\|x - y\|_2$.

(Check this identity!) Now, since $\bar{y}$ is a unit vector, (9.4) gives

$$\|X_{\bar{y}}\|_{\psi_2} \le CK^2.$$

Combining the two parts, we conclude that

$$\|X_x - X_y\|_{\psi_2} \le CK^2\big(\|x - \bar{y}\|_2 + \|\bar{y} - y\|_2\big). \tag{9.11}$$

At this point we might get nervous: we need to bound the right-hand side by $\|x - y\|_2$, but the triangle inequality would give the reverse bound! Nevertheless, looking at Figure 9.1 we may suspect that in our case the triangle inequality can be approximately reversed. The next exercise confirms this rigorously.

Exercise 9.1.7 (Reverse triangle inequality)☕☕ Consider vectors $x, y, \bar{y} \in \mathbb{R}^n$ satisfying (9.9) and (9.10). Show that

$$\|x - \bar{y}\|_2 + \|\bar{y} - y\|_2 \le \sqrt{2}\|x - y\|_2.$$

Using the result of this exercise, deduce from (9.11) the desired bound

$$\|X_x - X_y\|_{\psi_2} \le CK^2\|x - y\|_2.$$

Theorem 9.1.3 is completely proved. ■

Now that we have proved the matrix deviation inequality (Theorem 9.1.1), we can complement it with a high-probability version.

Exercise 9.1.8 (Matrix deviation inequality: tail bounds)☕ Show that, under the conditions of Theorem 9.1.1, we have the following. For any $u \ge 0$, the event

$$\sup_{x \in T} \Big| \|Ax\|_2 - \sqrt{m}\|x\|_2 \Big| \le CK^2 \left(w(T) + u \ \mathrm{rad}(T)\right) \tag{9.12}$$

holds with probability at least $1 - 2\exp(-u^2)$. Here $\mathrm{rad}(T)$ is the radius of T introduced in (8.47). ☞

Exercise 9.1.9☕ Argue that the right-hand side of (9.12) can be further bounded by $CK^2 u\gamma(T)$ for $u \ge 1$. Conclude that the bound in Exercise 9.1.8 implies Theorem 9.1.1.

Exercise 9.1.10 (Deviation of squares) Show that, under the conditions of Theorem 9.1.1, we have

$$\mathbb{E} \sup_{x \in T} \left| \|Ax\|_2^2 - m\|x\|_2^2 \right| \leq CK^4 \gamma(T)^2 + CK^2 \sqrt{m} \, \mathrm{rad}(T) \gamma(T).$$

Exercise 9.1.11 (Deviation of random projections) Prove a version of the matrix deviation inequality (Theorem 9.1.1) for random projections. Let P be an orthogonal projection in $\mathbb{R}^n$ onto an m-dimensional subspace uniformly distributed in the Grassmannian $G_{n,m}$. Show that, for any subset $T \subset \mathbb{R}^n$, we have

$$\mathbb{E} \sup_{x \in T} \left| \|Px\|_2 - \sqrt{\frac{m}{n}} \|x\|_2 \right| \leq \frac{CK^2 \gamma(T)}{\sqrt{n}}.$$

9.2 Random Matrices, Random Projections, and Covariance Estimation

The matrix deviation inequality has a number of important consequences, some of which we present in this and the next section.

9.2.1 Two-Sided Bounds on Random Matrices

First let us apply the matrix deviation inequality to the unit Euclidean sphere $T = S^{n-1}$. In this case, we can recover the bounds on random matrices that we proved in Section 4.6.

Indeed, the radius and Gaussian width of $T = S^{n-1}$ satisfy

$$\mathrm{rad}(T) = 1, \quad w(T) \leq \sqrt{n}.$$

(Recall (7.16).) The matrix deviation inequality in the form of Exercise 9.1.8 together with the triangle inequality imply that the event

$$\sqrt{m} - CK^2(\sqrt{n} + u) \leq \|Ax\|_2 \leq \sqrt{m} + CK^2(\sqrt{n} + u) \quad \forall x \in S^{n-1}$$

holds with probability at least $1 - 2\exp(-u^2)$.

We can interpret this event as a two-sided bound on the extreme singular values of A (recall (4.5)):

$$\sqrt{m} - CK^2(\sqrt{n} + u) \leq s_n(A) \leq s_1(A) \leq \sqrt{m} + CK^2(\sqrt{n} + u).$$

Thus we recover the result in Theorem 4.6.1.

9.2.2 Sizes of Random Projections of Geometric Sets

Another immediate application of the matrix deviation inequality is the bound on random projections of geometric sets that we gave in Section 7.7. In fact, the matrix deviation inequality yields a sharper bound:

Proposition 9.2.1 (Sizes of random projections of sets) *Consider a bounded set $T \subset \mathbb{R}^n$. Let A be an $m \times n$ matrix whose rows A_i are independent, isotropic, and sub-gaussian random vectors in $\mathbb{R}^n$. Then the scaled matrix*

$$P := \frac{1}{\sqrt{n}} A$$

(which is a "sub-gaussian projection") satisfies

$$\mathbb{E}\,\mathrm{diam}(PT) \le \sqrt{\frac{m}{n}}\,\mathrm{diam}(T) + CK^2 w_s(T).$$

Here $w_s(T)$ is the spherical width of T (recall Section 7.5.2) and $K = \max_i \|A_i\|_{\psi_2}$.

Proof Theorem 9.1.1 implies via the triangle inequality that

$$\mathbb{E} \sup_{x \in T} \|Ax\|_2 \le \sqrt{m} \sup_{x \in T} \|x\|_2 + CK^2 \gamma(T).$$

We can state this inequality in terms of the radii of the sets AT and T as

$$\mathbb{E}\,\mathrm{rad}(AT) \le \sqrt{m}\,\mathrm{rad}(T) + CK^2 \gamma(T).$$

Applying this bound for the difference set $T - T$ instead of T, we can write it as

$$\mathbb{E}\,\mathrm{diam}(AT) \le \sqrt{m}\,\mathrm{diam}(T) + CK^2 w(T).$$

(Here we have used (7.22) to pass from the Gaussian complexity to the Gaussian width.) Dividing both sides by $\sqrt{n}$ completes the proof. ■

Proposition 9.2.1 is more general and sharper than our previous bounds on random projections (Exercise 7.7.3). Indeed, it states that the diameter scales by the exact factor $\sqrt{m/n}$ without an absolute constant in front of it.

Exercise 9.2.2 (Sizes of projections: high-probability bounds)☕☕ Use the high-probability version of the matrix deviation inequality (Exercise 9.1.8) to obtain a high-probability version of Proposition 9.2.1. Namely, show that, for $\varepsilon > 0$, the bound

$$\mathrm{diam}(PT) \le (1+\varepsilon)\sqrt{\frac{m}{n}}\,\mathrm{diam}(T) + CK^2 w_s(T)$$

holds with probability at least $1 - \exp(-c\varepsilon^2 m/K^4)$.

Exercise 9.2.3☕☕☕ Deduce a version of Proposition 9.2.1 for the original model of P considered in Section 7.7, i.e., for a random projection P onto a random m-dimensional subspace $E \sim \mathrm{Unif}(G_{n,m})$. ☞

9.2.3 Covariance Estimation for Lower-Dimensional Distributions

Let us revisit the covariance estimation problem, which we studied in Section 4.7 for sub-gaussian distributions and in Section 5.6 in full generality. We found that the covariance matrix Σ of an n-dimensional distribution can be estimated from $m = O(n)$ sample points for sub-gaussian distributions, and from $m = O(n \log n)$ sample points in full generality.

An even smaller sample can be sufficient for covariance estimation when the distribution is approximately low dimensional, i.e., when $\Sigma^{1/2}$ has low stable rank,[1] which means that

[1] We introduced the notion of stable rank in Section 7.6.1.

the distribution tends to concentrate near a small subspace in $\mathbb{R}^n$. We should expect to do well with $m = O(r)$, where r is the stable rank of $\Sigma^{1/2}$. We noted this only for the general case in Remark 5.6.3, up to a logarithmic oversampling. Now let us address the sub-gaussian case, where no logarithmic oversampling is needed.

The following result extends Theorem 4.7.1 for approximately lower-dimensional distributions.

Theorem 9.2.4 (Covariance estimation for lower-dimensional distributions) *Let X be a sub-gaussian random vector in $\mathbb{R}^n$. More precisely, assume that there exists $K \geq 1$ such that*

$$\| \langle X, x\rangle \|_{\psi_2} \leq K \| \langle X, x\rangle \|_{L^2} \quad \textit{for any } x \in \mathbb{R}^n.$$

Then, for every positive integer m, we have

$$\mathbb{E}\, \|\Sigma_m - \Sigma\| \leq CK^4 \Big(\sqrt{\frac{r}{m}} + \frac{r}{m} \Big) \|\Sigma\|,$$

where $r = \operatorname{tr} \Sigma / \|\Sigma\|$ is the stable rank of $\Sigma^{1/2}$.

Proof We begin the proof exactly as in Theorem 4.7.1 by bringing the distribution to an isotropic position. Thus,

$$\begin{aligned}
\|\Sigma_m - \Sigma\| &= \|\Sigma^{1/2} R_m \Sigma^{1/2}\| \quad \Big(\text{where } R_m = \frac{1}{m} \sum_{i=1}^m Z_i Z_i^\mathsf{T} - I_n\Big) \\
&= \max_{x \in S^{n-1}} \big\langle \Sigma^{1/2} R_m \Sigma^{1/2} x, x \big\rangle \quad \text{(the matrix is positive-semidefinite)} \\
&= \max_{x \in T} \langle R_m x, x \rangle \quad \text{(if we define the ellipsoid } T := \Sigma^{1/2} S^{n-1}) \\
&= \max_{x \in T} \Big| \frac{1}{m} \sum_{i=1}^m \langle Z_i, x\rangle^2 - \|x\|_2^2 \Big| \quad \text{(by the definition of } R_m) \\
&= \frac{1}{m} \max_{x \in T} \big| \|Ax\|_2^2 - m \|x\|_2^2 \big|,
\end{aligned}$$

where in the last step A denotes the $m \times n$ matrix with rows Z_i. As in the proof of Theorem 4.7.1, the Z_i are mean-zero isotropic sub-gaussian random vectors with $\|Z_i\|_{\psi_2} \lesssim 1$. (For simplicity, let us hide the dependence on K within this argument.) This allows us to apply the matrix deviation inequality for A (in the form given in Exercise 9.1.10), which gives

$$\mathbb{E}\, \|\Sigma_m - \Sigma\| \lesssim \frac{1}{m} \big(\gamma(T)^2 + \sqrt{m}\, \mathrm{rad}(T) \gamma(T) \big).$$

The radius and Gaussian complexity of the ellipsoid $T = \Sigma^{1/2} S^{n-1}$ are easy to compute:

$$\mathrm{rad}(T) = \|\Sigma\|^{1/2} \quad \text{and} \quad \gamma(T) \leq (\operatorname{tr}(\Sigma))^{1/2}.$$

(Check!) This gives

$$\mathbb{E}\, \|\Sigma_m - \Sigma\| \lesssim \frac{1}{m} \Big(\operatorname{tr}(\Sigma) + \sqrt{m \|\Sigma\| \operatorname{tr}(\Sigma)} \Big).$$

Substitute $\operatorname{tr}(\Sigma) = r \|\Sigma\|$ and simplify the bound to complete the proof. ■

Exercise 9.2.5 (Tail bound) Prove a high-probability guarantee for Theorem 9.2.4 that is similar to the results of Exercises 4.7.3 and 5.6.4. Namely, check that, for any $u \geq 0$, we have

$$\|\Sigma_m - \Sigma\| \leq CK^4\Big(\sqrt{\frac{r+u}{m}} + \frac{r+u}{m}\Big)\,\|\Sigma\|$$

with probability at least $1 - 2e^{-u}$.

9.3 The Johnson–Lindenstrauss Lemma for Infinite Sets

Let us now apply the matrix deviation inequality for a finite set T. In this case, we recover the Johnson–Lindenstrauss lemma from Section 5.3, and more.

9.3.1 Recovering the Classical Johnson–Lindenstrauss Lemma

Let us check that the matrix deviation inequality contains the classical Johnson–Lindenstrauss lemma (Theorem 5.3.1). Let $\mathcal{X}$ be a set of N points in $\mathbb{R}^n$ and define T to be the set of normalized differences of $\mathcal{X}$, i.e.,

$$T := \Big\{\frac{x-y}{\|x-y\|_2} : x, y \in \mathcal{X} \text{ are distinct points}\Big\}.$$

The Gaussian complexity of T satisfies

$$\gamma(T) \leq C\sqrt{\log N} \tag{9.13}$$

(recall Exercise 7.5.10). The matrix deviation inequality (Theorem 9.1.1) now implies that the bound

$$\sup_{x,y\in\mathcal{X}} \Big|\frac{\|Ax - Ay\|_2}{\|x-y\|_2} - \sqrt{m}\Big| \lesssim \sqrt{\log N} \tag{9.14}$$

holds with high probability. To keep the calculation simple, we will be satisfied here with a probability 0.99, which can be obtained using Markov's inequality; Exercise 9.1.8 gives a better probability. Also, for simplicity we have suppressed the dependence on the sub-gaussian norm K.

Multiply both sides of (9.14) by $(1/\sqrt{m})\|x-y\|_2$ and rearrange the terms. We obtain that, with high probability, the scaled random matrix

$$Q := \frac{1}{\sqrt{m}}A$$

is an approximate isometry on $\mathcal{X}$, i.e.,

$$(1-\varepsilon)\|x-y\|_2 \leq \|Qx - Qy\|_2 \leq (1+\varepsilon)\|x-y\|_2 \quad \text{for all } x, y \in \mathcal{X},$$

where

$$\varepsilon \lesssim \sqrt{\frac{\log N}{m}}.$$

Equivalently, if we fix $\varepsilon > 0$ and choose the dimension m such that

$$m \gtrsim \varepsilon^{-2}\log N,$$

then, with high probability, Q is an ε-isometry on $\mathcal{X}$. Thus we recover the classical Johnson–Lindenstrauss lemma (Theorem 5.3.1).

Exercise 9.3.1 In the argument above, quantify the probability of success and dependence on K. In other words, use the matrix deviation inequality to give an alternative solution of Exercise 5.3.3.

9.3.2 Johnson–Lindenstrauss Lemma for Infinite Sets

The argument above does not really depend on $\mathcal{X}$ being a finite set. We used the fact that $\mathcal{X}$ is finite only to bound the Gaussian complexity in (9.13). This means that we can give a version of the Johnson–Lindenstrauss lemma for general, not necessarily finite, sets. Let us state such a version.

Proposition 9.3.2 (Additive Johnson–Lindenstrauss lemma) *Consider a set $\mathcal{X} \subset \mathbb{R}^n$. Let A be an $m \times n$ matrix whose rows A_i are independent, isotropic, and sub-gaussian random vectors in $\mathbb{R}^n$. Then, with high probability (say, 0.99), the scaled matrix*

$$Q := \frac{1}{\sqrt{m}} A$$

satisfies

$$\|x - y\|_2 - \delta \leq \|Qx - Qy\|_2 \leq \|x - y\|_2 + \delta \quad \textit{for all } x, y \in \mathcal{X},$$

where

$$\delta = \frac{CK^2 w(\mathcal{X})}{\sqrt{m}}$$

and $K = \max_i \|A_i\|_{\psi_2}$.

Proof Choose T to be the difference set, i.e., $T = \mathcal{X} - \mathcal{X}$, and apply the matrix deviation inequality (Theorem 9.1.1). It follows that, with high probability,

$$\sup_{x,y \in \mathcal{X}} \left| \|Ax - Ay\|_2 - \sqrt{m}\|x - y\|_2 \right| \leq CK^2 \gamma(\mathcal{X} - \mathcal{X}) = 2CK^2 w(\mathcal{X}).$$

(In the last step we used (7.22).) Dividing both sides by $\sqrt{m}$, we complete the proof. ■

Note that the error δ in Proposition 9.3.2 is additive, while the classical Johnson–Lindenstrauss lemma for finite sets (Theorem 5.3.1) has a multiplicative form of error. This may be a small difference, but in general it is necessary:

Exercise 9.3.3 (Additive error) Suppose that a set $\mathcal{X}$ has a non-empty interior. Check that, in order for the conclusion (5.10) of the classical Johnson-Lindenstrauss lemma to hold, one must have $m \geq n$, i.e., no dimension reduction is possible.

Remark 9.3.4 (Stable dimension) The additive version of the Johnson–Lindenstrauss lemma can be stated naturally in terms of the stable dimension of $\mathcal{X}$,

$$d(\mathcal{X}) \asymp \frac{w(\mathcal{X})^2}{\operatorname{diam}(\mathcal{X})^2},$$

which we introduced in Section 7.6. To see this, let us fix $\varepsilon > 0$ and choose the dimension m so that it *exceeds an appropriate multiple of the stable dimension*, namely

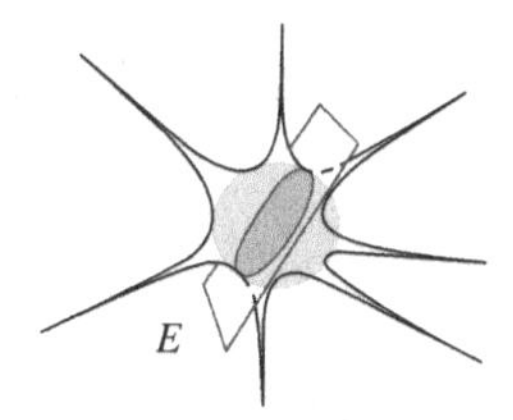

Figure 9.2 Illustration of an M^* bound: the intersection of a set T with a random subspace E.

$$m \geq (CK^4/\varepsilon^2)d(T).$$

Then, in Proposition 9.3.2 we have $\delta \leq \varepsilon \operatorname{diam}(\mathcal{X})$. This means that Q *preserves the distances in* $\mathcal{X}$ *to within a small fraction of the maximal distance*, which is the diameter of $\mathcal{X}$.

9.4 Random Sections: M^* Bound and Escape Theorem

Consider a set $T \subset \mathbb{R}^n$ and a random subspace E with given dimension. How large is the typical intersection of T and E? See Figure 9.2 for an illustration. There are two types of answer to this question. In Section 9.4.1 we give a general bound for the expected diameter of $T \cap E$; it is called the M^* bound. The intersection $T \cap E$ can even be empty; this is the content of the *escape theorem*, which we prove in Section 9.4.2. Both results are consequences of the matrix deviation inequality.

9.4.1 M^* Bound

First, it is convenient to realize the random subspace E as the kernel of a random matrix, i.e., to set

$$E := \ker A,$$

where A is a random $m \times n$ matrix. We always have

$$\dim(E) \geq n - m,$$

and for continuous distributions we have $\dim(E) = n - m$ almost surely.

Example 9.4.1 Suppose that A is a *Gaussian matrix*, i.e., it has independent $N(0, 1)$ entries. Rotation invariance implies that $E = \ker(A)$ is uniformly distributed in the Grassmannian:

$$E \sim \operatorname{Unif}(G_{n,n-m}).$$

Our main result is the following general bound on the diameters of random sections of geometric sets. For historic reasons, this result is called the M^* bound.

Theorem 9.4.2 (M^* bound) *Consider a set $T \subset \mathbb{R}^n$. Let A be an $m \times n$ matrix whose rows A_i are independent, isotropic, and sub-gaussian random vectors in $\mathbb{R}^n$. Then the random subspace $E = \ker A$ satisfies*

$$\mathbb{E}\,\mathrm{diam}(T \cap E) \le \frac{CK^2 w(T)}{\sqrt{m}},$$

where $K = \max_i \|A_i\|_{\psi_2}$.

Proof Apply Theorem 9.1.1 for $T - T$ and obtain

$$\mathbb{E} \sup_{x,y \in T} \Big| \|Ax - Ay\|_2 - \sqrt{m}\|x - y\|_2 \Big| \le CK^2 \gamma(T - T) = 2CK^2 w(T).$$

If we restrict the supremum to points x, y in the kernel of A then $\|Ax - Ay\|_2$ disappears, since $A(x - y) = 0$, and we have

$$\mathbb{E} \sup_{x,y \in T \cap \ker A} \sqrt{m}\|x - y\|_2 \le 2CK^2 w(T).$$

Dividing by $\sqrt{m}$ yields

$$\mathbb{E}\,\mathrm{diam}(T \cap \ker A) \le \frac{CK^2 w(T)}{\sqrt{m}},$$

which is the bound we claimed. ■

Exercise 9.4.3 (Affine sections) Check that the M^* bound holds not only for sections through the origin but for all affine sections as well:

$$\mathbb{E} \max_{z \in \mathbb{R}^n} \mathrm{diam}(T \cap E_z) \le \frac{CK^2 w(T)}{\sqrt{m}},$$

where $E_z = z + \ker A$.

Surprisingly, the random subspace E in the M^* bound is not low dimensional. On the contrary, $\dim(E) \ge n - m$ and we would typically choose $m \ll n$, so E has almost full dimension. This makes the M^* bound a strong and perhaps surprising statement.

Remark 9.4.4 (Stable dimension) It can be enlightening to look at the M^* bound through the lens of the notion of the stable dimension $d(T) \asymp w(T)^2/\mathrm{diam}(T)^2$, which we introduced in Section 7.6. Fix $\varepsilon > 0$. Then the M^* bound can be stated as

$$\mathbb{E}\,\mathrm{diam}(T \cap E) \le \varepsilon\ \mathrm{diam}(T)$$

as long as

$$m \ge C(K^4/\varepsilon^2)d(T). \tag{9.15}$$

In words, *the M^* bound becomes nontrivial – the diameter shrinks – as long as the codimension of E exceeds a multiple of the stable dimension of T.*

Equivalently, the dimension condition states that the sum of the dimension of E and a multiple of the stable dimension of T should be bounded by n. This condition should now make sense from the linear algebraic point of view. For example, if T is a centered Euclidean ball in some subspace $F \subset \mathbb{R}^n$ then a nontrivial bound $\operatorname{diam}(T \cap E) < \operatorname{diam}(T)$ is possible only if

$$\dim E + \dim F \le n.$$

(Why?)

Let us look at one remarkable application of the M^* bound.

Example 9.4.5 (The ℓ_1 ball) Let $T = B_1^n$, the unit ball of the ℓ_1 norm in $\mathbb{R}^n$. Since we proved in (7.18) that $w(T) \asymp \sqrt{\log n}$, the M^* bound (Theorem 9.4.2) gives

$$\mathbb{E} \operatorname{diam}(T \cap E) \lesssim \sqrt{\frac{\log n}{m}}.$$

For example, if $m = 0.1n$ then

$$\mathbb{E} \operatorname{diam}(T \cap E) \lesssim \sqrt{\frac{\log n}{n}}. \tag{9.16}$$

Comparing this with $\operatorname{diam}(T) = 2$, we see that the diameter shrinks by almost $\sqrt{n}$ as a result of intersecting T with a random subspace E that has almost full dimension (namely, $0.9n$).

For an intuitive explanation of this surprising fact, recall from Section 7.5.4 that the "bulk" of the octahedron $T = B_1^n$ is formed by the inscribed ball $(1/\sqrt{n})B_2^n$. Then it should not be surprising if a random subspace E tends to pass through the bulk and miss the "outliers" that lie closer to the vertices of T. This makes the diameter of $T \cap E$ essentially the same as the size of the bulk, which is $1/\sqrt{n}$.

This example indicates what makes a surprisingly strong and general result such as the M^* bound possible. Intuitively, the random subspace E tends to pass entirely through the bulk of T, which is usually a Euclidean ball with much smaller diameter than T, see Figure 9.2.

Exercise 9.4.6 (M^* bound with high probability)☕☕ Use the high-probability version of the matrix deviation inequality (Exercise 9.1.8) to obtain a high-probability version of the M^* bound.

9.4.2 Escape Theorem

In some circumstances, a random subspace E may completely miss a given set T in $\mathbb{R}^n$. This might happen, for example, if T is a subset of the sphere; see Figure 9.3. In this case, the intersection $T \cap E$ is typically empty under essentially the same conditions as in the M^* bound.

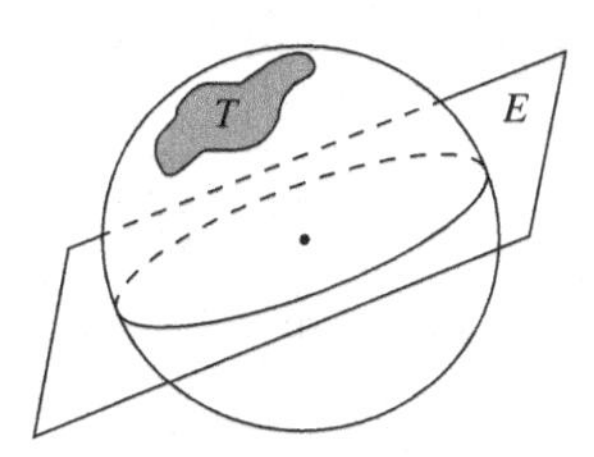

Figure 9.3 Illustration of the escape theorem: the set T has empty intersection with a random subspace E.

Theorem 9.4.7 (Escape theorem) *Consider a set $T \subset S^{n-1}$. Let A be an $m \times n$ matrix whose rows A_i are independent, isotropic, and sub-gaussian random vectors in $\mathbb{R}^n$. If*

$$m \geq CK^4 w(T)^2 \tag{9.17}$$

then the random subspace $E = \ker A$ satisfies

$$T \cap E = \emptyset$$

with probability at least $1 - 2\exp(-cm/K^4)$. Here $K = \max_i \|A_i\|_{\psi_2}$.

Proof Let us use the high-probability version of the matrix deviation inequality from Exercise 9.1.8. It states that the bound

$$\sup_{x \in T} \left| \|Ax\|_2 - \sqrt{m} \right| \leq C_1 K^2 (w(T) + u) \tag{9.18}$$

holds with probability at least $1 - 2\exp(-u^2)$. Suppose that this event indeed holds and $T \cap E \neq \emptyset$. Then for any $x \in T \cap E$ we have $\|Ax\|_2 = 0$, so our bound becomes

$$\sqrt{m} \leq C_1 K^2 (w(T) + u).$$

Choosing $u := \sqrt{m}/(2C_1 K^2)$, we simplify the bound to

$$\sqrt{m} \leq C_1 K^2 w(T) + \frac{\sqrt{m}}{2},$$

which yields

$$\sqrt{m} \leq 2C_1 K^2 w(T).$$

But this contradicts the assumption of the escape theorem, as long as we choose the absolute constant C large enough. This means that the event (9.18) with u chosen as above implies that $T \cap E = \emptyset$. The proof is complete. ■

Exercise 9.4.8 (Sharpness of escape theorem) ☕ Discuss the sharpness of the escape theorem for the example where T is the unit sphere of some subspace of $\mathbb{R}^n$.

Exercise 9.4.9 (Escape from a point set) ☕☕ Prove the following version of the escape theorem, with a rotation of a point set instead of a random subspace.

Consider a set $T \subset S^{n-1}$ and let $\mathcal{X}$ be a set of N points in $\mathbb{R}^n$. Show that if

$$\sigma_{n-1}(T) < \frac{1}{N}$$

then there exists a rotation $U \in O(n)$ such that

$$T \cap U\mathcal{X} = \emptyset.$$

Here σ_{n-1} denotes the normalized Lebesgue measure (of area) on S^{n-1}. ☞

9.5 Notes

The matrix deviation inequality (Theorem 9.1.1) and its proof were borrowed from [128]. Several important related results had been known before. In the partial case where A is Gaussian and T is a subset of the unit sphere, Theorem 9.1.1 can be deduced from Gaussian comparison inequalities. The upper bound on $\|Gx\|_2$ can be derived from the Sudakov–Fernique inequality (Theorem 7.2.11), while the lower bound can be obtained from Gordon's inequality (Exercise 7.2.14). G. Schechtman [177] proved a version of the matrix deviation inequality in the partial case of Gaussian random matrices A and for general norms (not necessarily Euclidean); we present this result in Section 11.1. For sub-gaussian matrices A, some earlier versions of the matrix deviation inequality can be found in [113, 141, 61]; see [128, Section 3] for a comparison with these results. Finally, a variant of the matrix deviation inequality for sparse matrices A (more precisely, for an A that is a sparse Johnson–Lindenstrauss transform) was obtained in [30].

A version of Proposition 9.2.1 is due to V. Milman [145]; see [11, Proposition 5.7.1]. Theorem 9.2.4 on covariance estimation for lower-dimensional distributions is due to V. Koltchinskii and K. Lounici [115]; they used a different approach that was also based on the majorizing measure theorem. R. van Handel showed in [207] how to derive Theorem 9.2.4 for Gaussian distributions from decoupling, conditioning, and the Slepian lemma. The bound in Theorem 9.2.4 can be reversed [115, 207].

A version of the Johnson–Lindenstrauss lemma for infinite sets that is similar to Proposition 9.3.2 can be found in [128].

The M^* bound, a version of which we proved in Section 9.4.1, is a useful result in geometric functional analysis; see [11, Sections 7.3, 7.4, and 9.3] and [85, 140, 217] for many known variants, proofs, and consequences of M^* bounds. The version we gave here, Theorem 9.4.2, is from [128].

The escape theorem from Section 9.4.2 is also called "escape from the mesh" in the literature. It was originally proved by Y. Gordon [85] for a Gaussian random matrix A and with a sharp constant factor in (9.17). The argument was based on Gordon's inequality from Exercise 7.2.14. Matching lower bounds for this sharp theorem are known for spherically convex sets [184, 9]. In fact, for a spherically convex set, the exact value of the hitting probability can be obtained by the methods of integral geometry [9]. Oymak and Tropp [159] proved that this sharp result is universal, i.e., it can be extended to non-Gaussian matrices. Our version of the escape theorem (Theorem 9.4.7), which is valid for even more general classes of random matrices but does not feature sharp absolute constants, is borrowed from [128]. As we will see in Section 10.5.1, the escape theorem is an important tool for signal recovery problems.

10

Sparse Recovery

In this chapter we focus entirely on applications of high-dimensional probability to data science. We will study basic signal recovery problems in compressed sensing and structured regression problems in high-dimensional statistics and develop algorithmic methods to solve them using convex optimization.

We introduce these problems in Section 10.1. Our first approach to them, which is very simple and general, is developed in Section 10.2 on the basis of the M^* bound. We then specialize this approach to two important problems. In Section 10.3 we study the sparse recovery problem, in which the unknown signal is sparse (i.e., has few nonzero coordinates). In Section 10.4 we study the low-rank matrix recovery problem, in which the unknown signal is a low-rank matrix. If instead of M^* bounds we use the escape theorem, it is possible to recover sparse signals *exactly* (without any error)! We prove this basic result in compressed sensing in Section 10.5. We first deduce it from the escape theorem and then study an important deterministic condition that guarantees sparse recovery: the restricted isometry property. Finally, in Section 10.6 we use the matrix deviation inequality to analyze Lasso, the most popular optimization method for sparse regression in statistics.

10.1 High-Dimensional Signal Recovery Problems

Mathematically, we model a *signal* as a vector $x \in \mathbb{R}^n$. Suppose that a priori we do not know x, but we have m random, linear, possibly noisy *measurements* of x. Such measurements can be represented as a vector $y \in \mathbb{R}^m$ with the following form:

$$y = Ax + w. \tag{10.1}$$

Here A is an $m \times n$ known measurement matrix, and $w \in \mathbb{R}^m$ is an unknown *noise* vector; see Figure 10.1. Our goal is to recover x from A and y as accurately as possible.

Note that the measurements $y = (y_1, \ldots, y_m)$ can be equivalently represented as

$$y_i = \langle A_i, x \rangle + w_i, \quad i = 1, \ldots, m, \tag{10.2}$$

where $A_i \in \mathbb{R}^n$ denotes a row of the matrix A. It is natural to assume that the A_i are independent, which makes the observations y_i independent too.

Example 10.1.1 (Audio sampling) In signal processing applications, x can be a digitized audio signal. The measurement vector y can be obtained by sampling x at m randomly chosen time points; see Figure 10.2.

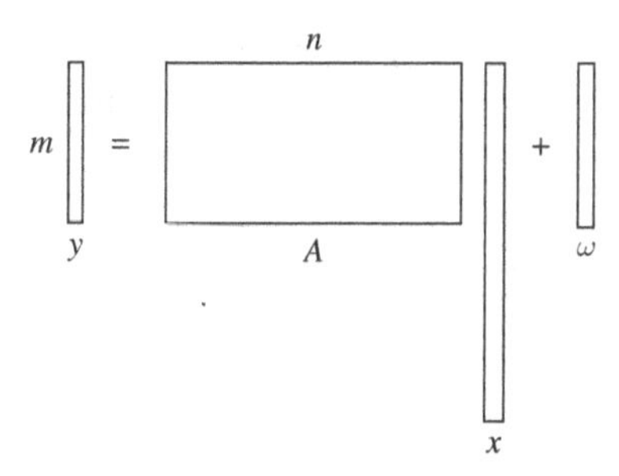

Figure 10.1 Signal recovery problem: recover a signal x from random linear measurements y.

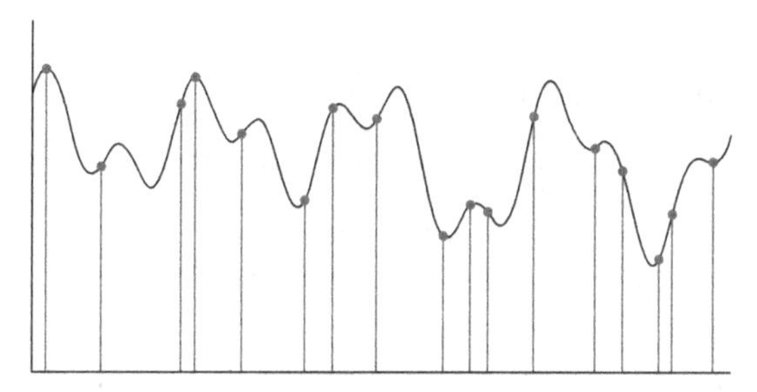

Figure 10.2 Signal recovery problem in audio sampling: recover an audio signal x from a sample of x taken at m random time points.

Example 10.1.2 (Linear regression) Linear regression is one of the major inference problems in statistics. Here we would like to model the relationship between n *predictor variables* and a *response variable*, using a sample of m observations. The regression problem is usually written as

$$Y = X\beta + w.$$

Here X is an $m \times n$ matrix that contains a sample of predictor variables, $Y \in \mathbb{R}^m$ is a vector that contains a sample of response variables, $\beta \in \mathbb{R}^n$ is a coefficient vector that specifies the relationship that we try to recover, and w is a noise vector.

For example, in genetics one could be interested in predicting a certain disease on the basis of genetic information. One thus performs a study on m patients, collecting the expressions of their n genes. The matrix X is defined by letting X_{ij} be the expression of gene j in patient i, and the coefficients Y_i of the vector Y can be set to quantify whether patient i has the disease (and to what extent). The goal is to recover the coefficients of β, which quantify how each gene affects the disease.

10.1.1 Incorporating Prior Information About the Signal

Many modern signal recovery problems operate in the regime where

$$m \ll n,$$

i.e., we have far fewer measurements than unknowns. For instance, in a typical genetic study like that described in Example 10.1.2, the number of patients is ~ 100 while the number of genes is $\sim 10\,000$.

In this regime the recovery problem (10.1) is *ill posed* even in the noiseless case where $w = 0$. It cannot be solved even approximately: the solutions form a linear subspace of dimension at least $n-m$. To overcome this difficulty, we can leverage some *prior information* about the signal x – something that we know, believe, or want to enforce about x. Such information can be mathematically expressed by assuming that

$$x \in T, \tag{10.3}$$

where $T \subset \mathbb{R}^n$ is a known set.

The smaller the set T, the fewer measurements m needed to recover x. For small T we can hope that signal recovery can be solved even in the ill-posed regime where $m \ll n$. We will see how this idea works in the following sections.

10.2 Signal Recovery Based on M^* Bound

Let us return to the recovery problem (10.1). For simplicity, we first consider the noiseless version of the problem, that is,

$$y = Ax, \quad x \in T.$$

To recap, here $x \in \mathbb{R}^n$ is the unknown signal, $T \subset \mathbb{R}^n$ is a known set that encodes our prior information about x, and A is a known $m \times n$ random measurement matrix. Our goal is to recover x from y.

Perhaps the simplest candidate for the solution would be *any* vector x' that is consistent both with the measurements and the prior, so we find

$$x' : \ y = Ax', \quad x' \in T. \tag{10.4}$$

If the set T is convex, this is a convex program (in the feasibility form), and many effective algorithms exist to solve it numerically.

This naïve approach actually works well. We now quickly deduce this from the M^* bound from Section 9.4.1.

Theorem 10.2.1 *Suppose that the rows A_i of A are independent, isotropic, and subgaussian random vectors. Then a solution $\widehat{x}$ of the program* (10.4) *satisfies*

$$\mathbb{E}\,\|\widehat{x} - x\|_2 \le \frac{CK^2 w(T)}{\sqrt{m}},$$

where $K = \max_i \|A_i\|_{\psi_2}$.

Proof Since $x, \widehat{x} \in T$ and $Ax = A\widehat{x} = y$, we have

$$x, \widehat{x} \in T \cap E_x,$$

where $E_x := x + \ker A$. (Figure 10.3 illustrates this situation.) Then the affine version of the M^* bound (Exercise 9.4.3) yields

$$\mathbb{E}\,\|\widehat{x} - x\|_2 \le \mathbb{E}\,\mathrm{diam}(T \cap E_x) \le \frac{CK^2 w(T)}{\sqrt{m}}.$$

This completes the proof. ∎

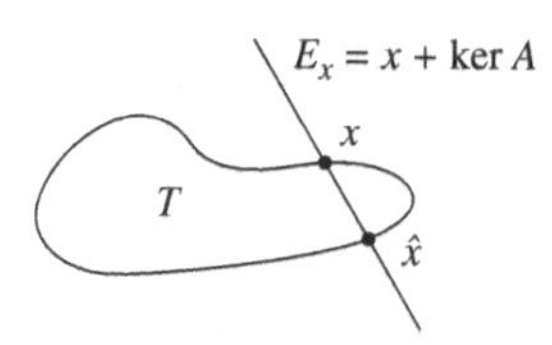

Figure 10.3 Signal recovery: the signal x and the solution $\widehat{x}$ lie in the prior set T and in the affine subspace E_x.

Remark 10.2.2 (Stable dimension) Arguing as in Remark 9.4.4, we obtain a nontrivial error bound

$$\mathbb{E}\,\|\widehat{x} - x\|_2 \le \varepsilon \text{ diam}(T),$$

provided that the number of measurements m is such that

$$m \ge C(K^4/\varepsilon^2)d(T).$$

In words, *the signal can be approximately recovered as long as the number of measurements m exceeds a multiple of the stable dimension $d(T)$ of the prior set T.*

Since the stable dimension can be much smaller than the ambient dimension n, the recovery problem may often be solved even in the high-dimensional, ill-posed, regime where

$$m \ll n.$$

We will see some concrete examples of this situation shortly.

Remark 10.2.3 (Convexity) If the prior set T is not convex, we can convexify it by replacing T with its convex hull $\text{conv}(T)$. This makes (10.4) a convex program and thus computationally tractable. At the same time, the recovery guarantees of Theorem 10.2.1 do not change since

$$w(\text{conv}(T)) = w(T),$$

by Proposition 7.5.2.

Exercise 10.2.4 (Noisy measurements)☕☕ Extend the recovery result (Theorem 10.2.1) for the noisy model $y = Ax + w$ we considered in (10.1). Namely, show that

$$\mathbb{E}\,\|\widehat{x} - x\|_2 \le \frac{CK^2 w(T) + \|w\|_2}{\sqrt{m}}.$$

☞

Exercise 10.2.5 (Mean squared error)☕☕☕ Prove that the error bound in Theorem 10.2.1 can be extended to the mean squared error

$$\mathbb{E}\,\|\widehat{x} - x\|_2^2.$$

☞

Exercise 10.2.6 (Recovery by optimization)☕☕ Suppose that T is the unit ball of some norm $\|\cdot\|_T$ in $\mathbb{R}^n$. Show that the conclusion of Theorem 10.2.1 holds also for the solution of the following optimization program:

$$\text{minimize } \|x'\|_T \text{ s.t. } y = Ax'.$$

10.3 Recovery of Sparse Signals

10.3.1 Sparsity

Let us give a concrete example of a prior set T. Very often we believe that x should be *sparse*, i.e., that most coefficients of x are zero, either exactly or approximately. For instance, in genetic studies like that described in Example 10.1.2, it is natural to expect that very few genes (~ 10) have a significant impact on a given disease, and we would like to find out which they are.

In some applications one needs to change basis so that the signals of interest are sparse. For instance, in the audio recovery problem considered in Example 10.1.1, we are typically dealing with *band-limited* signals x. Those are signals whose frequencies (the values of the Fourier transform) are constrained to some small set, such as a bounded interval. While the audio signal x itself is not sparse, as is apparent from Figure 10.2 the Fourier transform of x may be sparse. In other words, x may be sparse in the frequency but not the time domain.

To quantify the (exact) sparsity of a vector $x \in \mathbb{R}^n$, we consider the size of the support of x, which we denote

$$\|x\|_0 := |\operatorname{supp}(x)| = |\{i : x_i \neq 0\}| .$$

Assume that

$$\|x\|_0 = s \ll n. \tag{10.5}$$

This can be viewed as a special case of a general assumption (10.3) by putting

$$T = \left\{x \in \mathbb{R}^n : \|x\|_0 \le s\right\}.$$

Then a simple dimension count shows the recovery problem (10.1) could become well posed:

Exercise 10.3.1 (Sparse recovery problem is well posed) ☕☕☕ Argue that if $m \ge 2\|x\|_0$, the solution to the sparse recovery problem (10.1) is unique if it exists.

Even when the problem (10.1) is well posed, it could be computationally hard. It is easy if one knows the support of x (why?) but usually the support is unknown. An exhaustive search over all possible supports (subsets of a given size s) is impossible since the number of possibilities is exponentially large: $\binom{n}{s} \ge 2^s$.

Fortunately, there exist computationally effective approaches to high-dimensional recovery problems with general constraints (10.3), and to sparse recovery problems in particular. We cover these approaches next.

Exercise 10.3.2 (The "ℓ_p norms" for $0 \le p < 1$) ☕☕☕

(a) Check that $\|\cdot\|_0$ is not a norm on $\mathbb{R}^n$.
(b) Check that $\|\cdot\|_p$ is not a norm on $\mathbb{R}^n$ if $0 < p < 1$. Figure 10.4 illustrates the unit balls for various ℓ_p "norms".
(c) Show that, for every $x \in \mathbb{R}^n$,

$$\|x\|_0 = \lim_{p \to 0_+} \|x\|_p^p.$$

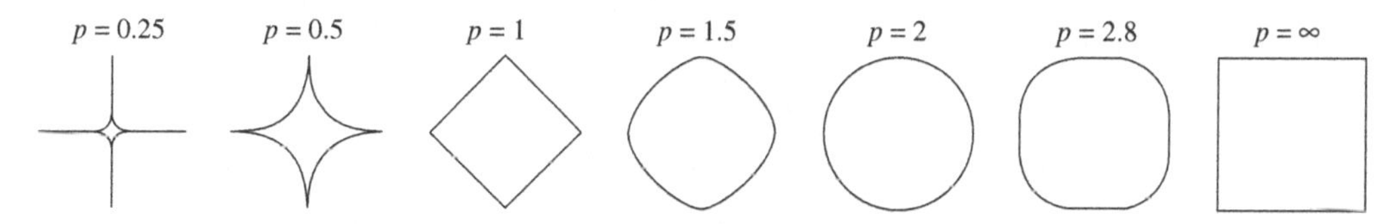

Figure 10.4 The unit balls of ℓ_p for various p in $\mathbb{R}^2$.

10.3.2 Convexifying the Sparsity by the ℓ_1 Norm, and Recovery Guarantees

Let us specialize the general recovery guarantees developed in Section 10.2 to the sparse recovery problem. To do this, we should choose the prior set T so that it promotes sparsity. In the previous section, we saw that the choice

$$T := \{x \in \mathbb{R}^n : \|x\|_0 \le s\}$$

does not allow for computationally tractable algorithms.

To make T convex, we may replace the "ℓ_0 norm" by the ℓ_p norm with the smallest exponent $p > 0$ that makes this a true norm. The exponent p is obviously 1, as we can see from Figure 10.4. So let us repeat this important heuristic: *we propose to replace the ℓ_0 "norm" by the ℓ_1 norm.*

Thus it makes sense to choose T to be a scaled ℓ_1 ball:

$$T := \sqrt{s} B_1^n.$$

The scaling factor $\sqrt{s}$ was chosen so that T can accommodate all s-sparse unit vectors:

Exercise 10.3.3☕ Check that

$$\{x \in \mathbb{R}^n : \|x\|_0 \le s,\ \|x\|_2 \le 1\} \subset \sqrt{s} B_1^n.$$

For this T, the general recovery program (10.4) becomes

$$\text{Find } x' : \; y = Ax', \quad \|x'\|_1 \le \sqrt{s}. \tag{10.6}$$

Note that this is a convex program and therefore is computationally tractable. And the general recovery guarantee, Theorem 10.2.1, specialized to our case, implies the following.

Corollary 10.3.4 (Sparse recovery: guarantees) *Assume that the unknown s-sparse signal $x \in \mathbb{R}^n$ satisfies $\|x\|_2 \le 1$. Then x can be approximately recovered from the random measurement vector $y = Ax$ by a solution $\hat{x}$ of the program* (10.6). *The recovery error satisfies*

$$\mathbb{E}\,\|\hat{x} - x\|_2 \le CK^2 \sqrt{\frac{s \log n}{m}}.$$

Proof Set $T = \sqrt{s} B_1^n$. The result follows from Theorem 10.2.1 and the bound (7.18) on the Gaussian width of the ℓ_1 ball:

$$w(T) = \sqrt{s}\, w(B_1^n) \le C\sqrt{s \log n}.$$

■

Remark 10.3.5 The recovery error guaranteed by Corollary 10.3.4 is small if

$$m \gtrsim s \log n$$

(and if the hidden constant here is appropriately large). In words, recovery is possible if *the number of measurements m is almost linear in the sparsity s*, while its dependence on the ambient dimension n is mild (logarithmic). This is good news. It means that, for sparse signals, one can solve recovery problems in the high-dimensional regime where

$$m \ll n,$$

i.e., with many fewer measurements than the dimension.

Exercise 10.3.6 (Sparse recovery by convex optimization)☕☕☕

(a) Show that an unknown s-sparse signal x (without restriction on the norm) can be approximately recovered by solving the convex optimization problem

$$\text{minimize } \|x'\|_1 \text{ s.t. } y = Ax'. \tag{10.7}$$

The recovery error satisfies

$$\mathbb{E}\,\|\widehat{x} - x\|_2 \le C\sqrt{\frac{s \log n}{m}}\,\|x\|_2.$$

(b) Argue that a similar result holds for approximately sparse signals. State and prove such a guarantee.

10.3.3 The Convex Hull of Sparse Vectors, and the Logarithmic Improvement

The replacement of s-sparse vectors by the octahedron $\sqrt{s}B_1^n$ that we made in Exercise 10.3.3 is almost sharp. In the following exercise, we show that the convex hull of the set of sparse vectors

$$S_{n,s} := \{x \in \mathbb{R}^n : \|x\|_0 \le s,\ \|x\|_2 \le 1\}$$

is approximately the truncated ℓ_1 ball

$$T_{n,s} := \sqrt{s}B_1^n \cap B_2^n = \{x \in \mathbb{R}^n : \|x\|_1 \le \sqrt{s},\ \|x\|_2 \le 1\}.$$

Exercise 10.3.7 (The convex hull of sparse vectors)☕☕☕

(a) Check that

$$\text{conv}(S_{n,s}) \subset T_{n,s}.$$

(b) To help prove a reverse inclusion, fix $x \in T_{n,s}$ and partition the support of x into disjoint subsets $I_1, I_2, \ldots$ so that I_1 indexes the s largest coefficients of x in magnitude, I_2 indexes the next s largest coefficients, and so on. Show that

$$\sum_{i \ge 1} \|x_{I_i}\|_2 \le 2,$$

where $x_I \in \mathbb{R}^T$ denotes the restriction of x onto a set I. ☞

(c) Deduce from part (b) that

$$T_{n,s} \subset 2\,\text{conv}(S_{n,s}).$$

Exercise 10.3.8 (Gaussian width of a set of sparse vectors) Use Exercise 10.3.7 to show that

$$w(T_{n,s}) \le 2w(S_{n,s}) \le C\sqrt{s\log(en/s)}.$$

Improve the logarithmic factor in the error bound for sparse recovery (Corollary 10.3.4) to

$$\mathbb{E}\,\|\widehat{x} - x\|_2 \le C\sqrt{\frac{s\log(en/s)}{m}}.$$

This shows that

$$m \gtrsim s\log(en/s)$$

measurements suffice for sparse recovery.

Exercise 10.3.9 (Sharpness) Show that

$$w(T_{n,s}) \ge w(S_{n,s}) \ge c\sqrt{s\log(2n/s)}.$$

☞

Exercise 10.3.10 (Garnaev–Gluskin theorem) Improve the logarithmic factor in the bound (9.4.5) on the sections of the ℓ_1 ball. Namely, show that

$$\mathbb{E}\,\mathrm{diam}(B_1^n \cap E) \lesssim \sqrt{\frac{\log(en/m)}{m}}.$$

In particular, this shows that the logarithmic factor in (9.16) is not needed. ☞

10.4 Low-Rank Matrix Recovery

In the following series of exercises, we establish a *matrix* version of the sparse recovery problem studied in Section 10.3. The unknown signal will now be a $d \times d$ matrix X instead of the signal $x \in \mathbb{R}^n$ considered previously.

There are two natural notions of sparsity for matrices. One is where most of the entries of X are zero, and it is quantified by the ℓ_0 "norm" $\|X\|_0$, which counts nonzero entries. For this notion, we can directly apply the analysis of sparse recovery from Section 10.3. Indeed, it is enough to vectorize the matrix X and think of it as a long vector in $\mathbb{R}^{d^2}$.

But, in this section, we consider an alternative and equally useful notion of sparsity for matrices: *low rank*. It is quantified by the rank of X, which we may think of as the ℓ_0 norm of the vector of the singular values of X, i.e.,

$$s(X) := (s_i(X))_{i=1}^d. \tag{10.8}$$

Our analysis of the low-rank matrix recovery problem will roughly go along the same lines as the analysis of sparse recovery, but will not be identical to it.

Let us set up a low-rank matrix recovery problem. We would like to recover an unknown $d \times d$ matrix from m random measurements of the form

$$y_i = \langle A_i, X\rangle, \quad i = 1, \ldots, m. \tag{10.9}$$

Here the A_i are independent $d \times d$ matrices, and $\langle A_i, X \rangle = \mathrm{tr}(A_i^\mathsf{T} X)$ is the canonical inner product of matrices (recall Section 4.1.3). In dimension $d = 1$, the matrix recovery problem (10.9) reduces to the vector recovery problem (10.2).

Since we have m linear equations in $d \times d$ variables, the matrix recovery problem is *ill posed* if

$$m < d^2.$$

To be able to solve it in this range, we make an additional assumption that X has low rank, i.e.,

$$\mathrm{rank}(X) \le r \ll d.$$

10.4.1 The Nuclear Norm

Like sparsity, the rank is not a convex function. To fix this, in Section 10.3 we replaced the sparsity (i.e. the ℓ_0 "norm") by the ℓ_1 norm. Let us try to do the same for the notion of rank. The rank of X is the ℓ_0 "norm" of the vector $s(X)$ of the singular values in (10.8). Replacing the ℓ_0 norm by the ℓ_1 norm, we obtain the quantity

$$\|X\|_* := \|s(X)\|_1 = \sum_{i=1}^{d} s_i(X) = \mathrm{tr}\,\sqrt{X^\mathsf{T} X}$$

which is called the *nuclear norm*, or *trace norm*, of X. (We omit the absolute values since the singular values are non-negative.)

Exercise 10.4.1 ☕☕☕ Prove that $\|\cdot\|_*$ is indeed a norm on the space of $d \times d$ matrices. ☞

Exercise 10.4.2 (Nuclear, Frobenius, and operator norms) ☕☕ Check that

$$\langle X, Y \rangle \le \|X\|_* \, \|Y\|. \tag{10.10}$$

Conclude that

$$\|X\|_F^2 \le \|X\|_* \, \|X\|.$$

☞

Now denote the unit ball of the nuclear norm by

$$B_* := \left\{ X \in \mathbb{R}^{d \times d} \colon \|X\|_* \le 1 \right\}.$$

Exercise 10.4.3 (Gaussian width of the unit ball of the nuclear norm) ☕ Show that

$$w(B_*) \le 2\sqrt{d}.$$

☞

The following is a matrix version of Exercise 10.3.3.

Exercise 10.4.4 ☕ Check that

$$\left\{ X \in \mathbb{R}^{d \times d} \colon \mathrm{rank}(X) \le r,\ \|X\|_F \le 1 \right\} \subset \sqrt{r} B_*.$$

10.4.2 Guarantees for Low-Rank Matrix Recovery

It makes sense to try to solve the low-rank matrix recovery problem (10.9) using the matrix version of the convex program (10.6), i.e.,

$$\text{find } X' : \; y_i = \langle A_i, X' \rangle \; \forall i = 1, \dots, m, \; \|X'\|_* \le \sqrt{r}. \tag{10.11}$$

Exercise 10.4.5 (Low-rank matrix recovery: guarantees) Suppose that the random matrices A_i are independent and have all independent sub-gaussian entries.[1] Assume the unknown $d \times d$ matrix X with rank r satisfies $\|X\|_F \le 1$. Show that X can be approximately recovered from the random measurements y_i by a solution $\widehat{X}$ of the program (10.11). Show that the recovery error satisfies

$$\mathbb{E}\,\|\widehat{X} - X\|_F \le CK^2\sqrt{\frac{rd}{m}}.$$

Remark 10.4.6 The recovery error becomes small if

$$m \gtrsim rd,$$

if the hidden constant here is appropriately large. This allows us to recover low-rank matrices even when the number of measurements m is too small, i.e., when

$$m \ll d^2$$

and the matrix recovery problem (without rank assumption) is ill posed.

Exercise 10.4.7 Extend the matrix recovery result for *approximately* low-rank matrices.

The following is a matrix version of Exercise 10.3.6.

Exercise 10.4.8 (Low-rank matrix recovery by convex optimization) Show that an unknown matrix X of rank r can be approximately recovered by solving the convex optimization problem

$$\text{minimize } \|X'\|_* \text{ s.t. } y_i = \langle A_i, X' \rangle \; \forall i = 1, \dots, m.$$

Exercise 10.4.9 (Rectangular matrices) Extend the matrix recovery result from quadratic to rectangular, $d_1 \times d_2$, matrices.

10.5 Exact Recovery and the Restricted Isometry Property

It turns out that the guarantees for sparse recovery that we have just developed can be dramatically improved: the recovery error for sparse signals x can actually be *zero*! We discuss two approaches to this remarkable phenomenon. First we deduce exact recovery from the escape theorem 9.4.7. Next we present a general deterministic condition on a matrix A which

[1] The independence of entries can be relaxed. How?

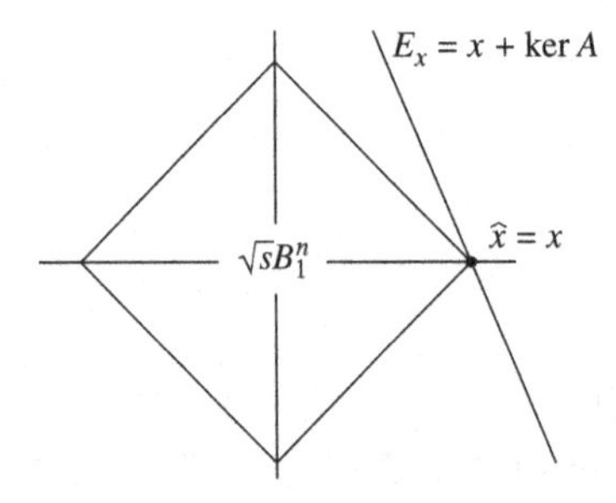

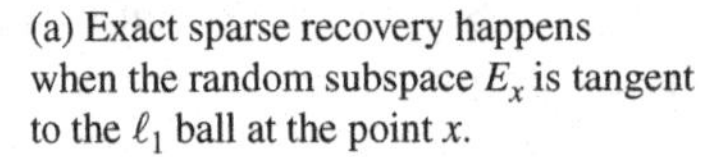
(a) Exact sparse recovery happens when the random subspace E_x is tangent to the ℓ_1 ball at the point x.

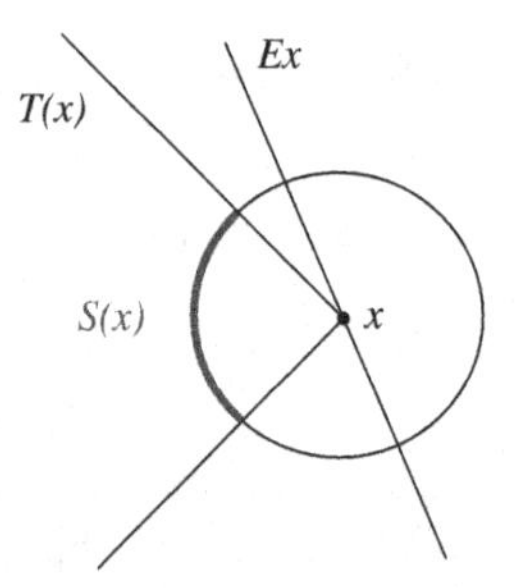

(b) The tangency occurs if and only if E_x is disjoint from the spherical part $S(x)$ of the tangent cone $T(x)$ of the ℓ_1 ball at point x.

Figure 10.5 Exact sparse recovery.

guarantees exact recovery; it is known as the restricted isometry property (RIP). We check that random matrices A satisfy RIP, which gives another approach to exact recovery.

10.5.1 Exact Recovery Based on the Escape Theorem

To see why exact recovery should be possible, let us look at the recovery problem from a geometric viewpoint, as illustrated by Figure 10.3. A solution $\widehat{x}$ of the program (10.6) must lie in the intersection of the prior set T, which in our case is the ℓ_1 ball $\sqrt{s}B_1^n$, and the affine subspace $E_x = x + \ker A$.

The ℓ_1 ball is a polytope, and the s-sparse unit vector x lies on the $(s-1)$-dimensional edge of that polytope; see Figure 10.5a.

It could happen with nonzero probability that the random subspace E_x is *tangent* to the polytope at the point x. If this does happen, x is the only point of intersection between the ℓ_1 ball and E_x. In this case, it follows that the solution $\widehat{x}$ to the program (10.6) is exact:

$$\widehat{x} = x.$$

To justify this argument, all we need to check is that the random subspace E_x is tangent to the ℓ_1 ball with high probability. We can do this using the escape theorem 9.4.7. To see a connection, look at what happens in a small neighborhood around the tangent point; see Figure 10.5b. The subspace E_x is tangent if and only if the *tangent cone* $T(x)$ (formed by all rays emanating from x toward the points in the ℓ_1 ball) intersects E_x at a single point x. Equivalently, this happens if and only if the *spherical part* $S(x)$ of the cone (the intersection of $T(x)$ with a small sphere centered at x) is disjoint from E_x. But this is exactly the conclusion of escape theorem 9.4.7!

Let us now formally state the exact recovery result. We shall consider the noiseless sparse recovery problem

$$y = Ax$$

and try to solve it using the optimization program (10.7), i.e.

$$\text{minimize } \|x'\|_1 \text{ s.t. } y = Ax'. \tag{10.12}$$

Theorem 10.5.1 (Exact sparse recovery) *Suppose that the rows A_i of A are independent, isotropic, and sub-gaussian random vectors, and let $K := \max_i \|A_i\|_{\psi_2}$. Then the event below happens with probability at least $1 - 2\exp(-cm/K^4)$.*

Assume that an unknown signal $x \in \mathbb{R}^n$ is s-sparse and the number of measurements m satisfies

$$m \geq CK^4 s \log n.$$

Then a solution $\widehat{x}$ of the program (10.12) *is exact, i.e.,*

$$\widehat{x} = x.$$

To prove the theorem, we would like to show that the recovery error

$$h := \widehat{x} - x$$

is zero. Let us examine the vector h more closely. First we show that h has more "energy" on the support of x than outside it.

Lemma 10.5.2 *Let $S := \text{supp}(x)$. Then*

$$\|h_{S^c}\|_1 \leq \|h_S\|_1.$$

Here $h_S \in \mathbb{R}^S$ denotes the restriction of the vector $h \in \mathbb{R}^n$ to a subset of coordinates $S \subset \{1, \ldots, n\}$.

Proof Since $\widehat{x}$ is the minimizer in the program (10.12), we have

$$\|\widehat{x}\|_1 \leq \|x\|_1. \tag{10.13}$$

But there is also a lower bound, as

$$\begin{aligned} \|\widehat{x}\|_1 = \|x + h\|_1 &= \|x_S + h_S\|_1 + \|x_{S^c} + h_{S^c}\|_1 \\ &\geq \|x\|_1 - \|h_S\|_1 + \|h_{S^c}\|_1, \end{aligned}$$

where the last line follows by the triangle inequality and we use $x_S = x$ and $x_{S^c} = 0$. Substitute this bound into (10.13) and cancel $\|x\|_1$ on both sides to complete the proof. ∎

Lemma 10.5.3 *The error vector satisfies*

$$\|h\|_1 \leq 2\sqrt{s}\|h\|_2.$$

Proof Using Lemma 10.5.2 and then the Cauchy-Schwarz inequality, we obtain

$$\|h\|_1 = \|h_S\|_1 + \|h_{S^c}\|_1 \leq 2\|h_S\|_1 \leq 2\sqrt{s}\|h_S\|_2.$$

Since trivially $\|h_S\|_2 \leq \|h\|_2$, the proof is complete. ∎

Proof of Theorem 10.5.1 Assume that the recovery is not exact, i.e.,

$$h = \widehat{x} - x \neq 0.$$

By Lemma 10.5.3, the normalized error $h/\|h\|_2$ lies in the set

$$T_s := \left\{ z \in S^{n-1} : \|z\|_1 \le 2\sqrt{s} \right\}.$$

Since also

$$Ah = A\widehat{x} - Ax = y - y = 0,$$

we have

$$\frac{h}{\|h\|_2} \in T_s \cap \ker A. \tag{10.14}$$

The escape theorem 9.4.7 states that this intersection is empty with high probability, as long as

$$m \ge CK^4 w(T_s)^2.$$

Now,

$$w(T_s) \le 2\sqrt{s}\, w(B_1^n) \le C\sqrt{s \log n}, \tag{10.15}$$

where we have used the bound (7.18) on the Gaussian width of the ℓ_1 ball. Thus, if $m \ge CK^4 s \log n$, the intersection in (10.14) is empty with high probability, which means that the inclusion in (10.14) cannot hold. This contradiction implies that our assumption that $h \ne 0$ is false with high probability. The proof is complete. ■

Exercise 10.5.4 (Improving the logarithmic factor) Show that the conclusion of Theorem 10.5.1 holds under a weaker assumption on the number of measurements, i.e.,

$$m \ge CK^4 s \log(en/s).$$ ☞

Exercise 10.5.5 Give a geometric interpretation of the proof of Theorem 10.5.1, using Figure 10.5b. What does the proof say about the tangent cone $T(x)$, and about its spherical part $S(x)$?

Exercise 10.5.6 (Noisy measurements) Extend the result on sparse recovery (Theorem 10.5.1) to noisy measurements, where

$$y = Ax + w.$$

You may need to modify the recovery program by making the constraint $y = Ax'$ approximate.

Remark 10.5.7 Theorem 10.5.1 shows that one can effectively solve *under-determined systems of linear equations* $y = Ax$ with $m \ll n$ equations in n variables, if the solution is sparse.

10.5.2 Restricted Isometries

This subsection is optional; further material is not based on it.

All the recovery results proved so far are probabilistic: they are valid for a random measurement matrix A and with high probability. We may wonder whether there exists a *deterministic* condition which can guarantee that a given matrix A can be used for sparse recovery. Such a condition is the restricted isometry property.

Definition 10.5.8 (RIP) An $m \times n$ matrix A satisfies the *restricted isometry property* (RIP) with parameters α, β, and s if the inequality

$$\alpha\|v\|_2 \le \|Av\|_2 \le \beta\|v\|_2$$

holds for all vectors $v \in \mathbb{R}^n$ such that[2] $\|v\|_0 \le s$.

In other words, a matrix A satisfies the RIP if the restriction of A to any s-dimensional coordinate subspace of $\mathbb{R}^n$ is an approximate isometry in the sense of (4.5).

Exercise 10.5.9 (RIP via singular values) Check that the RIP holds if and only if the singular values satisfy the inequality

$$\alpha \le s_s(A_I) \le s_1(A_I) \le \beta$$

for all subsets $I \subset [n]$ of size $|I| = s$. Here A_I denotes the $m \times s$ sub-matrix of A formed by selecting the columns indexed by I.

Now we will prove that RIP is indeed a sufficient condition for sparse recovery.

Theorem 10.5.10 (RIP implies exact recovery) *Suppose that an $m \times n$ matrix A satisfies the RIP with some parameters α, β, and $(1+\lambda)s$, where $\lambda > (\beta/\alpha)^2$. Then every s-sparse vector $x \in \mathbb{R}^n$ can be recovered exactly by solving the program* (10.12)*, i.e., the solution satisfies*

$$\widehat{x} = x.$$

Proof As in the proof of Theorem 10.5.1, we would like to show that the recovery error

$$h = \widehat{x} - x$$

is zero. To do this, we decompose h in a way similar to Exercise 10.3.7.

Step 1: Decomposing the support. Let I_0 be the support of x; let I_1 index the λs largest coefficients of $h_{I_0^c}$ in magnitude; let I_2 index the next λs largest coefficients of $h_{I_0^c}$ in magnitude, and so on. Finally, denote $I_{0,1} = I_0 \cup I_1$.

Since

$$Ah = A\widehat{x} - Ax = y - y = 0,$$

[2] Recall from Section 10.3.1 that by $\|v\|_0$ we denote the number of nonzero coordinates of v.

the triangle inequality yields

$$0 = \|Ah\|_2 \geq \|A_{I_{0,1}} h_{I_{0,1}}\|_2 - \|A_{I_{0,1}^c} h_{I_{0,1}^c}\|_2. \tag{10.16}$$

Next, we examine the two terms on the right-hand side.

Step 2: Applying RIP. Since $|I_{0,1}| \leq s + \lambda s$, RIP yields

$$\|A_{I_{0,1}} h_{I_{0,1}}\|_2 \geq \alpha \|h_{I_{0,1}}\|_2,$$

and the triangle inequality, followed by the RIP, also gives

$$\|A_{I_{0,1}^c} h_{I_{0,1}^c}\|_2 \leq \sum_{i \geq 2} \|A_{I_i} h_{I_i}\|_2 \leq \beta \sum_{i \geq 2} \|h_{I_i}\|_2.$$

Substituting into (10.16) gives

$$\beta \sum_{i \geq 2} \|h_{I_i}\|_2 \geq \alpha \|h_{I_{0,1}}\|_2. \tag{10.17}$$

Step 3: Summing up. Next, we bound the sum on the left in the same way as we did in Exercise 10.3.7. By the definition of I_i, each coefficient of h_{I_i} is bounded in magnitude by the average of the coefficients of $h_{I_{i-1}}$, i.e., by $\|h_{I_{i-1}}\|_1/(\lambda s)$ for $i \geq 2$. Thus

$$\|h_{I_i}\|_2 \leq \frac{1}{\sqrt{\lambda s}} \|h_{I_{i-1}}\|_1.$$

Summing up, we get

$$\begin{aligned}
\sum_{i \geq 2} \|h_{I_i}\|_2 &\leq \frac{1}{\sqrt{\lambda s}} \sum_{i \geq 1} \|h_{I_i}\|_1 = \frac{1}{\sqrt{\lambda s}} \|h_{I_0^c}\|_1 \\
&\leq \frac{1}{\sqrt{\lambda s}} \|h_{I_0}\|_1 \quad \text{(by Lemma 10.5.2)} \\
&\leq \frac{1}{\sqrt{\lambda}} \|h_{I_0}\|_2 \leq \frac{1}{\sqrt{\lambda}} \|h_{I_{0,1}}\|_2.
\end{aligned}$$

Putting this into (10.17) we conclude that

$$\frac{\beta}{\sqrt{\lambda}} \|h_{I_{0,1}}\|_2 \geq \alpha \|h_{I_{0,1}}\|_2.$$

This implies that $h_{I_{0,1}} = 0$, since $\beta/\sqrt{\lambda} > \alpha$ by assumption. By construction $I_{0,1}$ contains the largest coefficient of h. It follows that $h = 0$ as claimed. The proof is complete. ■

Unfortunately, it is unknown how to construct deterministic matrices A that satisfy the RIP with good parameters (i.e., with $\beta = O(\alpha)$ and with s as large as m, up to logarithmic factors). However, it is quite easy to show that random matrices A do satisfy the RIP with high probability:

Theorem 10.5.11 (Random matrices satisfy RIP) *Consider an $m \times n$ matrix A whose rows A_i of A are independent, isotropic, and sub-gaussian random vectors, and let $K := \max_i \|A_i\|_{\psi_2}$. Assume that*

$$m \geq C K^4 s \log(en/s).$$

Then, with probability at least $1 - 2\exp(-cm/K^4)$*, the random matrix* A *satisfies the RIP with parameters* $\alpha = 0.9\sqrt{m}$*,* $\beta = 1.1\sqrt{m}$*, and* s*.*

Proof By Exercise 10.5.9 it is enough to control the singular values of all $m \times s$ sub-matrices A_I. We will do this by using the two-sided bound from Theorem 4.6.1 and then taking the union bound over all sub-matrices.

Let us fix I. Theorem 4.6.1 yields

$$\sqrt{m} - r \le s_n(A_I) \le s_1(A_I) \le \sqrt{m} + r$$

with probability at least $1 - 2\exp(-t^2)$, where $r = C_0 K^2(\sqrt{s} + t)$. If we set $t = \sqrt{m}/(20C_0K^2)$ and use the assumption on m with appropriately large constant C, we can ensure that $r \le 0.1\sqrt{m}$. This yields

$$0.9\sqrt{m} \le s_n(A_I) \le s_1(A_I) \le 1.1\sqrt{m} \tag{10.18}$$

with probability at least $1 - 2\exp(-2cm^2/K^4)$, where $c > 0$ is an absolute constant.

It remains to take a union bound over all s-element subsets $I \subset [n]$; there are $\binom{n}{s}$ of them. We conclude that (10.18) holds with probability at least

$$1 - 2\exp(-2cm^2/K^4)\binom{n}{s} > 1 - 2\exp(-cm^2/K^4).$$

For the last inequality, recall that $\binom{n}{s} \le \exp(s\log(en/s))$ by (0.0.5) and use the assumption on m. The proof is complete. ■

The results we have just proved give us another approach to exact recovery with a random matrix A, Theorem 10.5.1.

Second proof of Theorem 10.5.1 By Theorem 10.5.11, A satisfies the RIP with $\alpha = 0.9\sqrt{m}$, $\beta = 1.1\sqrt{m}$, and $3s$. Thus, Theorem 10.5.10 for $\lambda = 2$ guarantees exact recovery. We conclude that Theorem 10.5.1 holds, and we even get the logarithmic improvement noted in Exercise 10.5.4. ■

An advantage of the RIP is it is often simpler to verify this property than to prove exact recovery directly. Let us give one example.

Exercise 10.5.12 (RIP for random projections) ☕☕☕ Let P be the orthogonal projection in $\mathbb{R}^n$ onto an m-dimensional random subspace uniformly distributed in the Grassmannian $G_{n,m}$.

(a) Prove that P satisfies the RIP with parameters that are similar to Theorem 10.5.11, up to a normalization.
(b) Conclude a version of Theorem 10.5.1 for exact recovery from random projections.

10.6 Lasso Algorithm for Sparse Regression

In this section we analyze an alternative method for sparse recovery. The operator in this method was originally developed in statistics for the equivalent problem of *sparse linear regression*, and it is called the Lasso ("least absolute shrinkage and selection operator").

10.6.1 Statistical Formulation

Let us recall the classical linear regression problem which we described in Example 10.1.2. It is

$$Y = X\beta + w, \tag{10.19}$$

where X is a known $m \times n$ matrix that contains a sample of predictor variables, $Y \in \mathbb{R}^m$ is a known vector that contains a sample of the values of the response variable, $\beta \in \mathbb{R}^n$ is an unknown coefficient vector that specifies the relationship between predictor and response variables, and w is a noise vector. We would like to recover β.

If we do not assume anything else, the regression problem can be solved by the method of *ordinary least squares*, which minimizes the ℓ_2 norm of the error over all candidates for β:

$$\text{minimize } \|Y - X\beta'\|_2 \text{ s.t. } \beta' \in \mathbb{R}^n. \tag{10.20}$$

Now let us make the extra assumption that β' is *sparse*, so that the response variable depends on only a few of the n predictor variables (e.g., the cancer depends on only a few genes). So, as in (10.5), we assume that

$$\|\beta\|_0 \le s$$

for some $s \ll n$. As we argued in Section 10.3, the ℓ_0 norm is not convex, and its convex proxy is the ℓ_1 norm. This prompts us to modify the ordinary least squares program (10.20) by including a restriction on the ℓ_1 norm which promotes sparsity in the solution:

$$\text{minimize } \|Y - X\beta'\|_2 \text{ s.t. } \|\beta'\|_1 \le R, \tag{10.21}$$

where R is a parameter which specifies the desired sparsity level of the solution. The program (10.21) is one formulation of Lasso, the most popular statistical method for sparse linear regression. It is a convex program, and therefore is computationally tractable.

10.6.2 Mathematical Formulation and Guarantees

It would be convenient to return to the notation we used for sparse recovery instead of using the statistical notation of the previous section. So let us restate the linear regression problem (10.19) as

$$y = Ax + w$$

where A is a known $m \times n$ matrix, $y \in \mathbb{R}^m$ is a known vector, $x \in \mathbb{R}^n$ is an unknown vector that we are trying to recover, and $w \in \mathbb{R}^m$ is the noise, which is either fixed or random and is independent of A. Then the Lasso program (10.21) becomes

$$\text{minimize } \|y - Ax'\|_2 \text{ s.t. } \|x'\|_1 \le R. \tag{10.22}$$

We prove the following guarantee of the performance of Lasso.

Theorem 10.6.1 (Performance of Lasso) *Suppose that the rows A_i of A are independent, isotropic, and sub-gaussian random vectors, and let $K := \max_i \|A_i\|_{\psi_2}$. Then the event below happens with probability at least* $1 - 2\exp(-s \log n)$.

Assume that an unknown signal $x \in \mathbb{R}^n$ is s-sparse and the number of measurements m satisfies

$$m \geq CK^4 s \log n. \tag{10.23}$$

Then a solution $\widehat{x}$ of the program (10.22) *with $R := \|x\|_1$ is accurate, namely*

$$\|\widehat{x} - x\|_2 \leq C\sigma\sqrt{\frac{s \log n}{m}},$$

where $\sigma = \|w\|_2/\sqrt{m}$.

Remark 10.6.2 (Noise) The quantity σ^2 is the *average squared noise per measurement*, since

$$\sigma^2 = \frac{\|w\|_2^2}{m} = \frac{1}{m}\sum_{i=1}^m w_i^2.$$

Then, if the number of measurements is

$$m \gtrsim s \log n,$$

Theorem 10.6.1 bounds the recovery error by the average noise per measurement σ. And if m is larger then the recovery error gets smaller.

Remark 10.6.3 (Exact recovery) In the noiseless model $y = Ax$ we have $w = 0$ and thus Lasso recovers x exactly, i.e.

$$\widehat{x} = x.$$

The proof of Theorem 10.6.1 will be similar to our proof of Theorem 10.5.1 on exact recovery, although instead of the escape theorem we use the matrix deviation inequality (Theorem 9.1.1) directly this time.

We would like to bound the norm of the error vector

$$h := \widehat{x} - x.$$

Exercise 10.6.4 ☕☕ Check that h satisfies the conclusions of Lemmas 10.5.2 and 10.5.3, so that we have

$$\|h\|_1 \leq 2\sqrt{s}\|h\|_2. \tag{10.24}$$

☞

In the case where the noise w is nonzero, we cannot expect to have $Ah = 0$ as in Theorem 10.5.1. (Why?) Instead, we can give upper and lower bounds for $\|Ah\|_2$.

Lemma 10.6.5 (Upper bound on $\|Ah\|_2$) *We have*

$$\|Ah\|_2^2 \leq 2\langle h, A^\mathsf{T} w\rangle. \tag{10.25}$$

Proof Since $\widehat{x}$ is the minimizer of the Lasso program (10.22), we have

$$\|y - A\widehat{x}\|_2 \le \|y - Ax\|_2.$$

Let us express both sides of this inequality in terms of h and w, using $y = Ax + w$ and $h = \widehat{x} - x$:

$$y - A\widehat{x} = Ax + w - A\widehat{x} = w - Ah,$$
$$y - Ax = w.$$

So, we have

$$\|w - Ah\|_2 \le \|w\|_2.$$

Square both sides:

$$\|w\|_2^2 - 2\langle w, Ah\rangle + \|Ah\|_2^2 \le \|w\|_2^2.$$

Simplifying this bound completes the proof. ■

Lemma 10.6.6 (Lower bound on $\|Ah\|_2$) *With probability that is at least* $1 - 2\exp(-4s\log n)$, *we have*

$$\|Ah\|_2^2 \ge \frac{m}{4}\|h\|_2^2.$$

Proof By (10.24), the normalized error $h/\|h\|_2$ lies in the set

$$T_s := \left\{z \in S^{n-1} : \|z\|_1 \le 2\sqrt{s}\right\}.$$

Use the matrix deviation inequality in its high-probability form (Exercise 9.1.8) with $u = 2\sqrt{s\log n}$. It yields that, with probability at least $1 - 2\exp(-4s\log n)$,

$$\begin{aligned}\sup_{z\in T_s}\left|\|Az\|_2 - \sqrt{m}\right| &\le C_1K^2\left(w(T_s) + 2\sqrt{s\log n}\right)\\ &\le C_2K^2\sqrt{s\log n} \quad \text{(recalling (10.15))}\\ &\le \frac{\sqrt{m}}{2} \quad \text{(by the assumption on } m).\end{aligned}$$

To obtain the last line, choose the absolute constant C in (10.23) large enough. By the triangle inequality this implies that

$$\|Az\|_2 \ge \frac{\sqrt{m}}{2} \quad \text{for all } z \in T_s.$$

Substituting $z := h/\|h\|_2$, we complete the proof. ■

The last piece of information that we need to prove Theorem 10.6.1 is an upper bound on the right-hand side of (10.25).

Lemma 10.6.7 *With probability at least* $1 - 2\exp(-4s\log n)$, *we have*

$$\left\langle h, A^{\mathsf{T}}w\right\rangle \le CK\|h\|_2\|w\|_2\sqrt{s\log n}. \tag{10.26}$$

Proof As in the proof of Lemma 10.6.6, the normalized error satisfies

$$z = \frac{h}{\|h\|_2} \in T_s.$$

So, dividing both sides of (10.26) by $\|h\|_2$, we see that it is enough to bound the supremum random process

$$\sup_{z \in T_s} \langle z, A^\mathsf{T} w \rangle$$

with high probability. We will use Talagrand's comparison inequality (Corollary 8.6.3). This result applies for random processes that have sub-gaussian increments, so let us check this latter condition first.

Exercise 10.6.8 Show that the random process

$$X_t := \langle t, A^\mathsf{T} w \rangle, \quad t \in \mathbb{R}^n,$$

has sub-gaussian increments and that

$$\|X_t - X_s\|_{\psi_2} \le CK \|w\|_2 \, \|t - s\|_2.$$

Now we can use Talagrand's comparison inequality in its high-probability form (Exercise 8.6.5) for $u = 2\sqrt{s \log n}$. We obtain that, with probability at least $1 - 2\exp(-4s \log n)$,

$$\begin{aligned}\sup_{z \in T_s} \langle z, A^\mathsf{T} w \rangle &\le C_1 K \|w\|_2 \left(w(T_s) + 2\sqrt{s \log n} \right) \\ &\le C_2 K \|w\|_2 \sqrt{s \log n} \quad \text{(recalling (10.15)).}\end{aligned}$$

This completes the proof of Lemma 10.6.7. ■

Proof of Theorem 10.6.1 Put together the bounds in Lemmas 10.6.5 and 10.6.6 and in (10.26). By the union bound, we have that, with probability at least $1 - 4\exp(-4s \log n)$,

$$\frac{m}{4} \|h\|_2^2 \le CK \|h\|_2 \|w\|_2 \sqrt{s \log n}.$$

Solving for $\|h\|_2$, we obtain

$$\|h\|_2 \le CK \frac{\|w\|_2}{\sqrt{m}} \sqrt{\frac{s \log n}{m}}.$$

This completes the proof of Theorem 10.6.1. ■

Exercise 10.6.9 (Improving the logarithmic factor) Show that Theorem 10.6.1 holds if $\log n$ is replaced by $\log(en/s)$, thus giving a stronger guarantee.

Exercise 10.6.10 Deduce the exact recovery guarantee (Theorem 10.5.1) directly from the Lasso guarantee (Theorem 10.6.1). The resulting probability could be a bit weaker.

Another popular form of the Lasso program (10.22) is the following *unconstrained version*:

$$\text{minimize } \|y - Ax'\|_2 + \lambda \|x'\|_1. \tag{10.27}$$

This is a convex optimization problem too. Here λ is a parameter which can be adjusted depending on the desired level of sparsity. The method of Lagrange multipliers shows that the constrained and unconstrained versions of Lasso are equivalent for appropriate R and λ. This, however, does not immediately tell us how to choose λ. The following exercise settles this question.

Exercise 10.6.11 (Unconstrained Lasso) Assume that the number of measurements satisfies

$$m \gtrsim s \log n.$$

Choose the parameter λ such that $\lambda \gtrsim \sqrt{\log n}\, \|w\|_2$. Then, with high probability, show that the solution $\widehat{x}$ of the unconstrained Lasso (10.27) satisfies

$$\|\widehat{x} - x\|_2 \lesssim \frac{\lambda\sqrt{s}}{m}.$$

10.7 Notes

The applications discussed in this chapter are drawn from two fields: signal processing (specifically, compressed sensing) and high-dimensional statistics (more precisely, high-dimensional structured regression). The tutorial [217] offers a unified treatment of these two kinds of problem, which we followed in this chapter. The survey [55] and the book [76] offer a deeper introduction to compressed sensing. The books [98, 41] discuss statistical aspects of sparse recovery.

Signal recovery using the M^* bound discussed in Section 10.2 is based on [217], which has various versions of Theorem 10.2.1 and Corollary 10.3.4. The Garnaev–Gluskin bound from Exercise 10.3.10 was first proved in [78]; see also [132] and [76, Chapter 10].

The survey [57] offers a comprehensive overview of the low-rank matrix recovery problem, which we discussed in Section 10.4. Our presentation was based on [217, Section 10].

The phenomenon of exact sparse recovery discussed in Section 10.5 goes back to the origins of compressed sensing; see [55] and the book [76] for its history and recent developments. Our presentation of exact recovery via the escape theorem in Section 10.5.1 partly follows [217, Section 9]; see also [51, 183] and especially [202] for applications of the escape theorem to sparse recovery. One can obtain very precise guarantees that give asymptotically sharp formulas (so-called phase transitions) for the number of measurements needed for signal recovery. The first such phase transitions were identified in [66] for sparse signals and uniform random projection matrices A; see also [65, 62–64]. More recent work has clarified phase transitions for general feasible sets T and more general measurement matrices [9, 158, 159].

The approach to exact sparse recovery based on the RIP presented in Section 10.5.2 was pioneered by E. Candes and T. Tao [45]; see [76, Chapter 6] for a comprehensive introduction. An early form of Theorem 10.5.10 appeared in [45]. The proof that we gave here was communicated to the author by Y. Plan; it is similar to the argument of [43]. The fact that random matrices satisfy the RIP (exemplified by Theorem 10.5.11) is a backbone of compressed sensing; see [76, Sections 9.1, 12.5] and [216, Section 5.6].

The Lasso algorithm for sparse regression that we studied in Section 10.6 was pioneered by R. Tibshirani [198]. The books [98, 41] offer a comprehensive introduction into statistical problems with sparsity constraints; these books discuss Lasso and its many variants. A version of Theorem 10.6.1 and some elements of its proof can be traced to the work of P. Bickel, Y. Ritov, and A. Tsybakov [21], although their argument was not based on the matrix deviation inequality. A theoretical analysis of Lasso is also presented in [98, Chapter 11] and [41, Chapter 6].

11

Dvoretzky–Milman Theorem

Here we extend the matrix deviation inequality from Chapter 9 to general norms on $\mathbb{R}^n$, and even to general sub-additive functions on $\mathbb{R}^n$. We use this result to prove the fundamental Dvoretzky–Milman theorem in high-dimensional geometry. It helps us to describe the shape of an m-dimensional random projection of an arbitrary set $T \subset \mathbb{R}^n$. The answer depends on whether m is larger or smaller than the critical dimension, which is the stable dimension $d(T)$. In the high-dimensional regime (where $m \gtrsim d(T)$), the additive Johnson–Lindenstrauss lemma, which we studied in Section 9.3.2, shows that such a random projection approximately preserves the geometry of T. In the low-dimensional regime (where $m \lesssim d(T)$), the geometry can no longer be preserved owing to "saturation". Instead, the Dvoretzky–Milman theorem shows that in this regime the projected set is approximately a *round ball*.

11.1 Deviations of Random Matrices with respect to General Norms

In this section we generalize the matrix deviation inequality from Section 9.1. We replace the Euclidean norm by any positive-homogeneous subadditive function.

Definition 11.1.1 Let V be a vector space. A function $f: V \to \mathbb{R}$ is called *positive-homogeneous* if

$$f(\alpha x) = \alpha f(x) \quad \text{for all } \alpha \geq 0 \text{ and } x \in V.$$

The function f is called *subadditive* if

$$f(x + y) \leq f(x) + f(y) \quad \text{for all } x, y \in V.$$

Note that despite being called positive-homogeneous, f is allowed to take negative values. ("Positive" here applies to the multiplier α in the definition.)

Example 11.1.2

(i) Any *norm* on a vector space is positive-homogeneous and subadditive. The subadditivity is nothing other than the triangle inequality in this case.

(ii) Clearly, any *linear functional* on a vector space is positive-homogeneous and subadditive. In particular, for any fixed vector $y \in \mathbb{R}^m$, the function $f(x) = \langle x, y \rangle$ is positive-homogeneous and subadditive on $\mathbb{R}^m$.

(iii) Consider a bounded set $S \subset \mathbb{R}^m$ and define the function

$$f(x) := \sup_{y \in S} \langle x, y \rangle, \quad x \in \mathbb{R}^m. \tag{11.1}$$

Then f is positive-homogeneous and subadditive on $\mathbb{R}^m$. This function is sometimes called the *support function* of S.

Exercise 11.1.3 Check that the function $f(x)$ in part (iii) of Example 11.1.2 is positive-homogeneous and subadditive.

Exercise 11.1.4 Let $f : V \to \mathbb{R}$ be a subadditive function on a vector space V. Show that

$$f(x) - f(y) \le f(x - y) \quad \text{for all } x, y \in V. \tag{11.2}$$

We are ready to state the main result of this section.

Theorem 11.1.5 (General matrix deviation inequality) *Let A be an $m \times n$ Gaussian random matrix with i.i.d. $N(0, 1)$ entries. Let $f\colon \mathbb{R}^m \to \mathbb{R}$ be a positive-homogeneous and subadditive function, and let $b \in \mathbb{R}$ be such that*

$$f(x) \le b\|x\|_2 \quad \textit{for all } x \in \mathbb{R}^n. \tag{11.3}$$

Then, for any subset $T \subset \mathbb{R}^n$, we have

$$\mathbb{E} \sup_{x \in T} \left| f(Ax) - \mathbb{E}\, f(Ax) \right| \le Cb\gamma(T).$$

Here $\gamma(T)$ is the Gaussian complexity introduced in Section 7.6.2.

This theorem generalizes the form of the matrix deviation inequality given in Exercise 9.1.2.

Exactly as in Section 9.1, Theorem 11.1.5 would follow from Talagrand's comparison inequality once we show that the random process $X_x := f(Ax) - \mathbb{E}\, f(Ax)$ has sub-gaussian increments. Let us do this now.

Theorem 11.1.6 (Sub-gaussian increments) *Let A be an $m \times n$ Gaussian random matrix with i.i.d. $N(0, 1)$ entries, and let $f\colon \mathbb{R}^m \to \mathbb{R}$ be a positive-homogeneous and subadditive function satisfying* (11.3). *Then the random process*

$$X_x := f(Ax) - \mathbb{E}\, f(Ax)$$

has sub-gaussian increments with respect to the Euclidean norm, namely

$$\|X_x - X_y\|_{\psi_2} \le Cb\|x - y\|_2 \quad \textit{for all } x, y \in \mathbb{R}^n. \tag{11.4}$$

Exercise 11.1.7 Deduce the general matrix deviation inequality (Theorem 11.1.5) from Talagrand's comparison inequality (in the form of Exercise 8.6.4) and Theorem 11.1.6.

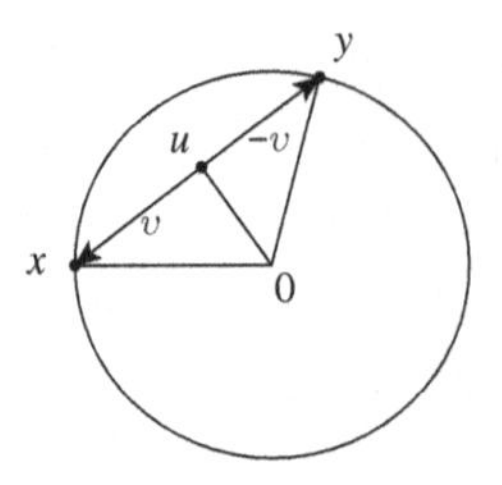

Figure 11.1 Creating a pair of orthogonal vectors u, v out of x, y.

Proof of Theorem 11.1.6 Without loss of generality we may assume that $b = 1$. (Why?) Just as in the proof of Theorem 9.1.3, let us first assume that

$$\|x\|_2 = \|y\|_2 = 1.$$

In this case, the inequality in (11.4) that we want to prove becomes

$$\|f(Ax) - f(Ay)\|_{\psi_2} \le C\|x - y\|_2. \tag{11.5}$$

Step 1. Creating independence. Consider the vectors

$$u := \frac{x+y}{2}, \quad v := \frac{x-y}{2}. \tag{11.6}$$

Then

$$x = u + v, \quad y = u - v$$

and thus

$$Ax = Au + Av, \quad Ay = Au - Av.$$

(See Figure 11.1.)

Since the vectors u and v are orthogonal (check!), the Gaussian random vectors Au and Av are independent. (Recall Exercise 3.3.6.)

Step 2. Using Gaussian concentration. Let us condition on $a := Au$ and study the conditional distribution of

$$f(Ax) = f(a + Av).$$

By rotation invariance $a + Av$ is a Gaussian random vector, which we can express as

$$a + Av = a + \|v\|_2\, g \quad \text{where} \quad g \sim N(0, I_m).$$

(Recall Exercise 3.3.3.) We claim that $f(a + \|v\|_2\, g)$ as a function of g is Lipschitz with respect to the Euclidean norm on $\mathbb{R}^m$, and so

$$\|f\|_{\mathrm{Lip}} \le \|v\|_2. \tag{11.7}$$

To check this, fix $t, s \in \mathbb{R}^m$ and note that

$$\begin{aligned} f(t) - f(s) &= f(a + \|v\|_2\, t) - f(a + \|v\|_2\, s) \\ &\le f(\|v\|_2\, t - \|v\|_2\, s) \quad \text{(by (11.2))} \\ &= \|v\|_2\, f(t - s) \quad \text{(by positive-homogeneity)} \\ &\le \|v\|_2\, \|t - s\|_2 \quad \text{(using (11.3) with } b = 1); \end{aligned}$$

then (11.7) follows.

Concentration in the Gauss space (Theorem 5.2.2) then yields

$$\|f(g) - \mathbb{E}\, f(g)\|_{\psi_2(a)} \le C\|v\|_2$$

or

$$\big\|f(a + Av) - \mathbb{E}_a\, f(a + Av)\big\|_{\psi_2(a)} \le C\|v\|_2, \tag{11.8}$$

where the index "a" reminds us that these bounds are valid for a conditional distribution with $a = Au$ fixed.

Step 3. Removing the conditioning. Since the random vector $a - Av$ has the same distribution as $a + Av$ (why?), it satisfies the same bound.

$$\big\|f(a - Av) - \mathbb{E}_a\, f(a - Av)\big\|_{\psi_2(a)} \le C\|v\|_2. \tag{11.9}$$

Subtract (11.9) from (11.8) and use the triangle inequality and the fact that the expectations are the same; this gives

$$\big\|f(a + Av) - f(a - Av)\big\|_{\psi_2(a)} \le 2C\|v\|_2.$$

This bound is for the conditional distribution, and it holds for any fixed realization of a random variable $a = Au$. Therefore, it holds for the original distribution, too:

$$\big\|f(Au + Av) - f(Au - Av)\big\|_{\psi_2} \le 2C\|v\|_2.$$

(Why?) Passing back to the x, y notation by means of (11.6), we obtain the desired inequality (11.5).

The proof is complete for the unit vectors x, y; Exercise 11.1.8 below extends it to the general case. ■

Exercise 11.1.8 (Non-unit x, y)☕ Extend the proof above to general (not necessarily unit) vectors x, y. ☞

Remark 11.1.9 It is an open question whether Theorem 11.1.5 holds for general sub-gaussian matrices A.

Exercise 11.1.10 (Anisotropic distributions)☕☕ Extend Theorem 11.1.5 to $m \times n$ matrices A whose columns are independent $N(0, \Sigma)$ random vectors, where Σ is a general covariance matrix. Show that

$$\mathbb{E} \sup_{x \in T} \Big| f(Ax) - \mathbb{E}\, f(Ax)\Big| \le Cb\gamma(\Sigma^{1/2}T).$$

Exercise 11.1.11 (Tail bounds)☕☕ Prove a high-probability version of Theorem 11.1.5. ☞

11.2 Johnson–Lindenstrauss Embeddings and Sharper Chevet Inequality

Like the original matrix deviation inequality from Chapter 9, the general theorem 9.1.1 has many consequences, which we now discuss.

11.2.1 Johnson–Lindenstrauss Lemma for General Norms

Using the general matrix deviation inequality as in Section 9.3, it should be quite straightforward to do the following exercises.

Exercise 11.2.1 State and prove a version of the Johnson–Lindenstrauss lemma for a general norm (as opposed to the Euclidean norm) on $\mathbb{R}^m$.

Exercise 11.2.2 (Johnson–Lindenstrauss lemma for ℓ_1 norm) Specialize the previous exercise to the ℓ_1 norm. Thus, let $\mathcal{X}$ be a set of N points in $\mathbb{R}^n$, let A be an $m \times n$ Gaussian matrix with i.i.d. $N(0, 1)$ entries, and let $\varepsilon \in (0, 1)$.

Suppose that

$$m \geq C(\varepsilon) \log N.$$

Show that with high probability the matrix $Q := \sqrt{\pi/2}\, m^{-1} A$ satisfies

$$(1-\varepsilon)\|x-y\|_2 \leq \|Qx - Qy\|_1 \leq (1+\varepsilon)\|x-y\|_2 \quad \text{for all } x, y \in \mathcal{X}.$$

This conclusion is very similar to that of the original Johnson–Lindenstrauss lemma (Theorem 5.3.1), except that the distance between the projected points is measured in the ℓ_1 norm.

Exercise 11.2.3 (Johnson–Lindenstrauss embedding into ℓ_∞) Use the same notation as in the previous exercise, but assume this time that

$$m \geq N^{C(\varepsilon)}.$$

Show that with high probability the matrix $Q := C(\log m)^{-1/2} A$, for some appropriate constant C, satisfies

$$(1-\varepsilon)\|x-y\|_2 \leq \|Qx - Qy\|_\infty \leq (1+\varepsilon)\|x-y\|_2 \quad \text{for all } x, y \in \mathcal{X}.$$

Note that in this case $m \geq N$, so Q gives an *almost isometric embedding* (rather than a projection) of the set $\mathcal{X}$ into ℓ_∞.

11.2.2 Two-Sided Chevet Inequality

The general matrix deviation inequality will help us to sharpen Chevet's inequality, which we originally proved in Section 8.7.

Theorem 11.2.4 (General Chevet inequality) *Let A be an $m \times n$ Gaussian random matrix with i.i.d. $N(0, 1)$ entries. Let $T \subset \mathbb{R}^n$ and $S \subset \mathbb{R}^m$ be arbitrary bounded sets. Then*

$$\mathbb{E} \sup_{x \in T} \left| \sup_{y \in S} \langle Ax, y\rangle - w(S)\|x\|_2 \right| \leq C\gamma(T)\, \mathrm{rad}(S).$$

Using the triangle inequality we can see that Theorem 11.2.4 is a sharper, two-sided, form of Chevet's inequality (Theorem 8.7.1).

Proof Let us apply the general matrix deviation inequality (Theorem 11.1.5) for the function f defined in (11.1), i.e., for

$$f(x) := \sup_{y \in S} \langle x, y \rangle .$$

To do this, we need to compute the b value for which (11.3) holds. Fix $x \in \mathbb{R}^m$ and use the Cauchy–Schwarz inequality to get

$$f(x) \leq \sup_{y \in S} \|x\|_2 \|y\|_2 = \text{rad}(S) \|x\|_2.$$

Thus (11.3) holds with $b = \text{rad}(S)$.

It remains to compute the expectation $\mathbb{E} f(Ax)$ appearing in the conclusion of Theorem 11.1.5. By the rotation invariance of the Gaussian distribution (see Exercise 3.3.3), the random vector Ax has the same distribution as $g\|x\|_2$, where $g \in N(0, I_m)$. Then

$$\begin{aligned}
\mathbb{E} f(Ax) &= \mathbb{E} f(g) \|x\|_2 \quad \text{(by positive-homogeneity)} \\
&= \mathbb{E} \sup_{y \in S} \langle g, y \rangle \, \|x\|_2 \quad \text{(by the definition of } f) \\
&= w(S) \|x\|_2 \quad \text{(by the definition of the Gaussian width)}.
\end{aligned}$$

Substituting this into the conclusion of Theorem 11.1.5, we complete the proof. ■

11.3 Dvoretzky–Milman Theorem

The Dvoretzky–Milman theorem is a remarkable result about the random projections of general bounded sets in $\mathbb{R}^n$. If the projection is onto a suitably low dimension, the convex hull of the projected set turns out to be *approximately a round ball* with high probability; see Figures 11.2 and 11.3.

11.3.1 Gaussian Images of Sets

It will be more convenient for us to work with Gaussian random projections than with ordinary projections. Here is a very general result that compares the Gaussian projection of a general set to a Euclidean ball.

Theorem 11.3.1 (Random projections of sets) *Let A be an $m \times n$ Gaussian random matrix with i.i.d. $N(0, 1)$ entries, and let $T \subset \mathbb{R}^n$ be a bounded set. Then the following holds with probability at least* 0.99*:*

$$r_- B_2^m \subset \text{conv}(AT) \subset r_+ B_2^m$$

where[1]

$$r_\pm := w(T) \pm C\sqrt{m} \, \text{rad}(T).$$

The left-hand inclusion holds only if r_- is non-negative; the right-hand inclusion holds always.

We will shortly deduce this theorem from the two-sided Chevet inequality. The following exercise will provide the link between the two results. It asks you to show that the support

[1] As before, $\text{rad}(T)$ denotes the radius of T, which we defined in (8.47).

function (11.1) of a general set S is the ℓ_2 norm if and only if S is the Euclidean ball; there is also a stability version of this equivalence.

Exercise 11.3.2 (Almost Euclidean balls and support functions)☕☕☕

(a) Let $V \subset \mathbb{R}^m$ be a bounded set. Show that $\operatorname{conv}(V) = B_2^m$ if and only if

$$\sup_{x \in V} \langle x, y \rangle = \|y\|_2 \quad \text{for all } y \in \mathbb{R}^m.$$

(b) Let $V \subset \mathbb{R}^m$ be a bounded set and $r_-, r_+ \geq 0$. Show that the inclusion

$$r_- B_2^m \subset \operatorname{conv}(V) \subset r_+ B_2^m$$

holds if and only if

$$r_- \|y\|_2 \leq \sup_{x \in V} \langle x, y \rangle \leq r_+ \|y\|_2 \quad \text{for all } y \in \mathbb{R}^m.$$

Proof of Theorem 11.3.1 Let us write the two-sided Chevet inequality in the following form:

$$\mathbb{E} \sup_{y \in S} \left| \sup_{x \in T} \langle Ax, y \rangle - w(T) \|y\|_2 \right| \leq C \gamma(S) \operatorname{rad}(T),$$

where $T \subset \mathbb{R}^n$ and $S \subset \mathbb{R}^m$. (To obtain this form use Theorem 11.2.4 with T and S swapped and for A^T instead of A – do this!)

Choose S to be the sphere S^{m-1} and recall that its Gaussian complexity $\gamma(S) \leq \sqrt{m}$. Then, by Markov's inequality, the following holds with probability at least 0.99:

$$\left| \sup_{x \in T} \langle Ax, y \rangle - w(T) \|y\|_2 \right| \leq C \sqrt{m} \operatorname{rad}(T) \quad \text{for every } y \in S^{m-1}.$$

Use the triangle inequality and recall the definition of $r_\pm$ to get

$$r_- \leq \sup_{x \in T} \langle Ax, y \rangle \leq r_+ \quad \text{for every } y \in S^{m-1}.$$

By homogeneity, this is equivalent to

$$r_- \|y\|_2 \leq \sup_{x \in T} \langle Ax, y \rangle \leq r_+ \|y\|_2 \quad \text{for every } y \in \mathbb{R}^m.$$

(Why?) Finally, note that

$$\sup_{x \in T} \langle Ax, y \rangle = \sup_{x \in AT} \langle x, y \rangle$$

and apply Exercise 11.3.2 for $V = AT$ to complete the proof. ∎

11.3.2 Dvoretzky–Milman Theorem

Theorem 11.3.3 (Dvoretzky–Milman theorem: Gaussian form) *Let A be an $m \times n$ Gaussian random matrix with i.i.d. $N(0,1)$ entries, let $T \subset \mathbb{R}^n$ be a bounded set, and let $\varepsilon \in (0,1)$. Suppose that*

$$m \leq c \varepsilon^2 d(T),$$

where $d(T)$ is the stable dimension of T introduced in Section 7.6. Then, with probability at least 0.99, we have

$$(1-\varepsilon)B \subset \mathrm{conv}(AT) \subset (1+\varepsilon)B,$$

where B is a Euclidean ball with radius $w(T)$.

Proof Translating T as necessary, we can assume that T contains the origin. Apply Theorem 11.3.1. All that remains to check is that $r_- \geq (1-\varepsilon)w(T)$ and $r_+ \leq (1+\varepsilon)w(T)$, which by definition would follow if

$$C\sqrt{m}\ \mathrm{rad}(T) \leq \varepsilon w(T). \tag{11.10}$$

To check this inequality, recall that by assumption and by Definition 7.6.2 we have

$$m \leq c\varepsilon^2 d(T) \leq \frac{\varepsilon^2 w(T)^2}{\mathrm{diam}(T)^2},$$

provided that the absolute constant $c > 0$ is chosen sufficiently small. Next, since T contains the origin, $\mathrm{rad}(T) \leq \mathrm{diam}(T)$. (Why?) This implies (11.10) and completes the proof. ∎

Remark 11.3.4 As is obvious from the above proof, if T contains the origin then the Euclidean ball B can be centered at the origin too. Otherwise the center of B could be chosen as x_0, where $x_0 \in T$ is any fixed point.

Exercise 11.3.5☕☕ State and prove a high-probability version of the Dvoretzky–Milman theorem.

Example 11.3.6 (Projections of the cube) Consider the cube

$$T = [-1, 1]^n = B_\infty^n.$$

Recall that

$$w(T) = \sqrt{\frac{2}{\pi}}\, n$$

and recall (7.17). Since $\mathrm{diam}(T) = 2\sqrt{n}$, the stable dimension of the cube is

$$d(T) \asymp \frac{w(T)^2}{\mathrm{diam}(T)^2} \asymp n.$$

Apply Theorem 11.3.3. If $m \leq c\varepsilon^2 n$ then with high probability we have

$$(1-\varepsilon)B \subset \mathrm{conv}(AT) \subset (1+\varepsilon)B,$$

where B is a Euclidean ball with radius $\sqrt{2/\pi}\, n$.

In words, *a random Gaussian projection of the cube onto a subspace of dimension $m \asymp n$ is close to a round ball.* Figure 11.2 illustrates this remarkable fact.

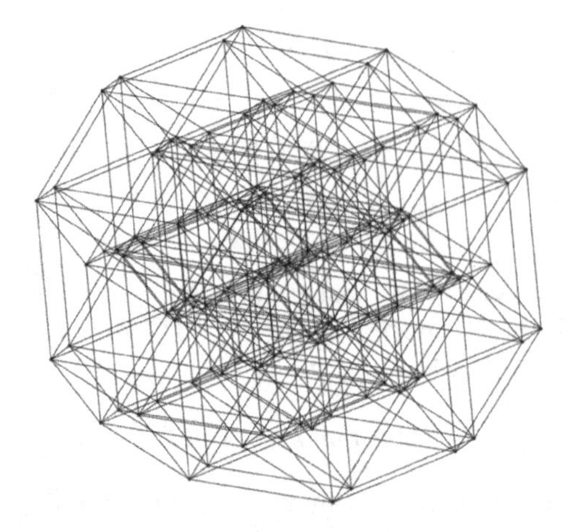

Figure 11.2 A random projection of a seven-dimensional cube onto the plane.

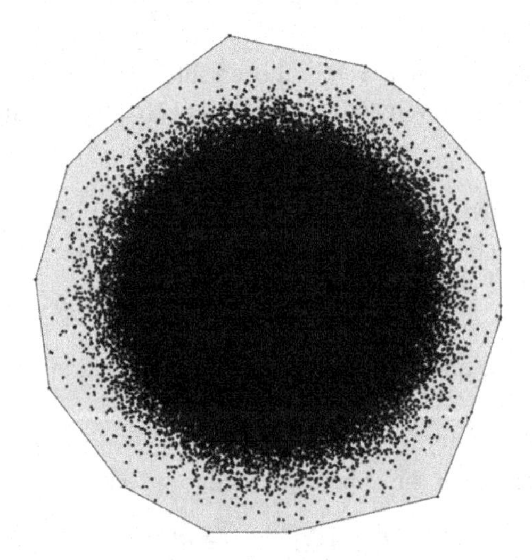

Figure 11.3 A Gaussian cloud of 10^7 points on the plane, and its convex hull.

Exercise 11.3.7 (Gaussian cloud) Consider a Gaussian cloud of n points in $\mathbb{R}^m$, which is formed by i.i.d. random vectors $g_1, \ldots, g_n \sim N(0, I_m)$. Suppose that

$$n \geq \exp(Cm)$$

with a large enough absolute constant C. Show that, with high probability, the convex hull of the Gaussian cloud is approximately a Euclidean ball with radius $\sim \sqrt{\log n}$. See Figure 11.3 for illustration. ☞

Exercise 11.3.8 (Projections of ellipsoids) Consider the ellipsoid $\mathcal{E}$ in $\mathbb{R}^n$ given as a linear image of the unit Euclidean ball, i.e.,

$$\mathcal{E} = S(B_2^n),$$

where S is an $n \times n$ matrix. Let A be the $m \times n$ Gaussian matrix with i.i.d. $N(0, 1)$ entries. Suppose that

$$m \gtrsim r(S),$$

where $r(S)$ is the stable rank of S (recall Definition 7.6.7). Show that with high probability the Gaussian projection $A(\mathcal{E})$ of the ellipsoid is almost a round ball with radius $\|S\|_F$:

$$A(\mathcal{E}) \approx \|S\|_F B_2^n.$$

☞

Exercise 11.3.9 (Random projection in the Grassmannian) ☕☕☕ Prove a version of the Dvoretzky–Milman theorem for a projection P onto a random m-dimensional subspace in $\mathbb{R}^n$. Under the same assumptions, the conclusion should be that

$$(1 - \varepsilon)B \subset \operatorname{conv}(PT) \subset (1 + \varepsilon)B$$

where B is a Euclidean ball with radius $w_s(T)$. (Recall that $w_s(T)$ is the spherical width of T, which we introduced in Section 7.5.2.)

Summary of Random Projections of Geometric Sets

It is useful to compare the Dvoretzky–Milman theorem with our earlier estimates on the diameter of random projections of geometric sets, which we developed in Sections 7.7 and 9.2.2. We found that a random projection P of a set T onto an m-dimensional subspace in $\mathbb{R}^n$ satisfies a phase transition. In the high-dimensional regime (where $m \gtrsim d(T)$), the projection shrinks the diameter of T by a factor of order $\sqrt{m/n}$, i.e.,

$$\operatorname{diam}(PT) \lesssim \sqrt{\frac{m}{n}} \operatorname{diam}(T) \quad \text{if } m \geq d(T).$$

Moreover, the additive Johnson–Lindenstrauss lemma from Section 9.3.2 shows that, in this regime, the random projection P approximately preserves the geometry of T (the distances between all points in T shrink roughly by the same scaling factor).

In the low-dimensional regime (where $m \lesssim d(T)$), surprisingly, the size of the projected set stops shrinking. All we can say is that

$$\operatorname{diam}(PT) \lesssim w_s(T) \asymp \frac{w(T)}{\sqrt{n}} \quad \text{if } m \leq d(T);$$

see Section 7.7.1.

The Dvoretzky–Milman theorem explains why the size of T stops shrinking for $m \lesssim d(T)$. Indeed, in this regime the projection PT is *approximately a round ball* of radius of order $w_s(T)$ (see Exercise 11.3.9), regardless of how small m is.

Let us summarize our findings. *A random projection of a set T in $\mathbb{R}^n$ onto an m-dimensional subspace approximately preserves the geometry of T if $m \gtrsim d(T)$. For smaller m, the projected set PT becomes approximately a round ball of diameter $\sim w_s(T)$ and its size does not shrink with m.*

11.4 Notes

The general matrix deviation inequality (Theorem 11.1.5) and its proof is due to G. Schechtman [177].

The original version of Chevet's inequality was proved by S. Chevet [53] and the constant factors there were improved by Y. Gordon [82]; see also [11, Section 9.4], [126, Theorem 3.20], and [199, 2]. The version of Chevet's inequality that we stated in Theorem 11.2.4 can be reconstructed from the work of Y. Gordon [82, 84]; see [126, Corollary 3.21].

The Dvoretzky–Milman theorem is a result with a long history in functional analysis. In proving a conjecture of A. Grothendieck, A. Dvoretzky [71, 72] also proved that any n-dimensional normed space has an m-dimensional almost Euclidean subspace, where $m = m(n)$ grows to infinity with n. V. Milman gave a probabilistic proof of this theorem and pioneered the study of the best possible dependence $m(n)$. Theorem 11.3.3 is due to V. Milman [144]. The stable dimension $d(T)$ is the critical dimension in the Dvoretzky–Milman theorem, i.e., its conclusion always fails for $m \gg d(T)$ owing to a result of V. Milman and G. Schechtman [147]; see [11, Theorem 5.3.3]. The tutorial [13] contains a light introduction to the Dvoretzky–Milman theorem. For a full exposition of the Dvoretzky–Milman theorem and many of its ramifications, see e.g. [11, Chapter 5 and Section 9.2], [126, Section 9.1], and the references therein.

An important question related to the Dvoretzky–Milman and central limit theorems regards the m-dimensional random projections (marginals) of a given probability distribution in $\mathbb{R}^n$; we may wonder whether such marginals are approximately normal. This question may be important in data science applications, where the "wrong" lower-dimensional random projections of data sets in $\mathbb{R}^n$ form a "Gaussian cloud". For log-concave probability distributions, a type of central limit theorem was first proved by B. Klartag [112]; see the history and more recent results in [11, Section 10.7]. For discrete sets, see E. Meckes [138] and the references there.

The phenomenon discussed in the summary at the end of Section 7.7 is due to V. Milman [145]; see [11, Proposition 5.7.1].

Hints for Exercises

Appetizer

0.0.5 To prove the upper bound, multiply both sides by the quantity $(m/n)^m$, replace this quantity by $(m/n)^k$ on the left-hand side, and use the binomial theorem.

0.0.6 The number of ways to choose k elements from an N-element set with repetitions is $\binom{N+k-1}{k}$. Simplify using Exercise 0.0.5.

Chapter 1

1.2.3 Use the integral identity for $|X|^p$ and change variables.

Chapter 2

2.4.1 Integrate by parts.

2.2.3 Compare the Taylor expansions of the two sides.

2.2.8 Apply Hoeffding's inequality for the case when the X_i are the indicators of the wrong answers.

2.2.9(a) Use the sample mean $\hat{\mu} := \frac{1}{N} \sum_{i=1}^{N} X_i$.

2.2.9(b) Use the median of the $O(\log(\delta^{-1}))$ weak estimates from part (a).

2.2.10 Rewrite the inequality $\sum X_i \leq \varepsilon N$ as $\sum(-X_i/\varepsilon) \geq -N$ and proceed as in the proof of Hoeffding's inequality. Use part (a) to bound the MGF.

2.3.3 Combine Chernoff's inequality with the Poisson limit theorem (Theorem 1.3.4).

2.3.5 Apply Theorem 2.3.1 and Exercise 2.3.2 for $t = (1 \pm \delta)\mu$ and analyze the bounds for small δ.

2.3.6 Combine Exercise 2.3.5 with the Poisson limit theorem (Theorem 1.3.4).

2.3.8 Derive this from the central limit theorem. Use the fact that the sum of independent Poisson distributions is a Poisson distribution.

2.4.2 Modify the proof of Proposition 2.4.1.

2.4.4 The principal difficulty is that the degrees d_i are not independent. To fix this, try to replace d_i by some d_i' that are independent. (Try to include not all the vertices in the counting.) Then use the Poisson approximation (2.9).

2.6.6 Use the following extrapolation trick. Prove the inequality $\|Z\|_2 \leq \|Z\|_1^{1/4} \|Z\|_3^{3/4}$ and use it for $Z = \sum a_i X_i$. Get a bound on $\|Z\|_3$ from Khintchine's inequality for $p = 3$.

2.6.7 Modify the extrapolation trick in Exercise 2.6.6.

2.8.5 Check that the numeric inequality

$$e^z \le 1 + z + \frac{z^2/2}{1 - |z|/3}$$

is valid provided that $|z| < 3$, apply it for $z = \lambda X$, and take expectations on both sides.

2.8.6 Follow the proof of Theorem 2.8.1.

Chapter 3

3.1.5 Use Exercise 3.1.4.

3.1.6 First check that $\mathbb{E}(\|X\|_2^2 - n)^2 \le K^4 n$ by expansion. This yields in a simple way that $\mathbb{E}(\|X\|_2 - \sqrt{n})^2 \le K^4$. Finally, replace $\sqrt{n}$ by $\mathbb{E}\,\|X\|_2$ arguing as in Exercise 3.1.4.

3.1.7 While this inequality does not follow from the result of Exercise 2.2.10 (why?), you can prove it by a similar argument.

3.3.4 Utilize a version of the Cramér–Wold theorem, which states that the totality of the distributions of one-dimensional marginals determines the distribution in $\mathbb{R}^n$ uniquely. More precisely, if X and Y are random vectors in $\mathbb{R}^n$ such that $\langle X, \theta \rangle$ and $\langle Y, \theta \rangle$ have the same distribution for each $\theta \in \mathbb{R}^n$ then X and Y have the same distribution.

3.3.6 Reduce the problem to the case where u and v are collinear with canonical basis vectors of $\mathbb{R}^n$.

3.3.9 Proceed similarly to the proof of Lemma 3.2.3.

3.5.3 Check and use the polarization identity $\langle Ax, y \rangle = \langle Au, u \rangle - \langle Av, v \rangle$ where $u = (x + y)/2$ and $v = (x - y)/2$.

3.5.5 Consider the *Gram matrix* of the vectors X_i, which is the $n \times n$ matrix with entries $\langle X_i, X_j \rangle$. Do not forget to describe how to translate a solution of (3.22) into a solution of (3.21).

3.5.7 First, express the objective function as $\frac{1}{2}\operatorname{tr}(\tilde{A} Z Z^{\mathsf{T}})$, where $\tilde{A} = \begin{bmatrix} 0 & A \\ A^{\mathsf{T}} & 0 \end{bmatrix}$, $Z = \begin{bmatrix} X \\ Y \end{bmatrix}$, and X and Y are the matrices with rows X_i^{T} and Y_j^{T}, respectively. Then express the set of matrices of the type ZZ^{T} with unit rows as the set of positive-semidefinite matrices whose diagonal entries equal 1.

3.6.4 Consider cutting G repeatedly. Give a bound on the expected number of experiments.

3.6.7 It will quickly follow once you show that the probability that $\langle g, u \rangle$ and $\langle g, v \rangle$ have opposite signs equals α/π, where $\alpha \in [0, \pi]$ is the angle between the vectors u and v. To check this, use rotation invariance to reduce the problem to $\mathbb{R}^2$. Once on the plane, a second use of rotation invariance will give the result.

3.7.5(a) Consider the direct sum $H = \mathbb{R}^{n \times n} \oplus \mathbb{R}^{n \times n \times n}$.

3.7.6 Construct Φ as in Exercise 3.7.5 but include the signs of a_k in the definition of Ψ.

Chapter 4

4.2.16 Adapt the volumetric argument by replacing volume by cardinality.

4.4.3(b) Proceed similarly to the proof of Lemma 4.4.1 and use the identity $\langle Ax, y \rangle - \langle Ax_0, y_0 \rangle = \langle Ax, y - y_0 \rangle + \langle A(x - x_0), y_0 \rangle$.

4.4.4 Assume without loss of generality that $\mu = 1$. Represent $\|Ax\|_2^2 - 1$ as a quadratic form $\langle Rx, x\rangle$ where $R = A^\mathsf{T}A - I_n$. Use Exercise 4.4.3 to compute the maximum of this quadratic form on a net.

4.4.7 Bound the operator norm of A below by the Euclidean norm of the first column and first row; use the concentration of the norm to complete the proof.

4.6.2 Use the integral identity from Lemma 1.2.1.

Chapter 5

5.1.9 If the conclusion in part (a) fails, the complement $B := (A_s)^c$ satisfies $\sigma(B) \ge 1/2$. Apply the blow-up lemma 5.1.7 for B.

5.1.13 To prove the upper bound, assume that $\|Z - \mathbb{E}\, Z\|_{\psi_2} \le K$ and use the definition of the median to show that $|M - \mathbb{E}\, Z| \le CK$.

5.1.14 First replace the expectation by the median. Then apply the assumption for the function $f(x) := \operatorname{dist}(x, A) = \inf\{d(x, y) \colon y \in A\}$, whose median is zero.

5.1.5 Construct the points $x_i \in S^{n-1}$ one at a time. Note that the set of points on the sphere that are almost orthogonal to a given point x_0 form a spherical cap. Show that the normalized area of that cap is exponentially small.

5.2.3 The ε-neighborhood of a half-space is still a half-space, and its Gaussian measure should be easy to compute.

5.3.4 Let $\mathcal{X}$ be an orthogonal basis and show that the projected set defines a packing.

5.4.5(d) Find 2×2 matrices such that $0 \preceq X \preceq Y$ but $X^2 \not\preceq Y^2$.

5.4.5(e) Using Courant-Fisher's min-max principle (4.2), show that $\lambda_i(X) \le \lambda_i(Y)$ for all i.

5.4.5(f) First consider the case where one of the matrices is the identity. Next, multiply the inequality $X \preceq Y$ by $Y^{-1/2}$ on the left and on the right.

5.4.5(g) Check and use the identity $\log x = \int_0^\infty (\frac{1}{1+t} - \frac{1}{x+t})\, dt$ and property (f).

5.4.11 Check that the matrix form of Bernstein's inequality implies that $\|\sum_{i=1}^N X_i\| \lesssim \|\sum_{i=1}^N \mathbb{E}\, X_i^2\|^{1/2}\sqrt{\log n + u} + K(\log n + u)$ with probability at least $1 - 2e^{-u}$. Then use the integral identity from Lemma 1.2.1.

5.4.12 Proceed as in the proof of Theorem 5.4.1. Instead of Lemma 5.4.10, check that $\mathbb{E} \exp(\lambda\varepsilon_i A_i) \le \exp(\lambda^2 A_i^2/2)$ just as in the proof of Hoeffding's inequality, Theorem 2.2.2.

5.4.15 Apply the matrix Bernstein inequality (Theorem 5.4.1) for the sum of $(m + n) \times (m + n)$ symmetric matrices $\begin{bmatrix} 0 & X_i^\mathsf{T} \\ X_i & 0 \end{bmatrix}$.

5.6.7 Think about the coordinate distribution from Section 3.3.4; argue as in Exercise 5.4.14.

5.6.8 Just as in the proof of Theorem 4.6.1, derive the conclusion from a bound on $m^{-1}A^\mathsf{T}A - I_n = m^{-1}\sum_{i=1}^m A_i A_i^\mathsf{T} - I_n$. Use Theorem 5.6.1.

Chapter 6

6.2.5 Use the singular value decomposition for A and the rotation invariance of $X \sim N(0, I_n)$ to simplify and control the quadratic form $X^\mathsf{T}AX$.

6.2.6(a) Argue as in the proof of the comparison lemma 6.2.3.

6.2.6(b) Argue as in the proof of Lemma 6.2.2.

6.2.7 The quadratic form in question can be represented as $X^\mathsf{T} A X$, as before, but now X is a $d \times n$ random matrix with columns X_i. Redo the computation for the MGF when X is Gaussian (Lemma 6.2.2), using the comparison lemma 6.2.3.

6.3.5 Use the bound on the MGF proved in Exercise 6.2.6.

6.4.6 Use the result of Exercise 6.4.5 with $F(x) = \exp(\lambda x)$ to bound the moment generating function or with $F(x) = \exp(cx^2)$.

6.5.2 Apply Theorem 6.5.1 for the $(m+n) \times (m+n)$ symmetric random matrix $\begin{bmatrix} 0 & A \\ A^\mathsf{T} & 0 \end{bmatrix}$.

6.6.2 Fix i and use Bernstein's inequality (Corollary 2.8.3) to get a tail bound for $\sum_{j=1}^n (\delta_{ij} - p)^2$. Conclude by taking a union bound over $i \in [n]$.

6.7.3 Use symmetrization, then the contraction principle (Theorem 6.7.1) conditioned on (X_i), and finish by applying symmetrization again.

6.7.7(b) To prove (6.17), condition on $\varepsilon_1, \ldots, \varepsilon_{n-1}$ and apply part (a).

6.7.8 Theorem 6.7.1 may help.

Chapter 7

7.1.9 Argue as in the proof of Lemma 6.4.2.

7.1.13 It might be simpler to think about increments $\|X_t - X_s\|_2$ instead of the covariance matrix.

7.2.4 Represent $X = \sigma Z$ for $Z \sim N(0,1)$, and apply Gaussian integration by parts.

7.2.6 Represent X as $\Sigma^{1/2} Z$ for $Z \sim N(0, I_n)$. Then

$$X_i = \sum_{k=1}^n (\Sigma^{1/2})_{ik} Z_k \quad \text{and} \quad \mathbb{E}\, X_i f(X) = \sum_{k=1}^n (\Sigma^{1/2})_{ik}\, \mathbb{E}\, Z_k f(\Sigma^{1/2} Z).$$

Apply univariate Gaussian integration by parts (Lemma 7.2.3) for $\mathbb{E}\, Z_k f(\Sigma^{1/2} Z)$ conditionally on all random variables except $Z_k \sim N(0,1)$, and simplify.

7.2.12 Differentiate f and check that

$$\frac{\partial f}{\partial x_i} = \frac{e^{\beta x_i}}{\sum_k e^{\beta x_k}} =: p_i(x) \quad \text{and} \quad \frac{\partial^2 f}{\partial x_i \partial x_j} = \beta \left(\delta_{ij} p_i(x) - p_i(x) p_j(x) \right),$$

where δ_{ij} is the Kronecker delta, which equals 1 if $i = j$ and 0 otherwise. Next, check the following numeric identity:

$$\text{if } \sum_{i=1}^n p_i = 1 \quad \text{then} \quad \sum_{i,j=1}^n \sigma_{ij} (\delta_{ij} p_i - p_i p_j) = \frac{1}{2} \sum_{i \neq j} (\sigma_{ii} + \sigma_{jj} - 2\sigma_{ij}) p_i p_j.$$

Use the Gaussian interpolation formula 7.2.7. Simplify the expression using the identity above with $\sigma_{ij} = \Sigma^X_{ij} - \Sigma^Y_{ij}$ and $p_i = p_i(Z(u))$. Deduce that

$$\frac{d}{du}\, \mathbb{E}\, f(Z(u)) = \frac{\beta}{4} \sum_{i \neq j} \left(\mathbb{E}(X_i - X_j)^2 - \mathbb{E}(Y_i - Y_j)^2 \right) \mathbb{E}\, p_i(Z(u))\, p_j(Z(u)).$$

By the assumptions, this expression is non-positive.

7.2.13 Use the Sudakov–Fernique inequality.

7.2.14 Use the Gaussian interpolation lemma 7.2.7 for $f(x) = \prod_i \left(1 - \prod_j h(x_{ij})\right)$, where $h(x)$ is an approximation to the indicator function $\mathbf{1}_{\{x \le \tau\}}$, as in the proof of Slepian's inequality.

7.3.4 Relate the smallest singular value to the min–max of a Gaussian process:

$$s_n(A) = \min_{u \in S^{n-1}} \max_{v \in S^{m-1}} \langle Au, v \rangle .$$

Apply Gordon's inequality (without the requirement of equal variances, as noted below Exercise 7.2.14) to show that

$$\mathbb{E}\, s_n(A) \ge \mathbb{E}\, \|h\|_2 - \mathbb{E}\, \|g\|_2 \quad \text{where} \quad g \sim N(0, I_n),\ h \sim N(0, I_m).$$

Combine this with the fact that $f(n) := \mathbb{E}\, \|g\|_2 - \sqrt{n}$ is increasing in dimension n. (Take this fact for granted; it can be proved by a tedious calculation.)

7.4.5 Use Proposition 4.2.12 and Corollary 7.4.4 and optimize in ε.

7.5.3 Use the rotation invariance of the Gaussian distribution.

7.5.4 Use the Sudakov–Fernique comparison inequality.

7.5.10 Argue as in the proof of Corollary 7.4.4.

7.6.1 Use Gaussian concentration to prove the upper bound.

7.7.2(a) Use the singular value decomposition of P.

7.7.2(b) It is enough to check the rotation invariance of the distribution of $Q^{\mathsf{T}} z$.

7.7.4 To obtain the bound $\mathbb{E}\, \mathrm{diam}(PT) \gtrsim w_s(T)$, reduce P to a one-dimensional projection by dropping terms from the singular value decomposition of P. To obtain the bound $\mathbb{E}\, \mathrm{diam}(PT) \ge \sqrt{m/n}\, \mathrm{diam}(T)$, base your argument on a pair of points in T.

7.7.5 Express the operator norm of PA in terms of the diameter of the ellipsoid $P(AB_2^k)$, and use Theorem 7.7.1 in part (a) and Exercise 7.7.3 in part (b).

Chapter 8

8.1.12(a) This should be straightforward from Exercise 2.5.10.

8.1.12(b) The first m vectors in T form a $(1/\sqrt{\log m})$-separated set.

8.2.6 Put a mesh on the square $[0, 1]^2$ with step ε. Given $f \in \mathcal{F}$, show that $\|f - f_0\|_\infty \le \varepsilon$ for some function f_0 whose graph follows the mesh; see Figure 8.5. The number of all mesh-following functions f_0 is bounded by $(1/\varepsilon)^{1/\varepsilon}$. Next, use the result of Exercise 4.2.9

8.2.7 Use that f is Lipschitz to find a better bound on the number of possible functions f_0.

8.3.15 Consider the set $\mathcal{F}$ of binary strings of length n with at most d ones. This set is called the *Hamming cube*.

8.3.17 Consider the Hamming cube from Exercise 8.3.15.

8.3.21 Argue as in Lemma 8.3.19 and then use the covering–packing relationship from Lemma 4.2.8.

8.3.24 Modify the proof of the symmetrization lemma 6.4.2.

8.3.25 Add the zero function to the class $\mathcal{F}$ and use Remark 8.1.5 to bound $|Z_f| = |Z_f - Z_0|$. Can the addition of one (zero) function significantly increase the VC dimension of $\mathcal{F}$?

8.3.29 Choose a subset $\Lambda \subset \Omega$ of arbitrarily large cardinality d that is shattered by $\mathcal{F}$, and let μ be the uniform measure on Λ, assigning probability $1/d$ to each point.

8.3.30 Proceed as in the proof of Theorem 8.3.23. Combine a concentration inequality with a union bound over the entire class $\mathcal{F}$. Control the cardinality of $\mathcal{F}$ using the Sauer–Shelah lemma.

8.4.6 Choose an $\varepsilon/4$-net $\{f_j\}_{j=1}^N$ of $\mathcal{F}$ and check that $\{(f_j - T)^2\}_{j=1}^N$ is an ε-net of $\mathcal{L}$.

8.4.9(b) Proceed as in the proof of Theorem 8.2.3.

8.5.2(a) Use the first 2^{2^k} vectors in T to define T_k.

8.6.4 Use Remark 8.5.4 and the majorizing measure theorem to get a bound in terms of the Gaussian width $w(T \cup \{0\})$, then pass to Gaussian complexity using Exercise 7.6.9.

8.6.5 Argue as in Exercise 8.6.4. Use Theorem 8.5.5 and Exercise 7.6.9.

8.7.2 Note that $\mathbb{E} \sup_{x \in T,\, y \in S} \langle Ax, y \rangle \ge \sup_{x \in T} \mathbb{E} \sup_{y \in S} \langle Ax, y \rangle$.

8.7.3 Use the result of Exercise 8.6.5.

8.7.4 Use the Sudakov–Fernique inequality (Theorem 7.2.11) instead of Talagrand's comparison inequality.

Chapter 9

9.1.2 Bound the difference between $\mathbb{E}\,\|Ax\|_2$ and $\sqrt{m}\|x\|_2$ using the concentration of the norm theorem 3.1.1.

9.1.5 In this range of s, the sub-gaussian tail will dominate in Bernstein's inequality. Do not forget to apply the inequality for $2K^2$ instead of K because of Lemma 9.1.4.

9.1.8 Use the high-probability version of Talagrand's comparison inequality from Exercise 8.6.5.

9.1.10 Reduce it to the original deviation inequality using the identity $a^2 - b^2 = (a-b)\times(a+b)$.

9.2.3 If $m \ll n$, the random matrix A in the matrix deviation inequality is an approximate projection: this follows from Section 4.6.

9.4.9 Consider a random rotation $U \in \mathrm{Unif}(SO(n))$ as in Section 5.2.5. Applying a union bound, show that the probability that there exists $x \in \mathcal{X}$ such that $Ux \in T$ is smaller than 1.

Chapter 10

10.2.4 Modify the argument that led to the M^* bound.

10.2.5 Modify the M^* bound accordingly.

10.3.7(b) Note that $\|x_{I_1}\|_2 \le 1$. Next, for $i \ge 2$, note that each coordinate of x_{I_i} is smaller in magnitude than the average coordinate of $x_{I_{i-1}}$; conclude that $\|x_{I_i}\|_2 \le (1/\sqrt{s})\|x_{I_{i-1}}\|_1$. Then sum up the bounds.

10.3.9 Construct a large ε-separated subset in $S_{n,s}$ and thus deduce a lower bound on the covering numbers of $S_{n,s}$. Then use Sudakov's minoration inequality (Theorem 7.4.1).

10.3.10 Fix $\rho > 0$ and apply the M^* bound for the truncated octahedron $T_\rho := B_1^n \cap \rho B_2^n$. Use Exercise 10.3.8 to bound the Gaussian width of T_ρ. Furthermore, note that if $\mathrm{rad}(T_\rho \cap E) \le \delta$ for some $\delta \le \rho$ then $\mathrm{rad}(T \cap E) \le \delta$. Finally, optimize in ρ.

10.4.1 This will follow upon checking the identity $\|X\|_* = \max\left\{ |\langle X, U \rangle| : U \in O(d) \right\}$, where $O(d)$ denotes the set of $d \times d$ orthogonal matrices. Prove the identity using the singular value decomposition of X.

10.4.2 Think of the nuclear norm $\|\cdot\|_*$, the Frobenius norm $\|\cdot\|_F$, and the operator norm $\|\cdot\|$ as matrix analogs of the ℓ_1 norm, ℓ_2 norm, and ℓ_∞ norms for vectors, respectively.

10.4.3 Use (10.10) followed by Theorem 7.3.1.

10.5.4 Use the result of Exercise 10.3.8.

10.6.4 The proofs of these lemmas are based on the fact that $\|\widehat{x}\|_1 \le \|x\|_1$, which holds for the present situation as well.

10.6.8 Recall the proof of the sub-gaussian Chevet inequality (Theorem 8.7.1).

10.6.9 Use the result of Exercise 10.3.8.

Chapter 11

11.1.8 Follow the argument in Section 9.1.4.

11.1.11 Follow Exercise 9.1.8.

11.3.7 Set T to be the canonical basis $\{e_1, \dots, e_n\}$ in $\mathbb{R}^n$, represent the points as $g_i = Te_i$, and apply Theorem 11.3.3.

11.3.8 First replace in Theorem 11.3.3 the Gaussian width $w(T)$ with the quantity $h(T) = (\mathbb{E}\sup_{t\in T}\langle g, t\rangle^2)^{1/2}$, which we discussed in (7.19) and which is easier to compute for ellipsoids.

References

[1] E. Abbe, A. S. Bandeira, G. Hall, Exact recovery in the stochastic block model, *IEEE Trans. Inform. Theory* 62 (2016), 471–487.

[2] R. Adamczak, R. Latala, A. Litvak, A. Pajor, N. Tomczak-Jaegermann, Chevet type inequality and norms of submatrices, *Studia Math.* 210 (2012), 35–56.

[3] R. J. Adler, J. E. Taylor, *Random Fields and Geometry.* Springer Monographs in Mathematics. Springer, New York, 2007.

[4] R. Ahlswede, A. Winter, Strong converse for identification via quantum channels, IEEE *Trans. Inform. Theory* 48 (2002), 568–579.

[5] F. Albiac, N. J. Kalton, *Topics in Banach Space Theory.* Second edition. With a foreword by Gilles Godefory. Graduate Texts in Mathematics, vol. 233. Springer, New York, 2016.

[6] S. Alesker, A remark on the Szarek–Talagrand theorem, *Combin. Probab. Comput.* 6 (1997), 139–144.

[7] N. Alon, A. Naor, Approximating the cut-norm via Grothendieck's inequality, *SIAM J. Comput.* 35 (2006), 787–803.

[8] N. Alon, J. H. Spencer, *The Probabilistic Method.* Fourth edition. Wiley Series in Discrete Mathematics and Optimization. John Wiley & Sons, Hoboken, NJ, 2016.

[9] D. Amelunxen, M. Lotz, M. B. McCoy, J. A. Tropp, Living on the edge: phase transitions in convex programs with random data, *Inform. Inference* 3 (2014), 224–294.

[10] A. Anandkumar, R. Ge, D. Hsu, S. Kakade, M. Telgarsky, Tensor decompositions for learning latent variable models, *J. Mach. Learn. Res.* 15 (2014), 2773–2832.

[11] S. Artstein-Avidan, A. Giannopoulos, V. Milman, *Asymptotic Geometric Analysis, Part I.* Mathematical Surveys and Monographs, vol. 202. American Mathematical Society, Providence, RI, 2015.

[12] D. Bakry, M. Ledoux, Lévy-Gromov's isoperimetric inequality for an infinite-dimensional diffusion generator, *Invent. Math.* 123 (1996), 259–281.

[13] K. Ball, Flavors of geometry, in: *An Elementary Introduction to Modern Convex Geometry,* pp. 1–58. Math. Sci. Res. Inst. Publ., vol. 31. Cambridge University Press, Cambridge, 1997.

[14] A. Bandeira, Ten lectures and forty-two open problems in the mathematics of data science. Lecture notes, 2016. Available at `www.cims.nyu.edu/~bandeira/TenLecturesFortyTwoProblems.pdf`.

[15] F. Barthe, B. Maurey, Some remarks on isoperimetry of Gaussian type, *Ann. Inst. H. Poincaré Probab. Statist.* 36 (2000), 419–434.

[16] F. Barthe, E. Milman, Transference principles for log-Sobolev and spectral-gap with applications to conservative spin systems, *Commun. Math. Phys.* 323 (2013), 575–625.

[17] P. Bartlett, S. Mendelson, Rademacher and Gaussian complexities: risk bounds and structural results, *J. Mach. Learn. Res.* 3 (2002), 463–482.

[18] M. Belkin, K. Sinha, Polynomial learning of distribution families, *SIAM J. Comput.* 44 (2015), 889–911.

[19] A. Blum, J. Hopcroft, R. Kannan, *Foundations of Data Science.* To appear.

[20] R. Bhatia, *Matrix Analysis.* Graduate Texts in Mathematics, vol. 169. Springer-Verlag, New York, 1997.

[21] P. J. Bickel, Y. Ritov, A. Tsybakov, Simultaneous analysis of Lasso and Dantzig selector, *Ann. Stat.* 37 (2009), 1705–1732.
[22] P. Billingsley, *Probability and Measure.* Third edition. Wiley Series in Probability and Mathematical Statistics. John Wiley & Sons, New York, 1995.
[23] S. G. Bobkov, An isoperimetric inequality on the discrete cube, and an elementary proof of the isoperimetric inequality in Gauss space, *Ann. Probab.* 25 (1997), 206–214.
[24] B. Bollobás, *Combinatorics: Set Systems, Hypergraphs, Families of Vectors, and Combinatorial Probability.* Cambridge University Press, Cambridge, 1986.
[25] B. Bollobás, *Random Graphs.* Second edition. Cambridge Studies in Advanced Mathematics, vol. 73. Cambridge University Press, Cambridge, 2001.
[26] C. Bordenave, M. Lelarge, L. Massoulie, Non-backtracking spectrum of random graphs: community detection and non-regular Ramanujan graphs, *Ann. Probab.*, to appear.
[27] C. Borell, The Brunn–Minkowski inequality in Gauss space, *Invent. Math.* 30 (1975), 207–216.
[28] J. Borwein, A. Lewis, *Convex Analysis and Nonlinear Optimization. Theory and Examples.* Second edition. CMS Books in Mathematics/Ouvrages de Mathématiques de la SMC, vol. 3. Springer, New York, 2006.
[29] S. Boucheron, G. Lugosi, P. Massart, *Concentration Inequalities. A Nonasymptotic Theory of Independence.* With a foreword by Michel Ledoux. Oxford University Press, Oxford, 2013.
[30] J. Bourgain, S. Dirksen, J. Nelson, Toward a unified theory of sparse dimensionality reduction in Euclidean Space, *Geom. Funct. Anal.* 25 (2015), 1009–1088.
[31] J. Bourgain, L. Tzafriri, Invertibility of "large" submatrices with applications to the geometry of Banach spaces and harmonic analysis, *Israel J. Math.* 57 (1987), 137–224.
[32] O. Bousquet, S. Boucheron, G. Lugosi, Introduction to statistical learning theory, in: *Advanced Lectures on Machine Learning*, Lecture Notes in Computer Science, vol. 3176, pp. 169–207. Springer Verlag, 2004.
[33] S. Boyd, L. Vandenberghe, *Convex Optimization.* Cambridge University Press, Cambridge, 2004.
[34] M. Braverman, K. Makarychev, Yu. Makarychev, A. Naor, The Grothendieck constant is strictly smaller than Krivine's bound, in: *Proc. 52nd Annual IEEE Symp. on Foundations of Computer Science (FOCS), 2011*, pp. 453–462.
[35] S. Brazitikos, A. Giannopoulos, P. Valettas, B.-H. Vritsiou, *Geometry of Isotropic Convex Bodies.* Mathematical Surveys and Monographs, vol. 196. American Mathematical Society, Providence, RI, 2014.
[36] S. Brooks, A. Gelman, G. Jones, Xiao-Li Meng, eds., *Handbook of Markov Chain Monte Carlo.* Chapman & Hall/CRC Handbooks of Modern Statistical Methods. Chapman and Hall/CRC, 2011.
[37] Z. Brzeźniak, T. Zastawniak, *Basic Stochastic Processes. A course Through Exercises.* Springer-Verlag, London, 1999.
[38] S. Bubeck, Convex optimization: algorithms and complexity, *Found. Trends Mach. Learn.* 8 (2015), 231–357.
[39] A. Buchholz, Operator Khintchine inequality in non-commutative probability, *Math. Ann.* 319 (2001), 1–16.
[40] A. Buchholz, Optimal constants in Khintchine type inequalities for fermions, Rademachers and q-Gaussian operators, *Bull. Pol. Acad. Sci. Math.* 53 (2005), 315–321.
[41] P. Bühlmann, S. van de Geer, *Statistics for High-Dimensional Data. Methods, Theory and Applications.* Springer Series in Statistics. Springer, Heidelberg, 2011.
[42] T. Cai, R. Zhao, H. Zhou, Estimating structured high-dimensional covariance and precision matrices: optimal rates and adaptive estimation, *Electron. J. Stat.* 10 (2016), 1–59.
[43] E. Candes, The restricted isometry property and its implications for compressed sensing, *C. R. Math. Acad. Sci. Paris* 346 (2008), 589–592.
[44] E. Candes, B. Recht, Exact matrix completion via convex optimization, *Found. Comput. Math.* 9 (2009), 717–772.
[45] E. Candes, T. Tao, Decoding by linear programming, *IEEE Trans. Inform. Theory* 51 (2005), 4203–4215.

[46] E. Candes, T. Tao, The power of convex relaxation: near-optimal matrix completion, *IEEE Trans. Inform. Theory* 56 (2010), 2053–2080.

[47] F. P. Cantelli, Sulla determinazione empirica delle leggi di probabilita, *Giorn. Ist. Ital. Attuari* 4 (1933), 221–424.

[48] B. Carl, Inequalities of Bernstein-Jackson-type and the degree of compactness of operators in Banach spaces, *Ann. Inst. Fourier (Grenoble)* 35 (1985), 79–118.

[49] B. Carl, A. Pajor, Gelfand numbers of operators with values in a Hilbert space, *Invent. Math.* 94 (1988), 479–504.

[50] P. Casazza, G. Kutyniok, F. Philipp, Introduction to finite frame theory, in: *Finite frames*, pp. 1–53. Applied and Numerical Harmonic Analysis Series. Birkhuser/Springer, New York, 2013.

[51] V. Chandrasekaran, B. Recht, P. A. Parrilo, A. S. Willsky, The convex geometry of linear inverse problems, *Found. Comput. Math.*, 12 (2012), 805–849.

[52] R. Chen, A. Gittens, J. Tropp, The masked sample covariance estimator: an analysis using matrix concentration inequalities, *Inform. Inference* 1 (2012), 2–20.

[53] S. Chevet, Séries de variables aléatoires gaussiennes à valeurs dans $E\hat{\otimes}_\varepsilon F$. Application aux produits d'espaces de Wiener abstraits, in: *Proc. Séminaire sur la Géométrie des Espaces de Banach (1977–1978)*, exp. no. 19, vol. 15. École Polytech., Palaiseau, 1978.

[54] P. Chin, A. Rao, and V. Vu, Stochastic block model and community detection in the sparse graphs: a spectral algorithm with optimal rate of recovery, preprint, 2015.

[55] M. Davenport, M. Duarte,Y. Eldar, G. Kutyniok, Introduction to compressed sensing, in: *Compressed Sensing*, pp. 1–64. Cambridge University Press, Cambridge, 2012.

[56] M. Davenport, Y. Plan, E. van den Berg, M. Wootters, 1-bit matrix completion, *Inform. Inference* 3 (2014), 189–223.

[57] M. Davenport, J. Romberg, An overview of low-rank matrix recovery from incomplete observations, preprint (2016).

[58] K. R. Davidson, S. J. Szarek, Local operator theory, random matrices and Banach spaces, in: *Handbook of the Geometry of Banach Spaces*, vol. I, pp. 317–366. Amsterdam, North-Holland, 2001.

[59] V. H. de la Peña, E. Giné, *Decoupling*. Probability and its Applications Series. Springer-Verlag, New York, 1999.

[60] V. H. de la Peña, S. J. Montgomery-Smith, Decoupling inequalities for the tail probabilities of multivariate U-statistics, *Ann. Probab.* 23 (1995), 806–816.

[61] S. Dirksen, Tail bounds via generic chaining, *Electron. J. Probab.* 20 (2015), art. no. 53, 29 pp.

[62] D. Donoho, M. Gavish, A. Montanari, The phase transition of matrix recovery from Gaussian measurements matches the minimax MSE of matrix denoising, *Proc. Natl. Acad. Sci. USA* 110 (2013), 8405–8410.

[63] D. Donoho, A. Javanmard, A. Montanari, Information-theoretically optimal compressed sensing via spatial coupling and approximate message passing, *IEEE Trans. Inform. Theory* 59 (2013), 7434–7464.

[64] D. Donoho, I. Johnstone, A. Montanari, Accurate prediction of phase transitions in compressed sensing via a connection to minimax denoising, *IEEE Trans. Inform. Theory* 59 (2013), 3396–3433.

[65] D. Donoho, A. Maleki, A. Montanari, The noise-sensitivity phase transition in compressed sensing, *IEEE Trans. Inform. Theory* 57 (2011), 6920–6941.

[66] D. Donoho, J. Tanner, Counting faces of randomly projected polytopes when the projection radically lowers dimension, *J. Amer. Math. Soc.* 22 (2009), 1–53.

[67] R. M. Dudley, The sizes of compact subsets of Hilbert space and continuity of Gaussian processes, *J. Funct. Anal.* 1 (1967), 290–330.

[68] R. M. Dudley, Central limit theorems for empirical measures, *Ann. Probab.* 6 (1978), 899–929.

[69] R. M Dudley, *Uniform Central Limit Theorems*. Cambridge University Press, 1999.

[70] R. Durrett, *Probability: Theory and Examples*. Fourth edition. Cambridge Series in Statistical and Probabilistic Mathematics, vol. 31. Cambridge University Press, Cambridge, 2010.

[71] A. Dvoretzky, A theorem on convex bodies and applications to Banach spaces, *Proc. Natl. Acad. Sci. USA* 45 (1959), 223–226.

[72] A. Dvoretzky, Some results on convex bodies and Banach spaces, in: *Proc. Symp. on Linear Spaces*, Jerusalem (1961), pp. 123–161.

[73] X. Fernique, *Regularité des trajectoires des fonctions aléatoires Gaussiens.* Lecture Notes in Mathematics, vol. 480, pp. 1–96. Springer, 1976.

[74] G. Folland, *A Course in Abstract Harmonic Analysis.* Studies in Advanced Mathematics. CRC Press, Boca Raton, FL, 1995.

[75] S. Fortunato, D. Hric, Community detection in networks: a user guide, *Phys. Rep.* 659 (2016), 1–44.

[76] S. Foucart, H. Rauhut, *A Mathematical Introduction to Compressive Sensing.* Applied and Numerical Harmonic Analysis Series. Birkhäuser/Springer, New York, 2013.

[77] P. Frankl, On the trace of finite sets, *J. Combin. Theory Ser. A* 34 (1983), 41–45.

[78] A. Garnaev, E. D. Gluskin, On diameters of the Euclidean sphere, *Dokl. A.N. USSR* 277 (1984), 1048–1052.

[79] A. Giannopoulos, V. Milman, Euclidean structure in finite dimensional normed spaces, in: *Handbook of the Geometry of Banach Spaces*, vol. I, pp. 707–779. North-Holland, Amsterdam, 2001.

[80] V. Glivenko, Sulla determinazione empirica della legge di probabilita, *Giorn. Ist. Ital. Attuari* 4 (1933), 92–99.

[81] M. Goemans, D. Williamson, Improved approximation algorithms for maximum cut and satisfiability problems using semidefinite programming, *J. ACM* 42 (1995), 1115–1145.

[82] Y. Gordon, Some inequalities for Gaussian processes and applications, *Israel J. Math.* 50 (1985), 265–289.

[83] Y. Gordon, Elliptically contoured distributions, *Prob. Theory Rel. Fields* 76 (1987), 429–438.

[84] Y. Gordon, Gaussian processes and almost spherical sections of convex bodies, *Ann. Probab.* 16 (1988), 180–188.

[85] Y. Gordon, On Milman's inequality and random subspaces which escape through a mesh in $\mathbb{R}^n$, in: *Proc. Conf. on Geometric Aspects of Functional Analysis* (1986/87), Lecture Notes in Mathematics, vol. 1317, pp. 84–106.

[86] Y. Gordon, Majorization of Gaussian processes and geometric applications, *Prob. Theory Rel. Fields* 91 (1992), 251–267.

[87] N. Goyal, S. Vempala, Y. Xiao, Fourier PCA and robust tensor decomposition, in: *Proc. Forty-sixth Annual ACM symp. on Theory of Computing*, pp. 584–593. New York, 2014.

[88] A. Grothendieck, Résumé de la théorie métrique des produits tensoriels topologiques, *Bol. Soc. Mat. Sao Paulo 8* (1953), 1–79.

[89] M. Gromov, Paul Lévy's isoperimetric inequality, Appendix C in: *Metric Structures for Riemannian and non-Riemannian Spaces.* Based on the 1981 French original. Progress in Mathematics, vol. 152. Birkhäuser Boston, 1999.

[90] D. Gross, Recovering low-rank matrices from few coefficients in any basis, *IEEE Trans. Inform. Theory* 57 (2011), 1548–1566.

[91] O. Guédon, Concentration phenomena in high-dimensional geometry. *J. MAS* (2012), 47–60. ArXiv: `https://arxiv.org/abs/1310.1204`.

[92] O. Guedon, R. Vershynin, Community detection in sparse networks via Grothendieck's inequality, *Probab. Theory Rel. Fields* 165 (2016), 1025–1049.

[93] U. Haagerup, The best constants in the Khintchine inequality, *Studia Math.* 70 (1981), 231–283.

[94] B. Hajek, Y. Wu, J. Xu, Achieving exact cluster recovery threshold via semidefinite programming, *IEEE Trans. Inform. Theory* 62 (2016), 2788–2797.

[95] D. L. Hanson, E. T. Wright, A bound on tail probabilities for quadratic forms in independent random variables, *Ann. Math. Statist.* 42 (1971), 1079–1083.

[96] L. H. Harper, Optimal numbering and isoperimetric problems on graphs, *Combin. Theory* 1 (1966), 385–393.

[97] T. Hastie, R. Tibshirani, J. Friedman, *The Elements of Statistical Learning.* Second edition. Springer Series in Statistics. Springer, New York, 2009.

[98] T. Hastie, R. Tibshirani, W. Wainwright, *Statistical Learning with Sparsity. The Lasso and Generalizations.* Monographs on Statistics and Applied Probability, vol. 143. CRC Press, Boca Raton, FL, 2015.

[99] D. Haussler, P. Long, A generalization of Sauer's lemma, *J. Combin. Theory Ser. A* 71 (1995), 219–240.

[100] T. Hofmann, B. Schölkopf, A. Smola, Kernel methods in machine learning, *Ann. Statist.* 36 (2008), 1171–1220.

[101] P. W. Holland, K. B. Laskey, S. Leinhardt, Stochastic blockmodels: first steps, *Social Networks* 5 (1983), 109–137.

[102] D. Hsu, S. Kakade, Learning mixtures of spherical Gaussians: moment methods and spectral decompositions, in: *Proc. 2013 ACM Conf. on Innovations in Theoretical Computer Science*, pp. 11–19. ACM, New York, 2013.

[103] F. W. Huffer, Slepian's inequality via the central limit theorem, *Canad. J. Statist.* 14 (1986), 367–370.

[104] G. James, D. Witten, T. Hastie, R. Tibshirani, *An Introduction to Statistical Learning, with Applications* in R. Springer Texts in Statistics, vol. 103. Springer, New York, 2013.

[105] S. Janson, T. Luczak, A. Rucinski, *Random Graphs.* Wiley-Interscience Series in Discrete Mathematics and Optimization. Wiley-Interscience, New York, 2000.

[106] A. Javanmard, A. Montanari, F. Ricci-Tersenghi, Phase transitions in semidefinite relaxations, *PNAS* 113 (2016), E2218–E2223.

[107] W. Johnson, J. Lindenstrauss, Extensions of Lipschitz mappings into a Hilbert space, *Contemp. Math.* 26 (1984), 189–206.

[108] J.-P. Kahane, Une inégalité du type de Slepian et Gordon sur les processus gaussiens, *Israel J. Math.* 55 (1986), 109–110.

[109] A. Kalai, A. Moitra, G. Valiant, Disentangling Gaussians, *Commun. ACM* 55 (2012), 113–120.

[110] S. Khot, G. Kindler, E. Mossel, R. O'Donnell, Optimal inapproximability results for MAX-CUT and other 2-variable CSPs?, *SIAM J. Computing* 37 (2007), 319–357.

[111] S. Khot, A. Naor, Grothendieck-type inequalities in combinatorial optimization, *Commun. Pure Appl. Math.* 65 (2012), 992–1035.

[112] B. Klartag, A central limit theorem for convex sets, *Invent. Math.* 168 (2007), 91–131.

[113] B. Klartag, S. Mendelson, Empirical processes and random projections, *J. Funct. Anal.* 225 (2005), 229–245.

[114] H. König, On the best constants in the Khintchine inequality for Steinhaus variables, *Israel J. Math.* 203 (2014), 23–57.

[115] V. Koltchinskii, K. Lounici, Concentration inequalities and moment bounds for sample covariance operators, *Bernoulli* 23 (2017), 110–133.

[116] I. Shevtsova, On the absolute constants in the Berry–Esseen type inequalities for identically distributed summands, preprint, 2012. arXiv:1111.6554.

[117] J. Kovacevic, A. Chebira, An introduction to frames, *Found. Trends Signal Proc.* 2 (2008), 1–94.

[118] J.-L. Krivine, Constantes de Grothendieck et fonctions de type positif sur les sphéres, *Adv. Math.* 31 (1979), 16–30.

[119] S. Kulkarni, G. Harman, *An Elementary Introduction to Statistical Learning Theory.* Wiley Series in Probability and Statistics. John Wiley & Sons, Hoboken, NJ, 2011.

[120] K. Larsen, J. Nelson, Optimality of the Johnson–Lindenstrauss lemma, submitted (2016). `https://arxiv.org/abs/1609.02094`.

[121] R. Latala, R. van Handel, P. Youssef, The dimension-free structure of nonhomogeneous random matrices, preprint (2017). `https://arxiv.org/abs/1711.00807`.

[122] M. Laurent, F. Vallentin, Semidefinite Optimization. Mastermath, 2012. Available at `http://page.mi.fu-berlin.de/fmario/sdp/laurentv.pdf`.

[123] G. Lawler, *Introduction to Stochastic Processes.* Second edition. Chapman & Hall/CRC, Boca Raton, FL, 2006.

[124] C. Le, E. Levina, R. Vershynin, Concentration and regularization of random graphs, *Random Struct. Algor.*, to appear.

[125] M. Ledoux, *The Concentration of Measure Phenomenon.* Mathematical Surveys and Monographs, vol. 89. American Mathematical Society, Providence, RI, 2001.

[126] M. Ledoux, M. Talagrand, *Probability in Banach spaces. Isoperimetry and Processes.* Ergebnisse der Mathematik und ihrer Grenzgebiete, vol. 3, p. 23. Springer-Verlag, Berlin, 1991.

[127] E. Levina, R. Vershynin, Partial estimation of covariance matrices, *Probab. Theory Rel. Fields* 153 (2012), 405–419.

[128] C. Liaw, A. Mehrabian, Y. Plan, R. Vershynin, A simple tool for bounding the deviation of random matrices on geometric sets, in: *Proc. Geometric Aspects of Functional Analysis: Israel Seminar (GAFA) 2014–2016*, B. Klartag, E. Milman, eds., pp. 277–299. Lecture Notes in Mathematics vol. 2169. Springer, 2017.

[129] J. Lindenstrauss, A. Pelczynski, Absolutely summing operators in L^p-spaces and their applications, *Studia Math.* 29 (1968), 275–326.

[130] F. Lust-Piquard, Inégalités de Khintchine dans C_p $(1 < p < \infty)$, *C. R. Math. Acad. Sci. Paris* 303 (1986), 289–292.

[131] F. Lust-Piquard, G. Pisier, Noncommutative Khintchine and Paley inequalities, *Ark. Mat.* 29 (1991), 241–260.

[132] Y. Makovoz, *A simple proof of an inequality in the theory of n-widths,* in: *Proc. conf. on Constructive Theory of Functions (Varna, 1987)*, pp. 305–308. Publ. House Bulgar. Acad. Sci., Sofia, 1988.

[133] J. Matoušek, *Geometric Discrepancy. An Illustrated Guide.* Algorithms and Combinatorics, vol. 18. Springer-Verlag, Berlin, 1999.

[134] J. Matoušek, *Lectures on Discrete Geometry.* Graduate Texts in Mathematics, vol. 212. Springer-Verlag, New York, 2002.

[135] B. Maurey, Construction de suites symétriques, *C.R.A.S., Paris* 288 (1979), 679–681.

[136] M. McCoy, J. Tropp, From Steiner formulas for cones to concentration of intrinsic volumes, *Discrete Comput. Geom.* 51 (2014), 926–963.

[137] F. McSherry, Spectral partitioning of random graphs, in: *Proc. 42nd FOCS (2001)*, pp. 529–537.

[138] E. Meckes, Projections of probability distributions: a measure-theoretic Dvoretzky theorem, in: Geometric aspects of functional analysis, pp. 317–326. Lecture Notes in Mathematics, vol. 2050. Springer, Heidelberg, 2012.

[139] S. Mendelson, A few notes on statistical learning theory, in: S. Mendelson, A.J. Smola, eds., *Advanced Lectures on Machine Learning*, LNAI, vol. 2600, pp. 1–40, 2003.

[140] S. Mendelson, A remark on the diameter of random sections of convex bodies, in: *Geometric Aspects of Functional Analysis (GAFA Seminar Notes)*, B. Klartag and E. Milman, eds., Lecture Notes in Mathematics, vol. 2116, pp. 3950, 2014.

[141] S. Mendelson, A. Pajor, N. Tomczak-Jaegermann, Reconstruction and subgaussian operators in asymptotic geometric analysis, *Geom. Funct. Anal.* 17 (2007), 1248–1282.

[142] S. Mendelson, R. Vershynin, Entropy and the combinatorial dimension, *Invent. Math.* 152 (2003), 37–55.

[143] F. Mezzadri, How to generate random matrices from the classical compact groups, *Not. Amer. Math. Soc.* 54 (2007), 592–604.

[144] V. D. Milman, New proof of the theorem of Dvoretzky on sections of convex bodies, *Funct. Anal. Appl.* 5 (1971), 28–37.

[145] V. D. Milman, A note on a low M^*-estimate, in: P. F. Muller and W. Schachermayer, eds., *Geometry of Banach Spaces*: LMS Lecture Note Series, vol. 158, pp. 219–229. Cambridge University Press, 1990.

[146] V. D. Milman, G. Schechtman, *Asymptotic Theory of Finite-Dimensional Normed Spaces.* With an appendix by M. Gromov. Lecture Notes in Mathematics, vol. 1200. Springer-Verlag, Berlin, 1986.

[147] V. D. Milman, G. Schechtman, Global versus local asymptotic theories of finite-dimensional normed spaces, *Duke Math. J.* 90 (1997), 73–93.

[148] M. Mitzenmacher, E. Upfal, Probability and computing. *Randomized Algorithms and Probabilistic Analysis*. Cambridge University Press, Cambridge, 2005.

[149] A. Moitra, Algorithmic aspects of machine learning, Preprint, MIT Special Subject in Mathematics, 2014.

[150] A. Moitra, G. Valiant, Settling the polynomial learnability of mixtures of Gaussians, in: *Proc. 2010 IEEE 51st Annual Symp. on Foundations of Computer Science*, pp. 93–102. IEEE Computer Society, Los Alamitos, CA, 2010.

[151] S. J. Montgomery-Smith, The distribution of Rademacher sums, *Proc. Amer. Math. Soc.* 109 (1990), 517–522.

[152] P. Mörters, Y. Peres, *Brownian Motion.* Cambridge University Press, Cambridge, 2010.

[153] E. Mossel, J. Neeman, A. Sly, Belief propagation, robust reconstruction and optimal recovery of block models, *Ann. Appl. Probab.* 26 (2016), 2211–2256.

[154] M. E. Newman, *Networks. An Introduction.* Oxford University Press, Oxford, 2010.

[155] R. I. Oliveira, Sums of random Hermitian matrices and an inequality by Rudelson, *Electron. Commun. Probab.* 15 (2010), 203–212.

[156] R. I. Oliveira, Concentration of the adjacency matrix and of the Laplacian in random graphs with independent edges, unpublished manuscript, 2009. arXiv: 0911.0600.

[157] S. Oymak, B. Hassibi, New null space results and recovery thresholds for matrix rank minimization, in: *Proc. ISIT 2011.* ArXiv: https://arxiv.org/abs/1011.6326.

[158] S. Oymak, C. Thrampoulidis, B. Hassibi, The squared-error of generalized LASSO: a precise analysis, in: *Proc. 51st Annual Allerton Conf on Communication, Control and Computing,* 2013. ArXiv: https://arxiv.org/abs/1311.0830.

[159] S. Oymak, J. Tropp, Universality laws for randomized dimension reduction, with applications, *Inform. Inference*, to appear (2017).

[160] A. Pajor, *Sous espaces ℓ_1^n des espaces de Banach.* Hermann, Paris, 1985.

[161] D. Petz, A survey of certain trace inequalities, in: *Proc. Conf. on Functional Analysis and Operator Theory* (Warsaw, 1992), pp. 287–298. Polish Acad. Sci. Inst. Math., Warsaw, 1994.

[162] G. Pisier, Remarques sur un résultat non publié de B. Maurey, in: *Proc. Seminar on Functional Analysis, 1980–1981*, exp. no. V, 13 pp. École Polytechnique, Palaiseau, 1981.

[163] G. Pisier, *The Volume of Convex Bodies and Banach Space Geometry.* Cambridge Tracts in Mathematics, vol. 94. Cambridge University Press, Cambridge, 1989.

[164] G. Pisier, Grothendieck's theorem, past and present, *Bull. Amer. Math. Soc. (NS)* 49 (2012), 237–323.

[165] Y. Plan, R. Vershynin, Robust 1-bit compressed sensing and sparse logistic regression: a convex programming approach, *IEEE Trans. Inform. Theory* 59 (2013), 482–494.

[166] Y. Plan, R. Vershynin, E. Yudovina, High-dimensional estimation with geometric constraints, *Inform. Inference* 6 (2016), 1–40.

[167] D. Pollard, *Empirical Processes: Theory and Applications.* NSF-CBMS Regional Conference Series in Probability and Statistics, vol. 2. Institute of Mathematical Statistics, Hayward, CA; American Statistical Association, Alexandria, VA, 1990.

[168] H. Rauhut, *Compressive sensing and structured random matrices*, in: M. Fornasier, ed., *Theoretical Foundations and Numerical Methods for Sparse Recovery*, pp. 1–92. Radon Series on Computational and Applied Mathematics, vol. 9. de Gruyter, Berlin, 2010.

[169] B. Recht, A simpler approach to matrix completion, *J. Mach. Learn. Res.* 12 (2011), 3413–3430.

[170] P. Rigollet, High-dimensional statistics. Lecture notes, Massachusetts Institute of Technology, 2015. Available at MIT Open CourseWare.

[171] M. Rudelson, Random vectors in the isotropic position, *J. Funct. Anal.* 164 (1999), 60–72.

[172] M. Rudelson, R. Vershynin, Combinatorics of random processes and sections of convex bodies, *Ann. Math.* 164 (2006), 603–648.

[173] M. Rudelson, R. Vershynin, Sampling from large matrices: an approach through geometric functional analysis, *J. ACM* (2007), art. no. 21, 19 pp.

[174] M. Rudelson, R. Vershynin, On sparse reconstruction from Fourier and Gaussian measurements, *Commun. Pure Appl. Math.* 61 (2008), 1025–1045.

[175] M. Rudelson, R. Vershynin, Hanson–Wright inequality and sub-gaussian concentration, *Electroni. Commun. Probab.* 18 (2013), 1–9.

[176] N. Sauer, On the density of families of sets, *J. Comb. Theor.* 13 (1972), 145–147.

[177] G. Schechtman, Two observations regarding embedding subsets of Euclidean spaces in normed spaces, *Adv. Math.* 200 (2006), 125–135.

[178] R. Schilling, L. Partzsch, *Brownian Motion. An Introduction to Stochastic Processes.* Second edition. De Gruyter, Berlin, 2014.

[179] S. Shelah, A combinatorial problem: stability and order for models and theories in infinitary languages, *Pacific J. Math.* 41 (1972), 247–261.

[180] M. Simonovits, How to compute the volume in high dimension?, in: *Proc. ISMP, 2003 (Copenhagen), Math. Program. Ser. B*, 97 (2003), nos. 1–2, 337–374.

[181] D. Slepian, The one-sided barrier problem for Gaussian noise, *Bell. System Tech. J.* 41 (1962), 463–501.

[182] D. Slepian, *On the zeroes of Gaussian noise,* in: M. Rosenblatt, ed., *Time Series Analysis*, pp. 104–115. Wiley, New York, 1963.

[183] M. Stojnic, Various thresholds for ℓ_1-optimization in compressed sensing, unpublished manuscript, 2009. ArXiv: `https://arxiv.org/abs/0907.3666`.

[184] M. Stojnic, Regularly random duality, unpublished manuscript, 2013. ArXiv: `https://arxiv.org/abs/1303.7295`.

[185] V. N. Sudakov, Gaussian random processes and measures of solid angles in Hilbert spaces, *Soviet Math. Dokl.* 12 (1971), 412–415.

[186] V. N. Sudakov, B. S. Cirelson, Extremal properties of half-spaces for spherically invariant measures (in Russian); *LOMI* 41 (1974), 14–24.

[187] V. N. Sudakov, Gaussian random processes and measures of solid angles in Hilbert space, *Dokl. Akad. Nauk. SSR* 197 (1971), 4345; English translation in *Soviet Math. Dokl.* 12 (1971), 412–415.

[188] V. N. Sudakov, Geometric problems in the theory of infinite-dimensional probability distributions, *Trud. Mat. Inst. Steklov* 141 (1976); English translation in *Proc. Steklov Inst. Math* 2, American Mathematical Society.

[189] S. J. Szarek, On the best constants in the Khinchin inequality, *Studia Math.* 58 (1976), 197–208.

[190] S. Szarek, M. Talagrand, An "isomorphic" version of the Sauer–Shelah lemma and the Banach–Mazur distance to the cube, in: *Proc. conf. on Geometric Aspects of Functional Analysis (1987–88)*, pp. 105–112, Lecture Notes in Mathematics, 1376. Springer, Berlin, 1989.

[191] S. Szarek, M. Talagrand, On the convexified Sauer–Shelah theorem, *J. Combin. Theory Ser. B* 69 (1997), 1830–192.

[192] M. Talagrand, A new look at independence, *Ann. Probab.* 24 (1996), 1–34.

[193] M. Talagrand, *The Generic Chaining. Upper and Lower Bounds of Stochastic Processes.* Springer Monographs in Mathematics. Springer-Verlag, Berlin, 2005.

[194] C. Thrampoulidis, E. Abbasi, B. Hassibi, Precise error analysis of regularized M-estimators in high-dimensions, preprint. ArXiv: `https://arxiv.org/abs/1601.06233`.

[195] C. Thrampoulidis, B. Hassibi, Isotropically random orthogonal matrices: performance of LASSO and minimum conic singular values, ISIT 2015. ArXiv: `https://arxiv.org/abs/503.07236`.

[196] C. Thrampoulidis, S. Oymak, B. Hassibi, Simple error bounds for regularized noisy linear inverse problems, ISIT 2014. ArXiv: `https://arxiv.org/abs/1401.6578`.

[197] C. Thrampoulidis, S. Oymak, B. Hassibi, The Gaussian min–max theorem in the presence of convexity, 2014. ArXiv: `https://arxiv.org/abs/1408.4837`.

[198] R. Tibshirani, Regression shrinkage and selection via the lasso, *J. Roy. Statist. Soc. Ser. B* 58 (1996), 267–288.

[199] N. Tomczak-Jaegermann, *Banach–Mazur Distances and Finite-Dimensional Operator Ideals.* Pitman Monographs and Surveys in Pure and Applied Mathematics, vol. 38. Longman Scientific & Technical, Harlow; John Wiley & Sons, New York, 1989.

[200] J. Tropp, User-friendly tail bounds for sums of random matrices, *Found. Comput. Math.* 12 (2012), 389–434.

[201] J. Tropp, An introduction to matrix concentration inequalities, *Found. Trends Mach. Learn.* 8 (2015) 1–230.

[202] J. Tropp, Convex recovery of a structured signal from independent random linear measurements, in: G. Pfander, ed. *Sampling Theory, a Renaissance: Compressive Sampling and Other Developments.* Series on Applied and Numerical Harmonic Analysis. Birkhäuser, Basel, 2015.

[203] J. Tropp, The expected norm of a sum of independent random matrices: an elementary approach, in: C. Houdre, D. M. Mason, P. Reynaud-Bouret, J. Rosinski, eds. *High-Dimensional Probability VII: The Cargese Volume.* Series on Progress in Probability, vol. 71. Birkhäuser, Basel, 2016.

[204] S. van de Geer, *Applications of Empirical Process Theory.* Cambridge Series in Statistical and Probabilistic Mathematics, vol. 6. Cambridge University Press, Cambridge, 2000.

[205] A. van der Vaart, J. Wellner, *Weak Convergence and Empirical Processes, with Applications to Statistics.* Springer Series in Statistics. Springer-Verlag, New York, 1996.

[206] R. van Handel, Probability in high dimension, Lecture notes. Available at `www.princeton.edu/~rvan/APC550`.

[207] R. van Handel, Structured random matrices, in: IMA volume *Discrete Structures: analysis and Applications*, Springer, to appear.

[208] R. van Handel, Chaining, interpolation, and convexity, *J. Eur. Math. Soc.*, to appear, 2016.

[209] R. van Handel, Chaining, interpolation, and convexity II: the contraction principle, preprint, 2017.

[210] J. H. van Lint, *Introduction to Coding Theory.* Third edition. Graduate Texts in Mathematics, vol. 86. Springer-Verlag, Berlin, 1999.

[211] V. N. Vapnik, A. Ya. Chervonenkis, The uniform convergence of frequencies of the appearance of events to their probabilities, *Teor. Verojatnost. i Primenen* 16 (1971), 264–279.

[212] S. Vempala, Geometric random walks: a survey, in: *Combinatorial and Computational Geometry*, pp. 577–616. Math. Sci. Res. Inst. Publ., vol. 52, Cambridge University Press, Cambridge, 2005.

[213] R. Vershynin, Integer cells in convex sets, *Adv. Math.* 197 (2005), 248–273.

[214] R. Vershynin, A note on sums of independent random matrices after Ahlswede–Winter, unpublished manuscript, 2009, available at `www.math.uci.edu/~rvershyn/papers/ahlswede-winter.pdf`.

[215] R. Vershynin, Golden–Thompson inequality, unpublished manuscript, 2009, available at `www.math.uci.edu/~rvershyn/papers/golden-thompson.pdf`.

[216] R. Vershynin, Introduction to the non-asymptotic analysis of random matrices, in: *Compressed Sensing*, pp. 210–268. Cambridge University Press, Cambridge, 2012.

[217] R. Vershynin, Estimation in high dimensions: a geometric perspective, in: *Sampling Theory, a Renaissance*, pp. 3–66. Birkhauser, Basel, 2015.

[218] C. Villani, *Topics in Optimal Transportation.* Graduate Studies in Mathematics, vol. 58. American Mathematical Society, Providence, RI, 2003.

[219] A. Wigderson, D. Xiao, Derandomizing the Ahlswede–Winter matrix-valued Chernoff bound using pessimistic estimators, and applications, *Theory Comput.* 4 (2008), 53–76.

[220] E. T. Wright, A bound on tail probabilities for quadratic forms in independent random variables whose distributions are not necessarily symmetric, *Ann. Probab.* 1 (1973), 1068–1070.

[221] H. Zhou, A. Zhang, Minimax rates of community detection in stochastic block models, *Ann. Statist.*, to appear.

Index

Printed in the United States
By Bookmasters